FLIGHT TO THE FUTURE

HUMAN FACTORS IN AIR TRAFFIC CONTROL

Christopher D. Wickens, Anne S. Mavor, and James P. McGee, editors

Panel on Human Factors in Air Traffic Control Automation

Committee on Human Factors

Commission on Behavioral and Social Sciences and Education
National Research Council

NATIONAL ACADEMY PRESS
Washington, D.C. 1997

NATIONAL ACADEMY PRESS • 2101 Constitution Avenue, NW • Washington, DC 20418

NOTICE: The project that is the subject of this report was approved by the Governing Board of the National Research Council, whose members are drawn from the councils of the National Academy of Sciences, the National Academy of Engineering, and the Institute of Medicine. The members of the committee responsible for the report were chosen for their special competences and with regard for appropriate balance.

This report has been reviewed by a group other than the authors according to procedures approved by a Report Review Committee consisting of members of the National Academy of Sciences, the National Academy of Engineering, and the Institute of Medicine.

This work is sponsored by the Federal Aviation Administration and funded under Grant No. 94-G-042. The views, opinions, and findings contained in this report are those of the author(s) and should not be construed as an official Department of Transportation position, policy, or decision, unless so designated by other official documentation.

Library of Congress Cataloging-in-Publication Data

The human factors of air traffic control / Panel on Human Factors in
    Air Traffic Control Automation, Commission on Behavioral and Social
    Sciences and Education, National Research Council ; Christopher D.
    Wickens, Anne S. Mavor, and James P. McGee, editors.
        p.   cm
    Includes bibliographical references (p.) and index.
    ISBN 0-309-05637-3
    1. Air traffic control—United States—Automation. 2. Air traffic
control—United States—Safety measures. 3. Aeronautics—Human
factors. I. Wickens, Christopher D. II. Mavor, Anne S.
III. McGee, J. (James), 1950- . IV. National Research Council
(U.S.). Panel on Human Factors in Air Traffic Control Automation.
TL725.3.T7H865   1997
629.136'6—dc21   96-37616
                                                                                                                                            CIP

Additional copies of this report are available from:

National Academy Press
2101 Constitution Avenue, N.W., Box 285
Washington, D.C. 20418
800-624-6242 or 202-334-3313 (in Washington Metropolitan Area).
http://www.nap.edu.

Copyright 1997 by the National Academy of Sciences. All rights reserved.

Printed in the United States of America

PANEL ON HUMAN FACTORS IN AIR TRAFFIC CONTROL AUTOMATION

CHRISTOPHER D. WICKENS (*Chair*), Aviation Research Laboratory, University of Illinois, Savoy
CHARLES B. AALFS, Federal Aviation Administration/Air Traffic Control Service (retired), Fountain Valley, CA
TORA K. BIKSON, Rand Corporation, Santa Monica, CA
MARVIN S. COHEN, Cognitive Technologies, Inc., Arlington, VA
DIANE DAMOS, Human Factors Department, University of Southern California
JAMES DANAHER, National Transportation Safety Board, Washington, DC
ROBERT L. HELMREICH, National Aeronautics and Space Adminstration/University of Texas Aerospace Crew Research Project, Austin
V. DAVID HOPKIN, Centre for Aviation/Aerospace Research, Embry-Riddle Aeronautical University, Daytona Beach, FL
TODD R. LaPORTE, Department of Political Science, University of California, Berkeley
RAJA PARASURAMAN, Department of Psychology, Catholic University
JOSEPH O. PITTS, VITRO, Rockville, MD
THOMAS B. SHERIDAN, Engineering and Applied Psychology, Massachusetts Institute of Technology
PAUL STAGER, Department of Psychology, York University, Toronto
RICHARD B. STONE, Bountiful, UT
EARL L. WIENER, Department of Psychology and Industrial Engineering, University of Miami
LAURENCE R. YOUNG, Department of Aeronautics and Astronautics, Massachusetts Institute of Technology

ANNE S. MAVOR, Study Director
JAMES P. McGEE, Senior Research Associate
JERRY KIDD, Senior Adviser
SUSAN R. McCUTCHEN, Senior Project Assistant
THERESA NOONAN, Senior Project Assistant

COMMITTEE ON HUMAN FACTORS

WILLIAM B. ROUSE (*Chair*), Enterprise Support Systems, Norcross, Georgia
TERRY CONNOLLY, Department of Management and Policy, College of Business and Public Administration, University of Arizona, Tucson
PAUL S. GOODMAN, Center for Management of Technology, Graduate School of Industrial Administration, Carnegie Mellon University
ROBERT L. HELMREICH, National Aeronautics and Space Administration/University of Texas Aerospace Crew Research Project, Austin
WILLIAM C. HOWELL, American Psychological Association Science Directorate, Washington, DC
ROBERTA L. KLATZKY, Department of Psychology, Carnegie Mellon University
TOM B. LEAMON, Liberty Mutual Research Center, Hopkinton, MA
ANN MAJCHRZAK, Human Factors Department, Institute of Safety and Systems Management, University of Southern California
DAVID C. NAGEL, AT&T Laboratories, Basking Ridge, NJ
BENJAMIN SCHNEIDER, Department of Psychology, University of Maryland
LAWRENCE W. STARK, School of Optometry, University of California, Berkeley
EARL L. WIENER, Department of Management Science, University of Miami
GREG L. ZACHARIAS, Charles River Analytics, Cambridge, MA

ANNE S. MAVOR, Study Director
JERRY KIDD, Senior Adviser
SUSAN R. McCUTCHEN, Senior Project Assistant

Contents

PREFACE ix

EXECUTIVE SUMMARY 1

PART I: BASELINE SYSTEM DESCRIPTION

1 OVERVIEW 17
    Air Traffic Control Operations, 19
    Safety and Efficiency, 21
    The Pilot's Perspective, 23
    Key Historical Events, 25
    Scope and Organization of the Report, 30

2 TASKS IN AIR TRAFFIC CONTROL 32
    Air Traffic Control Organization, 32
    The Tower, 34
    The TRACON, 37
    The En Route Center, 45
    TRACON and En Route: Similarities and Differences, 48
    Central Flow Control, 48
    Flight Service Stations, 51
    Summary, 52

3 PERFORMANCE ASSESSMENT, SELECTION, AND TRAINING   54
Performance Assessment, 56
Selection, 63
Training, 69
Summary, 74

4 AIRWAY FACILITIES   76
Scope of Responsibilities, 77
Equipment Supporting Supervisory Control Operations, 78
Operations, 79
Staffing, 82
Summary, 87

PART II: HUMAN FACTORS AND AUTOMATION ISSUES

5 COGNITIVE TASK ANALYSIS OF AIR TRAFFIC CONTROL   91
Cognitive Model of the Controller's Task, 92
Cognitive Vulnerabilities in the Controller's Task, 98
Moderating Factors, 105
Conclusions, 109

6 WORKLOAD AND VIGILANCE   112
Mental Workload, 113
Modeling Workload, 116
Vigilance, 125
Work-Rest Schedules, Shift Work, and Sleep Disruption, 130
Conclusions, 133

7 TEAMWORK AND COMMUNICATIONS   135
Team Performance Issues, 137
Team-Related Research in Air Traffic Control, 140
Team Training for the Flight Deck, 143
Team Training in Air Traffic Control, 145
Implications of Automation for Teamwork and Communications, 148
Conclusions, 150

8 SYSTEMS MANAGEMENT   152
Approaches to Describing the Context of Air Traffic Control Tasks, 153
Assessing Safety and Efficiency, 155
Formal Organizational Context Variables, 158
Informal Organizational Context Variables, 166
Coordinating Human Factors Research Activities, 172
Conclusions, 174

9 HUMAN FACTORS IN AIRWAY FACILITIES — 177
Effects of Increased Automation, 177
Operations, 183
Staffing, 188
Human Factors Research, 191
Conclusions, 195

10 STRATEGIES FOR RESEARCH — 197
Human Engineering Databases and Literature, 199
Analysis of Controller Responses, 201
Modeling and Computer Simulation, 210
Design Prototyping, 215
Real-Time Simulation, 216
Field Studies, 220
Combining Sources of Human Factors Data, 222
Measurement in Complex Systems, 223
Conclusions, 225

11 HUMAN FACTORS AND SYSTEM DEVELOPMENT — 226
History, Orientation, and Rationale, 227
Formal Arrangements for Incorporating Human Factors, 229
Undertakings with Respect to Air Traffic Control, 231
The Implementation of Innovations, 235
Conclusions, 239

12 AUTOMATION — 241
Forms of Automation, 243
Functional Characteristics, 247
Human Factors Aspects of Automation, 265
Human-Centered Automation, 280
Conclusions, 288

REFERENCES — 290

APPENDIXES

A Aviation and Related Acronyms — 339
B Contributors to the Report — 343
C Biographical Sketches — 345

INDEX — 353

The National Academy of Sciences is a private, nonprofit, self-perpetuating society of distinguished scholars engaged in scientific and engineering research, dedicated to the furtherance of science and technology and to their use for the general welfare. Upon the authority of the charter granted to it by the Congress in 1863, the Academy has a mandate that requires it to advise the federal government on scientific and technical matters. Dr. Bruce M. Alberts is president of the National Academy of Sciences.

The National Academy of Engineering was established in 1964, under the charter of the National Academy of Sciences, as a parallel organization of outstanding engineers. It is autonomous in its administration and in the selection of its members, sharing with the National Academy of Sciences the responsibility for advising the federal government. The National Academy of Engineering also sponsors engineering programs aimed at meeting national needs, encourages education and research, and recognizes the superior achievements of engineers. Dr. William A. Wulf is interim president of the National Academy of Engineering.

The Institute of Medicine was established in 1970 by the National Academy of Sciences to secure the services of eminent members of appropriate professions in the examination of policy matters pertaining to the health of the public. The Institute acts under the responsibility given to the National Academy of Sciences by its congressional charter to be an adviser to the federal government and, upon its own initiative, to identify issues of medical care, research, and education. Dr. Kenneth I. Shine is president of the Institute of Medicine.

The National Research Council was organized by the National Academy of Sciences in 1916 to associate the broad community of science and technology with the Academy's purposes of furthering knowledge and advising the federal government. Functioning in accordance with general policies determined by the Academy, the Council has become the principal operating agency of both the National Academy of Sciences and the National Academy of Engineering in providing services to the government, the public, and the scientific and engineering communities. Dr. Bruce M. Alberts and Dr. William A. Wulf are chairman and interim vice chairman, respectively, of the National Research Council.

Preface

This report is the work of the Panel on Human Factors in Air Traffic Control Automation, which was established in fall 1994 at the request of the Federal Aviation Administration (FAA). The panel was appointed to conduct a two-phase study of the human factors aspects of the nation's air traffic control system, of the national airspace system of which it is a part, and of proposed future automation issues in terms of the human's role in the system. The impetus for the study grew out of a concern by members of the Subcommittee on Aviation of the Public Works and Transportation Committee of the U.S. House of Representatives that efforts to modernize and further automate the air traffic control system should not compromise safety and efficiency by marginalizing the human controller's ability to effectively monitor the process, intervene as spot failures in the software or environmental disturbances require, or assume manual control if the automation becomes untrustworthy. Panel members represent expertise in human factors, decision making, cognitive psychology, organization structure and culture, training and simulation, system design, controller operations, and pilot operations. The primary focus of the study is the relationship between the human and the tools provided to assist in accomplishment of system tasks.

The panel's charge calls for two phases. The first phase focuses on the current air traffic control system and its development and operation within the national airspace system from a human factors perspective. The specific purposes are to understand the complexities of the current system that automation is intended to address, characterize the manner in which some levels of automation have already been implemented, and provide a baseline of human factors knowledge as it relates to the functions of the air traffic controller in the system and the organizational context within which these functions are performed. The second phase is to assess future automation alternatives and the role of the human operator in ensuring safety and efficiency in the air traffic control system. A critical aspect of this second phase is to examine the interaction between the automation and the controller on the ground and the automation and the pilot in the cockpit. Specifically, we plan to project future tasks and examine the consequences of automation on them, assess possible changes in the pattern of controller work and the potential effects on performance, and evaluate procedures needed for the smooth evolution of the national airspace system.

This report provides the results of the panel's deliberations during the first phase. The first part of the report presents a baseline description of the air traffic control system, the selection, training, and assessment of controllers, and the operations associated with keeping the systems and equipment functioning. The second part of the report discusses current knowledge about human factors as it relates to the air traffic controller. We begin this part with human factors principles and findings concerning the cognitive and workload characteristics of the job of the controller, working both as an individual and as part of a team. Then we examine system management, human factors considerations in Airway Facilities, and the integration of human factors research and development into the organization. Finally, the discussion of automation issues serves as a bridge to our work in the second phase. The panel's recommendations concerning human factors considerations appear in the executive summary.

We hope the readers of this report will encompass a broad audience, including those interested in the air traffic control system and its operation and policy as well as those interested in general issues of aviation psychology research and air safety. We direct the attention of our policy readers to the executive summary with our conclusions and recommendations, the chapters on system management and automation, and the final sections of each chapter that contain a brief discussion of the major points covered. Our readers from the research community are directed to the chapter details in Part II.

Many individuals have made contributions to the panel's thinking and to various sections of this report by serving as presenters, advisers, and liaisons to useful sources of information. A complete list of contributors and their affiliations is presented in Appendix B. Although all of these individuals provided us with valuable information, a few played a more direct role in the coordination of information used in the preparation of this volume, and they deserve special mention. We extend our gratitude to several individuals in the FAA and NASA: to Mark Hofmann for frequent and detailed updates on human factors issues and activities within the FAA and for consistent support of the panel's activities; to David Cherry for consistently helpful and timely responses to numerous requests from the panel for documentation and for arranging visits to FAA facilities and discussions with FAA subject-matter experts; to Carol Manning for similarly responsive requests for information and for coordinating presentations to the panel by staff at the Civil Aeromedical Institute and the Mike Monroney training facility; and to Kevin Corker for coordinating presentations to the panel by staff at NASA Ames. We are especially grateful to Neil Planzer, director of Air Traffic Plans and Requirements for the FAA, for informative perspectives on air traffic control historical developments, challenges, and future plans and concepts.

Although this report is the collective product of the entire panel, each member took an active role in drafting sections of chapters, leading discussions, and/ or reading and commenting on successive drafts. In particular, Raja Parasuraman assumed major responsibility for the chapters on automation and on workload and vigilance, Robert Helmreich for the chapter on teamwork and communication, and Paul Stager for the chapter on human factors research methodology. Charles Aalfs, Joseph Pitts, and Richard Stone provided materials reflecting operational expertise that were especially critical for the development of the chapters that describe the air traffic control system and the tasks of air traffic controllers. Tora Bikson contributed sections for the chapters on system management and system development, Marvin Cohen for the chapter on cognitive task analysis, and David Hopkin and Thomas Sheridan for the chapter on automation. Todd LaPorte provided critical conceptual and detailed considerations with respect to the issues of system reliability and organizational context, James Danaher with respect to system safety and efficiency, Earl Wiener with respect to automation and to the management of human factors activities, Diane Damos with respect to the selection of air traffic controllers, and Laurence Young with respect to historical developments in related domains.

Staff at the National Research Council made important contributions to our work in many ways. We would like to express our appreciation to Alexandra Wigdor, director of the Division on Education, Labor, and Human Performance, for her valuable insight, guidance, and support; to Theresa Noonan and Susan McCutchen, the panel's administrative assistants, who were indispensable in organizing meetings, arranging travel, compiling agenda materials, and managing the exchange of documentation across the panel. We are also indebted to Christine McShane, who edited and significantly improved the report, and to Gary Baldwin, who generously shared his wealth of knowledge and experience with respect to FAA organization, policies, procedures, and information sources.

> Christopher D. Wickens, *Chair*
> Anne S. Mavor, *Study Director*
> James P. McGee, *Senior Research Associate*
> Panel on Human Factors in Air Traffic
>   Control Automation

FLIGHT TO THE FUTURE

HUMAN FACTORS IN AIR TRAFFIC CONTROL

Executive Summary

The nation's air traffic control system, which is part of the national airspace system, is responsible for managing a complex mixture of air traffic from commercial, general, corporate, and military aviation. Despite the strong safety record achieved over the last several decades, the system does suffer occasional serious disruption, often the result of outdated and failed equipment. When equipment failures occur, system safety relies on the skills of controllers and pilots. Under these circumstances, safety is maintained by reducing the number of aircraft in the air. Pressures to provide the capacity to handle a greater number of flights in the future and to maintain high levels of safety and efficiency have led to proposals to provide more reliable and powerful equipment and at the same time increase the level of automation in air traffic control facilities—that is, to use advances in technology to take over tasks that are currently performed by humans. Such proposals have raised concern from members of the Subcommittee on Aviation of the Public Works and Transportation Committee of the U.S. House of Representatives that automation not compromise the safety or efficiency of the system by marginalizing the human controller's ability to provide the necessary backup when disruptions occur.

As a result, the Panel on Human Factors in Air Traffic Control Automation was convened at the request of the Federal Aviation Administration (FAA) for the purposes of gaining an understanding of, and providing recommendations on, the human factors characteristics of the current air traffic control system, the national airspace system, and future automation alternatives in terms of the human's role in the system. The panel's charge divides the tasks into two phases. The first focuses on the current system and its development as a means to:

(1) understand the complexities of and problems with the current air traffic control system that automation is intended to address; (2) describe the manner in which some levels of automation have already been implemented; and (3) provide a baseline of human factors knowledge as it relates to the functions of the air traffic controller in the system. The second phase is to assess future automation alternatives and the role of the human operator in ensuring safety and efficiency in the air traffic control system.

This report provides the results of the panel's work during the first phase.

SYSTEM RELIABILITY

The goal of the air traffic control system is to satisfy and balance the two critical goals of safety and efficiency. Human participants in the system must make continuous adjustments in flight scheduling and flight paths to maximize efficiency without compromising safety. The many redundant components in the system, and the smooth communications between its operators (both on the ground, and between ground and air) have generally allowed it to recover gracefully from failures, without accident. Because perfect system reliability can never be assumed, it is important that planners not change the system in ways that will destroy these critical failure-recovery aspects.

Conclusions Despite the complexity of the national airspace system and of its many semiautonomous air traffic control facilities, the system has operated with a remarkably good safety record. However, the skills of air traffic controllers are being increasingly relied on to compensate for the limited capacity and declining reliability of aging equipment. Although new procedures and technologies that represent increases in automation are being considered as means of meeting projected increases in air traffic, human controllers are expected to maintain responsibility for the safe and efficient flow of air traffic for the forseeable future. No matter how well the system is engineered and tested, some level of system unreliability (or some degree of system failure) is inevitable. And some level of human error is also inevitable, so long as human operators remain in the system. Such errors are not likely to be damaging to system performance if they can be caught and corrected by error-tolerant systems.

Recommendation The panel recommends that the FAA expand current efforts to reduce errors by employing good human factors in design and by adopting a fault-resistant and fault-tolerant philosophy of system design. Such a philosophy would render the system less susceptible to catastrophic failures in the case of errors by the human operator (controller, pilot, maintenance specialist) or failures of equipment.

MODELS FOR ASSESSING SAFETY AND EFFICIENCY

The FAA is considering increased application of automation to air traffic control as a means of achieving minimum accidents and maximum efficiency of air travel. However, these goals can potentially be in conflict; the FAA has expressed its resolve that, under such circumstances, safety will remain the number one priority.

Conclusions Unclear definitions of efficiency factors, efficiency measures, and acceptable levels of efficiency, as well as lack of knowledge about the effects of proposed improvements on the cognitive tasks of controllers, inhibit predictive assessment of whether and to what extent proposed improvements to equipment and to procedures—including automated features—will actually contribute to safety or to efficiency.

Recommendations The panel recommends that the FAA specify acceptability criteria for safety and efficiency and foster the development and application of models of the controller and the national airspace system that: (a) clearly identify indicators and measures and make use of the levels of acceptability for safety and for efficiency set by FAA management; (b) assess the interaction of safety and efficiency factors; (c) take into account the cognitive tasks of controllers in balancing the pressures of both safety and efficiency in tactical and strategic decisions and behavior; and (d) take into account the variables associated with different air traffic control options, regions, and facilities. The developed models should be applied to the evaluation of proposed changes in equipment, software, and procedures.

SELECTION AND TRAINING OF CONTROLLERS

Controllers are trained for their duties by a combination of formal classroom instruction and on-the-job training. The selection and screening criteria have varied over the years, with current emphasis on a "train for success" philosophy, designed to reduce training program attrition. As a consequence, the major component of both training and selection takes place within the facilities where developmental controllers (i.e., trainees) receive on-the-job training from full-performance-level controllers, who serve as instructors. Much of this training is received while developmental controllers handle live traffic, closely supervised and evaluated by other controllers.

Job-Related Criteria

Conclusions The FAA has recognized that, in order to select new controllers effectively, the agency must validate its selection procedures against on-the-job performance. Detailed job performance criteria are necessary to enable effective selection, training, and performance assessment of controllers. The FAA has commenced an assessment program with the goal of establishing clear definitions of the tasks and criteria that characterize effective performance.

Recommendations The panel recommends that efforts be continued to develop job-related performance criteria as prerequisites for both personnel selection and evaluation of training procedures. We further encourage the FAA to reexamine controller job tasks and performance criteria when new air traffic control technology, including automation, is introduced; and to coordinate the development of performance criteria with the development of a comprehensive selection battery for controllers, as well as with continued study of the relationship between performance and the personality and demographic characteristics of controllers.

Controller Training

Conclusion Due to reductions in staffing, full-performance-level controllers may not have sufficient time to leave their assigned facilities to receive refresher training or training in use of new or upgraded software or equipment introduced into the current air traffic control system.

Recommendations The panel recommends the accelerated development and the use of computer-based training simulations that incorporate both performance assessment and the particular characteristics of a facility's airspace. These simulations should include a capability to provide augmented feedback for training. SATORI (situation assessment through re-creation of incidents) is a good example of such a capability; it provides a graphic display of data along with synchronized voice that can be used to review performance on a simulation exercise.

COGNITIVE TASK ANALYSIS

Controllers in all types of air traffic control facilities develop strategic plans for traffic flow, monitor these plans with visual inputs to update their "big picture" of the traffic flow, and communicate heavily with pilots and other controllers to ensure continued safety and efficiency. Controllers in the towers depend heavily on direct visual sightings of traffic at the airport, while those in the TRACON (terminal radar approach control around airports) and in the en route environments are supported by computer-based, partially automated radar displays. All controllers must be prepared to deal with unanticipated events—for example, equipment failure, weather emergency, or pilot noncompliance with instructions—in a flexible manner that preserves safety even if it temporarily disrupts efficiency.

Conclusions The technique of cognitive task analysis has revealed several strengths of the skilled air traffic controller, along with a number of vulnerabilities inherent in human information processing. The strengths include the ability to bring experiences stored in long-term memory to bear in solving novel unexpected problems. Weaknesses include vulnerability in detecting subtle and infrequent events, in predicting events occurring in a three-dimensional space, and in temporarily storing and sometimes communicating information. Considerable human factors knowledge exists as to how these vulnerabilities can be addressed by design and training.

Recommendations The panel recommends that changes to air traffic control systems should include not only efforts to retain and capitalize on the controller's cognitive strengths, but also efforts to compensate for weaknesses. Such compensation includes making subtle and infrequent events more prominent, providing explicit (and reliable) predictive displays whenever possible, providing redundant communications and visual backup for working memory when errors can be critical, providing visible feedback for state changes, and using display techniques to improve individual and shared situation awareness, both among controllers and between controllers and pilots.

WORKLOAD AND VIGILANCE

Workload is one of the most critical characteristics of the controller's task. It is driven by objective characteristics of the air traffic control system (e.g., number of aircraft, complexity of sector routes, quality of displays) and is experienced by the controller (e.g., measurable by behavioral or physiological indices). Skilled controllers adapt their performance strategies in handling aircraft as workload increases, in order to prevent excessive workload or loss of safety.

Conclusions Increases in air traffic density and complexity have led to substantial demands on the mental workload of controllers. Very high workload can lower performance and set an upper limit on traffic-handling capacity. Very low workload may result in boredom and reduced alertness, with consequent implications for handling emergencies. Factors influencing controller mental workload include airspace variables, display factors, work team dynamics, and controller-pilot communications. Most controllers use various adaptive strategies to manage their performance and subjective perceptions of task involvement. When evidence relating air traffic control operational errors to performance and workload has been found, the errors have been linked to both low and high task load conditions. Such conditions increase demands on controller monitoring and vigilance, and they could increase in the future as the system becomes more automated. Current work-rest schedules have not been documented to have a negative impact on controller performance, although subjective complaints of fatigue occur. However, shift work and the consequent disruption of circadian rhythms and sleep loss continue to be a major source of concern.

Recommendations The panel recommends that workload assessment span the entire range of air traffic control workload, from low to high (underload to overload). Current performance measures are not adequate to provide indices of performance potential—and hence safety. These must be accompanied by measures of controller workload. We therefore recommend developing predictive air traffic control workload models similar to those used for flight control and management tasks, initiating additional studies to rectify the relative paucity of studies of underload and boredom in air traffic control, and encouraging the scheduling of controller shift and work-rest cycles that will be consistent with the state of the art in research on fatigue, circadian rhythms, and sleep loss. We also recommend that the FAA discourage current shift-work patterns (e.g., phase-advanced shifts and compressed work weeks), which may be associated with degraded performance.

TEAMWORK AND COMMUNICATION

The individual controller is part of an interlocking set of teams. Communication is vital to these team functions. Much communication can be analyzed from an information processing perspective, and much of it depends on both sender and receiver sharing the same mental model or awareness of the situation. But several critical aspects of team communication relate more to the discipline of social and personality psychology. Over the last 20 years, these have been revealed through the study of cockpit resource management on the flight deck. Breakdowns in flight deck crew resource management that resulted in accidents have spawned a series of training programs for pilots that have been successfully implemented in many airlines. These programs have a record of improved safety and improved attitudes toward teamwork. A corresponding program, called air traffic teamwork enhancement (ATTE), has been developed for air traffic controllers.

Conclusions Teamwork, reflected in communication among controllers and their supervisors and between controllers and flight crews, is a critical component of air traffic control. Cockpit resource management (CRM) team training has proved effective in improving team coordination, communications, and task management for aircraft flight crews. Similar training for air traffic controllers, their supervisors, and their trainers has the potential to provide similar enhancement of teamwork. The air traffic teamwork enhancement program, a team training program developed for controllers based on CRM principles, contains positive features, but it does not provide the necessary recurrent training or hands-on practice and reinforcement of team skills. In general, there is a severe lack of research pertaining to teamwork aspects of air traffic control, including teamwork aspects of selection, training, performance appraisal, communication, cognitive behavior, shared situation awareness, workload, and system design.

Recommendations The panel recommends that the FAA initiate a systematic effort to reinforce the value of teamwork within its organizational culture. Ways to do this are to define team coordination as part of task descriptions, to include evaluation of team skills as part of performance assessment, and to consider team factors during investigation of operational errors. We further recommend that an improved program of team training for controllers, their supervisors, and on-the-job training instructors should be a centrally funded program required at all air traffic facilities. The program should include training in controller-to-supervisor interface issues and team-related automation issues. The training should be refined on the basis of empirical data derived from analysis of team issues in operational errors, surveys of controllers' attitudes regarding team issues, behavioral measures of team performance based on simulation, and evaluations of the program by participants. Team training should provide for recurrent training, hands-on practice, and reinforcement of team skills. We further recommend that the FAA initiate efforts to fill gaps in knowledge of teamwork aspects of air traffic control, including research on teamwork aspects of selection, training, performance appraisal, communication, cognitive behavior and performance, workload, and system design.

SYSTEMS MANAGEMENT

Broader than simply teams, the air traffic control system incorporates a much larger organizational context within which the controller works, a context that is characterized by procedures, regulations, a labor-management structure, and performance-based rewards and penalties. All of these factors contribute, in a complex but poorly understood way, to the performance of the controller both as an individual and as a team member. In addition to the organizational structure, which can be documented by formal written procedures, different facilities are also characterized by a less formal, but equally potent organizational culture, defined by attitudes, ways of doing things, and informal delegations of responsibility that may be quite different from formal responsibilities. These differences in culture may have dramatic influences on the ways in which new technologies are received within a facility and the ways in which emergencies are handled.

Effects of Organizational Context

Conclusions Both formal and informal (cultural) organizational context factors contribute to safety and efficiency and are implicated in the successful introduction of new technologies. From this perspective, key aspects of the FAA's formal organizational context include: policies governing safety and efficiency; allocation of authority and responsibility; procedures for selecting, training, managing, and evaluating the workforce; legal liability; labor-management relations; and processes for procuring and implementing new technologies and for introducing changes. The informal cultural context includes such important safety- and efficiency-related factors as: informal rules and norms, subculture differences, job satisfaction, and attitudes toward change. Organizational culture affects and is affected by organizational structure.

There is a lack of research data that would permit identification of the specific mechanisms by which formal and informal organizational contexts within the FAA interact and how they affect organizational climate and controller performance. Studies to date have shown, however, that during high-tempo conditions (represented by such events as responding to very high traffic loads, large-scale outages, and weather contingencies) informal teaming arrangements, leadership roles, and procedures are often invoked; these informal responses may help to account for the remarkable safety record demonstrated to date during such conditions.

Recommendations The panel recommends that the FAA initiate a systematic and comprehensive research program to study the effects of introducing new air traffic control technology, including automation, on the accomplishment of safety and efficiency goals. This program should include attention to the ways in which both formal and cultural organizational context factors contribute to or detract from the effective introduction of the new technology and, conversely, the ways in which the FAA's organizational context may be modified as a consequence of the new technology. The FAA should conduct further research to clarify, within the context of air traffic control, the ways in which formal and informal organizational context factors affect one another and the ways in which they both affect the performance of controllers. Specific study should be undertaken of informal teaming and procedures during high-tempo operations with the goal of eliciting recommendations for ways in which formal procedures and organizational structures may be improved.

Job Satisfaction of Air Traffic Employees

Conclusions The results of the FAA's biennial employee attitude survey (EAS) provide information about organizational culture and climate, including employees' perception of the extent to which the formal structure and practices meet their needs. The results of the most recent survey (1995) indicate that air traffic employees generally consider their jobs to be satisfying, but they are generally dissatisfied or only moderately satisfied with management practices and with the organizational context within which they perform their jobs. Results also indicate that a significant proportion of FAA employees are reluctant to express dissatisfaction or disagreement to their management; communication difficulties may therefore be influencing other indicators of job satisfaction.

Recommendations The panel recommends that the FAA utilize its employee attitude survey as a useful source of supporting information for studies of how formal and informal FAA organizational contexts affect one another and how they can affect controller performance. The surveys could be used to support detailed study of discrepancies between managers' and controllers' perceptions of organizational context factors; ways in which controller-management communications could be improved; ways in which different air traffic control options and geographic locations may form different subcultures with different perceptions; and ways in which the introduction of new technology, including automation, is perceived by controllers.

Managing Human Factors Activities

Conclusions Human factors activities within the FAA, including both research and practice activities, are extensive but fragmented. Research activities are not adequately coordinated across research centers, research is not always systematically performed to support system design needs, and research findings are not systematically applied. System design activities are sometimes uncoordinated and rely heavily on subcontractor efforts that are not managed by FAA human factors professionals.

Recommendations The panel recommends that the FAA focus the overall management of human factors research and development activities for the agency and provide the necessary authority, responsibility, and resources to ensure that such activities are systematically conducted and applied; that they are adequately coordinated across research centers (including government research organizations such as the National Aeronautics and Space Administration, the Department of Defense, and the MITRE Corporation); and that human factors contractor efforts are effectively overseen by FAA human factors professionals. We understand that the issue of human factors management within the agency is currently being addressed by a subcommittee of the FAA's Research, Engineering, and Development Advisory Council.

HUMAN FACTORS PERTAINING TO AIRWAY FACILITIES

Airway Facilities specialists are critical partners of air traffic controllers with respect to supervisory control and restoration of the air traffic control system. They are responsible for such critical tasks as system monitoring, diagnostics, certification, maintenance, and restoration of equipment, systems, and services after outages.

Conclusions To date the introduction of new equipment has involved far more automation of Airway Facilities functions than of air traffic control functions. By comparison with the human factors efforts that have been devoted to air traffic controllers, attention to human factors characteristics of Airway Facilities operations, personnel, software, and equipment has been meager. Airway Facilities personnel are currently faced with a bewildering array of equipment and workstation devices that are provided by different vendors, apply different levels of automation, and present different computer-human interaction procedures and characteristics.

The trend toward centralized maintenance control centers has not alleviated this concatenation of equipment and workstation devices, which does not reflect the effective application of human factors analysis, design support, or evaluation. The decreasing reliability of national airspace system equipment, an associated requirement to develop creative workarounds, the lack of workstations designed using human factors principles, and a progression of large numbers of Airway

Facilities personnel toward near-term eligibility for retirement threaten to overwhelm efforts to continue to maintain the operation of the system. The recent creation of the GS-2101 automation specialist job classification acknowledges that Airway Facilities specialists are increasingly required to serve as systems engineers, but the job classification has not been accompanied by the development of validated selection or assignment procedures, performance assessment procedures tied to clear job criteria, or tailored training programs. Although the FAA has recently produced a human factors design guide applicable to Airway Facilities, the paucity of research and analysis to support continued maintenance and updating of knowledge bases for human factors applications to Airway Facilities is of deep concern.

Recommendations The panel recommends that the FAA significantly expand its application of human factors research, analysis, design, and test and evaluation activities to Airway Facilities. These activities should be directed toward the development of integrated workstations, teamwork and communication strategies, and procedures for selecting, assigning, training, managing, and assessing the performance of Airway Facilities specialists.

STRATEGIES FOR RESEARCH

Human factors research should support long-range innovation by evaluating developmental concepts and should also serve in the solution of immediate design problems by assessing specific design options. To fulfill these imperatives, human factors researchers need to employ a wide array of information resources and study methods. The invention of methodological refinements and the precise determination of which methods to marshal for each research question as it arises confront the human factors research community as continuing challenges.

Conclusions The inherent complexity of the air traffic control system and the unpredictability of system error mean that all potential sources of human factors engineering data must be used in research and system design activities. In many instances, different methods or approaches to data collection must be combined in an integrated collection process, with one method, set of tools, or data source often complementing another. Most methods have inherent constraints or limitations. For example, databases of design guidelines do not always address future design issues; interpretation of accident analyses may be ambiguous; the databases of reporting systems are sometimes difficult to access and integrate and are not always available in user-friendly form; user opinions, either from surveys or from rapid prototype evaluations, are often biased; many air traffic control models are not validated, and existing models focus more on system performance than on human performance; studies using simulations must trade off precision and complexity with expense; and field studies impose particular constraints on the experimental control of variables. In all behavioral analysis techniques (i.e., all of the above except for modeling), there is a need to assess representative samples of users across all levels of expertise.

The representativeness of participating controllers, operational scenarios, and determinant variables ultimately determines the validity of the assessment data that are collected in air traffic control research. Contemporary advances in computer simulation have made it possible to undertake economical rapid prototyping in order to address most design questions, although such prototyping should augment, rather than substitute for, comprehensive real-time simulation with actual performance data.

Recommendations In order to overcome the current constraints on the usability of air traffic control reporting systems, the panel recommends accelerated efforts to provide user-friendly access to the aviation safety reporting system and other reporting systems. Human performance models should be developed and validated to support research and design activities in air traffic control system development. Part of the modeling effort should be to address the articulation of universally recognized quantifiable dimensions of controller performance, including dependent variables that define performance across the range of operational contexts.

The panel recommends systematic work to formalize the role and enhance the contribution of rapid prototyping in determining the characteristics of human-computer interaction. Whenever feasible, a cost-effective simulation capability should be included within design programs that will enable progressive acceptance testing and assessments of the risk of poor design, which may engender operational problems and necessitate costly redesign, to be carried well beyond preliminary rapid prototyping and into design validation activities. Early field testing should be conducted as a means of further mitigation of the risk of poor design. Additional human engineering standards and guidelines should be developed for design validation during system development to ensure that determinants of controller (and system) reliability are adequately addressed prior to system implementation.

SYSTEM DEVELOPMENT

Successful human factors programs closely link research and development activities to ensure that research activities are responsive to developmental needs and that research findings are applied in product development, testing and modification. Several factors are critical: extensive user input into the design process at all stages, often employing rapid prototyping; extensive involvement of human factors practitioners, who are knowledgeable about human factors design guidelines and about appropriate assessment techniques that capitalize on users' expertise; frequent opportunities for behavioral testing (not just expert opinion) of interfaces, and for refinement of those interfaces at several points throughout the design cycle; and sensitivity to the special needs, wishes, and organizational cultures of different facilities at which technology will be ultimately introduced.

Conclusions Although specifications for new systems typically detail the functional and performance criteria for equipment, the human factors design specifications and guidelines, traditionally applied as part of the definition of system requirements, do not detail the functional and performance criteria for the human controllers who also constitute critical elements of the system. The evolution of human-computer interaction characteristics of new systems therefore often relies on user evaluations of these characteristics as the design progresses. User participation, however, is not a substitute for the expert knowledge of the human factors specialist; rather, the two should function as complementary teammates during the acquisition process. One example of an apparently successful form of this team relationship is the center-TRACON automation system (CTAS), in which a key feature is the continuous involvement of customers and users in both the design and evaluation process. The result has been a useful set of tools for projecting and automatically sequencing aircraft approaches to airports that has been well accepted by controllers. Human factors analysis, test, and evaluation activities are important for both developmental and nondevelopmental items, including commercial-off-the-shelf items; the application of such items may involve both trade-offs of capabilities and transition issues that affect controller tasks.

Recommendations The panel recommends that the FAA include both representative users (e.g., controllers and maintainers) and human factors specialists on product acquisition and development teams. Also, they should systematize user inputs to the design process according to human factors procedures for performing analyses and for conducting prototype and pilot trials to inform user requirements and assess the likely impacts of design features on controller task performance and workload.

Prototyping, simulation, and modeling exercises—carefully designed by human factors specialists and supported by user participation—should be applied to assist in making design trade-offs and to distinguish between user preference and usability. The FAA should use such studies to refine and tailor existing research databases and design standards and to coordinate studies and results with operational data and recommendations derived from the FAA's safety databases and systems effectiveness databases. Human factors analysis, test, and evaluation activities should be applied, for a given application, during the acquisition process for nondevelopmental and commercial-off-the-shelf items to ensure that they are compatible with the capabilities and limitations of users.

AUTOMATION

There has recently been a great deal of discussion of the concept of human-centered automation, which is viewed by many as a critical issue to the successful introduction of automation. Unfortunately, however, there are many different attributes that researchers have identified with this concept, and not all of these are consistent with one another. Furthermore, not all of them are necessarily consistent with the goal of attaining the best (i.e., safe and expeditious) system performance. Detailed consideration of the human centered automation concept as it applies to air traffic control will form the core of the second phase of our work.

Conclusions A number of components of automation have been introduced into the air traffic control system over the past decades in the areas of sensing, warning, prediction, and information exchange. These automated systems have provided a number of system benefits, and the attitudes of controllers have generally been positive. There are also a series of lessons that have been learned from other domains about the appropriate and inappropriate implementation of automation as it affects the human user or supervisor of that automation. Of particular concern is the human being's possible loss of alertness and awareness of automated functions and system functioning, which may become critical if sudden manual intervention is necessary. Humans may distrust the automation because they fail to understand its complexities, and it is possible that reliance on automation may lead to a loss of human proficiency in the skills that the automation replaces.

Recommendations The panel recommends that lessons learned from other domains be carefully heeded in the further introduction of air traffic control automation and that research be pursued to establish the generalizability of the research findings to the air traffic control domain.

Part I

BASELINE SYSTEM DESCRIPTION

1

Overview

The American airspace system is impressive in both its capacity and its safety. Still, during such events as severe weather and recent power outages at the air traffic control centers around New York City, Chicago, and Pittsburgh, we realize the vulnerability of the system's capacity and its complete dependence for safety on the skilled coordination of air traffic control and flight deck personnel. This sense of vulnerability is heightened by the outdated technology underlying much of the physical equipment that controllers must use (Stix, 1994) and the chronic shortage of personnel at many facilities. Nevertheless, there are severe pressures to stress the system still further by pushing for more capacity and to fly in even more degraded weather conditions, while still maintaining and improving the current standards of safety.

In order to meet these demands, many have argued that the level of automation in air traffic control facilities should be increased, to keep pace with the rapid development of automation in the flight deck and with the developing availability of satellite-based navigational technology. But as we have learned from other domains, automation—the replacement of human functioning by that of machines—is a mixed blessing (Wiener, 1995). It can sometimes create more problems than it can solve, and these issues lie very much at the heart of this report.

Public safety considerations raise particular concern about the possibility that automation will marginalize the controllers' tasks to a point at which they can no longer effectively monitor the process or intervene when system failures or environmental disturbances occur. As a consequence of these concerns, members of the Subcommittee on Aviation of the Public Works and Transportation Committee of the U.S. House of Representatives suggested that the Federal Aviation Administration (FAA) ask the National Academy of Sciences/National Research Council to undertake an independent assessment of the automation in the current and future air traffic control system from a human factors perspective. In fall 1994, the National Research Council established the Panel on Human Factors in Air Traffic Control Automation.

The panel's charge is to conduct a two-phase study to describe the current system and assess future automation alternatives in terms of the human's role in the system and its effects on total system performance. The first phase involves a review and analysis of the development and operation of the current system; the second phase focuses on the system of the future. Each phase includes two distinct tasks: one concerns the operation of the air traffic control system, and the other concerns the integration of the system with the larger national airspace system. Consideration of the organizational issues and institutional adjustments that could strengthen the role of human factors science in the automation of the system is also included in each phase.

The panel adopts a specific interpretation of the term *automation*. As noted above, we define it to mean the replacement by machines (usually computers) of tasks previously done by humans. We explicitly distinguish this term from that of *modernization*, which includes any and all upgrades of air traffic control technology, including those, such as the installation of new radar systems and computers, that do not substantially alter the controller's job (except perhaps by offering more reliable data). When we use the term *human-centered automation*, we are referring to a philosophy that guides the design of automated systems in a way that both enhances system safety and efficiency and optimizes the contribution of human operators.

A key concept underlying the panel's work, as well as recent events that have occurred within the air traffic control system, is that of *system reliability*. Formally, this concept can be expressed either as a time measure (mean time between failures) for continuously operating systems or as a probability measure (1 – number of failures/number of opportunities) for systems or components associated with discrete events. In this report, we make the important distinction between *reliability* and *trust*.

Designers continuously seek ways to make system reliability as high as possible, often by striving to meet specific FAA procurement requirements. Whereas this appears to be an admirable goal, absolute reliance on specified reliability should be treated with some caution, for three reasons. First, like any estimate, a reliability number has both an expected value (mean) and an estimated variance. Yet the variance is often ill defined and hard to estimate. When it is left unstated, it is tempting to read the offered reliability figure (e.g., $r = .999$) as a firm promise rather than the midpoint of a range.

Second, objective data of past system performance reveal ample evidence of systems whose promised level of reliability greatly overestimated the actual reliabilities. For example, a recent outage of the newly installed voice switching and control system (VSCS) used hundreds of years of its specified nonavailability (C. Grundmann, personal communication, 1995). There is little reason to doubt that this projected level of overconfidence will persist, given the inherent human tendencies toward overconfidence in forecast estimation (Fischhoff, 1982).

Third, experience also reveals that it is next to impossible to forecast all inevitable circumstances that may lead a well-designed system to "fail," even given the near boundless creativity of the system engineer. One example of this is the failure of the triply redundant hydraulics system in the United Airlines Sioux City crash in 1989. Another example is the near impossibility of debugging the entire set of software codes underlying functioning of the Airbus A320.

In short, we believe that it is impossible to bring the reliability of any system up to infinity, or even to accurately estimate that level (without variability) when it is quite high, and this has profound implications for the introduction of automation. That is, one must introduce automation under the assumption that somewhere, sometime, the system may fail; system design must therefore accommodate the human response to system failure.

In contrast to system and human reliability, *trust* refers to people's belief in the reliability of a system. Hence trust may accurately correspond to reliability or not. Miscalibration can involve either *overtrust* (complacency) or *undertrust*. We further distinguish two sorts of trust: the trust of the air traffic control specialist in the components of the system under his or her control and supervision, and the trust of the flying public in the ability of the system to transport safely and efficiently.

AIR TRAFFIC CONTROL OPERATIONS

The task of air traffic control includes several phases: ground operations from the gate to the taxiway to the runway, takeoff and climb operations to reach a cruising altitude, cross-country flight to the destination, approach and landing operations at the destination, and finally, taxi back to the gate (or other point of unloading). Figure 1.1 is a generic representation of these phases. The traffic to be controlled includes not only commercial flights, but also corporate, military, and general aviation flights (some in the latter class may choose, in good weather, to fly in unrestricted regions of the airspace without the benefit of positive air traffic control).

Control is accomplished by three general classes of controllers, each resident in different sorts of control facilities. First, ground and local controllers (both referred to as tower controllers) handle aircraft on the taxiways and runways, through takeoffs and landings. Second, radar controllers handle aircraft from their takeoff to their cruising path at the origin (departure control) and return them through their approach at the destination (approach control), through the busy airspace surrounding airport facilities. This region is referred to as a terminal radar control area or TRACON. Third, en route controllers working at the air

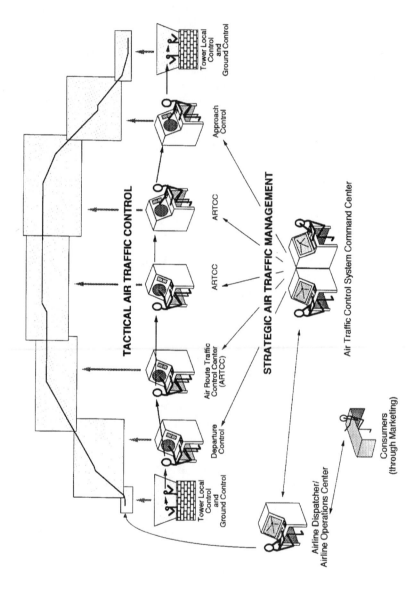

FIGURE 1.1 Phases in air traffic control operations. Source: Adapted from Billings (1996a). Reprinted by permission

route traffic control center (ARTCC) manage the flow of traffic along the airways between the TRACON areas. Overall flow of aircraft across the entire United States is managed by the Air Traffic Control System Command Center in Herndon, Virginia. Separate elements of the air traffic control system are also represented by oceanic control for overseas flights and by military controllers when military aircraft are flying within special-use airspace.

SAFETY AND EFFICIENCY

The stated goal of the air traffic control system is to accomplish *the safe, efficient flow of traffic from origin to destination*. The joint goals of safety and efficiency are accomplished by controllers through an intricate series of procedures, judgments, plans, decisions, communications, and coordinated activities. The communications and coordinations between the pilot and the controller are most familiar to the public. However, every bit as critical are the coordinations that take place within and between the air traffic control facilities themselves. Controllers must hand off aircraft as they pass from one controller's sector of responsibility to another. This handoff communication is sometimes done within a facility and sometimes between them. Also, hierarchical communications flow from the most global, national perspectives to more regional and local ones. That is, the Air Traffic Control System Command Center in Virginia considers national weather patterns and traffic needs each day and establishes national traffic patterns. The constraints established there are passed downward and outward throughout the system. Hour by hour, traffic patterns are monitored in the en route systems and may be used to identify bottlenecks, which in turn may give rise to specific instructions to hold aircraft from proceeding from one sector to the next.

The two goals of safety and efficiency are to some extent partially contradictory, and each is subject to tremendous pressures. We describe each in detail below.

Safety is ensured, in large part, by guaranteeing minimum separation between aircraft, a separation defined by altitude and lateral dimensions, creating a sort of "hockey puck" of space around each aircraft. These dimensions have different values in different regions of the airspace. The pressures for safety obviously come from the traveling community and are increased by reports of very rare midair collisions (Wiener, 1989) and somewhat less rare near-midair collisions (Office of Technology Assessment, 1988).

Of course, to ensure total safety, aircraft would never fly; to ensure a greater safety level than we have today, separations between aircraft would be greater than is currently the practice. However, that would compromise the second goal: efficiency. Two forces put strong pressures on the system for efficient flow: consumers and pilots. The traveling public, whose wishes are generally expressed by airline management, is understandably impatient with overbooked flights, ground holds, and delayed arrivals. Pilots, too, are understandably anxious to make the flight time from gate to gate as short as possible, provided that safety is not compromised. As shown in Figure 1.1, within the national airspace system, this pressure is directly expressed to the air traffic control system via the airline dispatchers who are heavily responsible for adhering to published arrival and departure times.

The specific manner in which the goal of efficiency is met, thereby maximizing the capacity of the current airspace, is a more complex and constraining process than meeting the goal of safety (minimum separation). The limiting factor to capacity maximization is usually the rate of arrivals at an airport, particularly at the large hubs. The constraints at the hubs are dictated jointly by the number of gates, the number of runways, and the speed with which aircraft can exit the runways. Arrivals, more than departures, represent the limiting factor.

Every airport has a capacity in terms of number of aircraft it can receive per unit of time. The goal of the air traffic control system is to meet that capacity (to optimize flow) by delivering airplanes, at regularly spaced intervals, to line up for the final approach. Several factors conspire to prevent this goal from being achieved, causing the system to underutilize its maximum capacity.

(1) Controllers cannot normally "stack" aircraft at the arriving TRACON, to be delivered as soon as a slot is available. So schedule departures and speed changes must be scheduled far upstream in strategic plans, in anticipation that the capacity limits will be realized when the aircraft approach their destination. But weather, head winds, and other uncertain conditions may influence a flight schedule well before the airplane reaches its scheduled point near the final approach. Optimization is limited by the limited ability of all elements within the air traffic control system to predict the future.

(2) Wake vortices, particularly following the passage of heavy aircraft, force the controller to maintain greater separation on the final approach for some aircraft.

(3) Sudden changes in weather may force changes in the configuration of airports—for example, closing or reversing runways, slowing taxiing.

A host of system design efforts at all levels are intended to counteract these bottlenecks to efficiency (Federal Aviation Administration, 1996). On the ground, although it is unlikely that newer, larger-capacity airports will be built in the near future, more realistic possibilities exist for expanding the capacity of existing airports by building added runways. Efforts are also under way to allow the more efficient use of existing runways by expanding the opportunities to use parallel or converging runways for landing. The National Aeronautics and Space Administration has a program of research on terminal area productivity (Eckhardt et al., 1996), to allow more rapid exits from runways after landing and more rapid taxiing to the gates, particularly in bad weather. Increased developments of heads-up displays and automatic landing (autoland) systems are designed to allow more aircraft to land in poor visibility. Perhaps the biggest pressure for efficiency has been exerted on the air traffic control system itself, to increase its efficiency through equipment upgrades, in a manner that may challenge the cognitive capacities of the individual controller to the utmost. This important issue is a key theme in the next three chapters.

The typical controller is able to address the sometimes-conflicting pressures for safety and efficiency in two ways: (1) by adhering to a well-developed and extensive set of FAA procedures that have evolved over the years and (2) by being able to augment them with skilled problem solving on the infrequent occasions when following procedures fails to specify the appropriate actions. Understanding how the system came to evolve as it is today and appreciating the current pressures can be facilitated by considering the pilot's perspective, as well as some key historical events that have occurred in the evolution of the national airspace system.

THE PILOT'S PERSPECTIVE

Because of the somewhat differing perspectives of pilots and controllers, their views on the best tactics to achieve their mutual goals of safety and efficiency are not always the same. Most critically, the air traffic controller has a number of aircraft to deal with, whereas the pilot is concerned with only one. The air traffic control system, of which an individual controller is only one part, is spread over a large area and must be managed so that aircraft cross over or under each other safely. The commercial pilot, generally reflecting the goals of the airline dispatcher, would like to fly the aircraft in the most efficient manner by choosing the most direct route (a straight course or a great circle arc) and at the aircraft's optimal altitude. This ideal course is not always compatible with the constrained routes available. This situation had led the FAA to allow greater flexibility for commercial aircraft to fly preferred routings at high altitudes and to fly great circle routes under its expanded national route plan (NRP) program.

Another difference is that the pilot's goal of efficiency is not always in harmony with that of the controller. The controller's goal is to maintain the maximum evenly spaced flow of all aircraft from airport to airport, even if this means that a given aircraft must slow down or do an extra turn.

Aircraft automation is also a potential source of conflict. Many newer aircraft employ very sophisticated systems. The flight management system (FMS) is one such system; based on temperature, wind, and the weight of the aircraft, the system can select the best altitude for the aircraft to fly. Although many aircraft may prefer the same altitude, the controller assigns it to the first aircraft departing along that route. If it were permissible, the aircraft's flight management system would constantly change altitudes throughout the flight in order to select the best altitude: following takeoff, the aircraft would commence a slow climb as fuel was burned off, climb until it reached its optimal altitude, cruise while seeking the optimal altitude to minimize head winds and maximize tail winds, and then start a steep descent near its destination. However, the most efficient method for operating one jet aircraft is not necessarily compatible with the controller's need to maintain safe and expeditious flow control across all the aircraft aloft.

In addition to the flight management system, other forms of automation can amplify the differences in perspective between pilots and controllers. For example, the traffic alert and collision avoidance system (TCAS), described more fully in Chapter 12, allows pilots to initiate a traffic avoidance maneuver without direction from air traffic control. If not carefully implemented and if prior notification is not given to the controller, such a maneuver can severely disrupt the overall traffic flow plan (Mellone and Frank, 1993).

The goals and tactics of pilots and controllers generally coincide as the destination airport is approached, although sometimes a controller's request that the pilot make a steep, last-minute descent or a visual approach can conflict with the pilot's sense of safety, comfort, or both (Monan, 1987). As aircraft near their destination, it becomes necessary for the controller to become more involved in maintaining their precise trajectory. This is usually accomplished by directing the aircraft to follow a set of orders. The controller uses a radar display to adjust the distance between aircraft so that a stream of aircraft flows toward the runway that is being used for landing.

Although aircraft automation may provide some sources of tension between pilots and controllers, there is no doubt that, on the whole, both the flight management system and the traffic alert and collision avoidance system have been well received by both communities and can be viewed as safety enhancing. Pilots greatly appreciate automation that provides them with more integrated information, although they are less pleased with some features of the more complex forms of automation that involve extensive programming and reprogramming (Wiener, 1989). Before-flight programming has added significantly more time to the preflight actions required by the flight crew, as much as an additional 15-20 minutes. And in-flight reprogramming of the system can take precious minutes in an already high-workload, dynamic situation.

Complex cockpit automation such as the flight management system may also "hide" the logic behind its control of the aircraft's trajectory in ways that pilots do not always understand, leading them to feel that they are "out of the loop" (Sarter and Woods, 1995). Finally, there is the very real danger that automation can be trusted too much, leading to a sense of pilot complacency (and resulting failure to monitor the automated device). Overtrust and overreliance may even result in a potential loss of manual flying skills.

In a very different manner, flight deck automation has the capability of further improving national airspace system safety if lessons (both good and bad) that pilots have learned from their automated devices can be transferred effectively to the implementation of air traffic control automation. Examples of these lessons are discussed more fully in Chapter 12.

These and other human factors concerns regarding the national airspace have evolved from events and developments that have taken place over the last half century, some of which are described below.

KEY HISTORICAL EVENTS

Since the introduction of radar in the late 1940s, drastic changes have occurred in the national airspace system. Some changes have resulted from technological developments (e.g., the introduction of radar, sensor technology or global positioning system). Others have been more abrupt, resulting in part from analysis of catastrophic accidents. Had these accidents never happened, the changes might never have taken place. The list below is an approximate chronology of key events, most of them accidents, that have transformed the national airspace system since the 1950s to its current state. Some, but not all, have direct implications for air traffic control. Others have broader implications for air safety in the national airspace system. Many of the advances in air traffic control technology that have led to the evolution of the national airspace system are described in subsequent chapters. Figure 1.2 represents these events on a time line, along with certain key developments or policy changes that have been instigated as a result.

1. On June 30, 1956, a United Airlines DC-7 and a Trans World Airlines (TWA) Constellation collided over the Grand Canyon. The TWA aircraft was operating at 21,000 feet with an "on top" clearance, and the United one was cleared under instrument flight rules at 21,000 feet. Because of this accident, a radar positive control system was implemented with the requirement that all air carriers operate under instrument flight rules. It has been said that this accident stimulated more action to modernize the air traffic control system than any other single occurrence (Luffsey, 1990).

2. During an approach to Miami International Airport, all three members of the flight crew of Eastern Airlines L-1011 were distracted by a landing gear warning light and none recognized that the autopilot had disconnected and the aircraft was descending until only a few seconds before the crash. The accident served as the first prominent example of two critical problems that had great impact on later human factors developments in aviation: the problem of complacency with an automated system and the problem of crew resource management (Wiener, 1977). As we discuss in subsequent chapters, both issues have direct relevance for the evolution of air traffic control.

3. In 1974, a TWA flight on approach to Dulles International Airport crashed into a mountainside, the flight crew unaware of the mountain's presence in the forward flight path. The accident led the FAA to greater concern for problems in communication ambiguity between ground and air. In the air traffic control

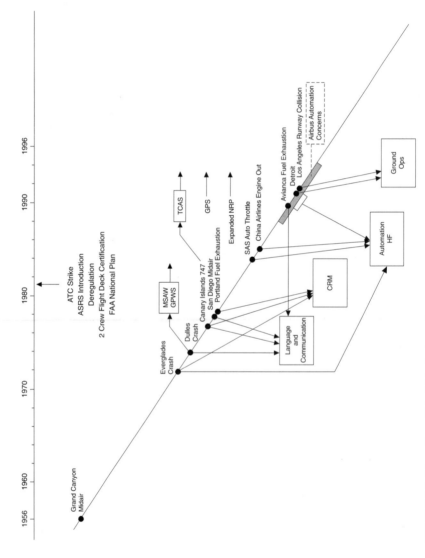

FIGURE 1.2 Time line of key historical events.

facility, it provided a trigger for the introduction of the minimum safe altitude warning, an early form of automation. The accident also provided a major impetus for the introduction of the ground proximity warning system into the cockpits of all transport aircraft, a procedural change that had a profound and positive influence on air safety.

4. In 1977, two Boeing 747s collided on a runway at Tenerife in the Canary Islands, the worst commercial aircraft disaster in history. As with any major accident, several factors were involved, but a critical one was the misunderstanding in voice communications between the controller and the captain of one of the aircraft. As a consequence of this disaster, major efforts were made to standardize communications procedures in international airspace.

5. The San Diego midair collision of a PSA B-727 and a Cessna 172 on September 25, 1978, again illustrated ground-air communications problems and also heightened the need for an airborne collision avoidance system for pilots. (The ground-based conflict alert system for controllers had been introduced in 1976.) The Cessna was climbing under air traffic control and the B-727 was making a visual approach to the San Diego airport. The B-727 crew stated they had the Cessna in sight but had misidentified the target in the busy southern California airspace. The visibility was limited due to hazy conditions. The National Transportation Safety Board found that the PSA pilots did not maintain adequate visual separation, and the controller was cited for allowing an aircraft to use visual separation only. A consequence of this accident in part led to accelerated development, testing, and implementation of the traffic alert and collision avoidance system, now a feature on all commercial aircraft.

6. A United Airlines DC-8 crashed on an approach to the Portland, Oregon, airport on December 28, 1978. The cause of the accident was fuel exhaustion. Because of a malfunction of the landing gear warning system, preparation for an emergency landing preoccupied the captain and, despite warnings by other crew members about the low fuel state, he delayed the landing. The poor use of resources by the captain instigated a reevaluation of the organization of the cockpit. Researchers identified and analyzed other accidents that were mainly related to the chain of command, crew coordination, management style, and team-building elements of airline cockpit (Cooper et al., 1980; Murphy, 1980; Foushee, 1984:Chapter 7). This effort gave rise in the 1980s to a new requirement by the FAA to promulgate optimal cockpit management. It is now generally described as cockpit resource management or crew resource management and has become a training course for pilots. The essence of resource management training has found its way into flight attendant, air traffic controller, and maintenance training as well as operations outside the airline industry. Its positive benefits on air safety have been well documented (Diehl, 1991).

7. During the period around 1980, four nonaccident events took place that have had a profound influence on the national airspace system. (a) In 1978, deregulation of the airline industry created both a decrease in ticket prices (lead-

ing to a much greater demand for flying) and a change in flying routes, as airlines formed hubs at some major cities. These factors combined to create much greater traffic congestion in certain regions of the airspace. (b) In 1981, an air traffic controller strike led to the firing of much of the air traffic control workforce, creating a shortage of trained personnel to manage the increased traffic and exacerbating the stresses on the system. This shortage remains today. (c) In 1980, the aviation safety reporting system was developed at NASA Ames and implemented by the FAA, providing valuable insight into the nature of incidents in both the flight deck and the air traffic control facility. This database has provided valuable indicators of airspace hazards. (d) In 1981, the FAA certified the two-person flight deck on a new generation of Boeing and McDonald Douglas aircraft, triggering a modest revolution in aerospace automation and further development of the flight management system (Billings, 1996b).

The introduction of higher levels of flight deck automation in the 1980s has produced its own set of accidents that have revealed certain human factors problems, previewed by the 1972 Everglades crash—number 2 above (Wiener and Curry, 1980). Although none of these accidents directly influenced the national airspace system or air traffic control procedures, all have led to the evolution in thinking about the role of humans and automation in flight safety.

8. The downing of Korean Airlines flight 007 demonstrates a major concern with automation. On September 3, 1983, a Soviet fighter shot down this B-747 over Sakhalin Island. The aircraft had transgressed the boundary of the Soviet Union without permission. The Korean flight crew had inadvertently left the inertial navigation system in a heading mode, which allowed the autoflight system to maintain a constant heading rather than a programmed track. Complacent in their belief that the "dumb and dutiful" automation system was correctly doing its job, they apparently failed to monitor their position in the airspace.

9. On February 28, 1984, a Scandinavian Airlines DC-10 skidded off the runway at John F. Kennedy Airport in New York. The autothrottle system, a means of automatically controlling the speed of the aircraft, was in use during the landing approach. Although the system had a history of malfunctioning, the crew did not override the overspeed of the aircraft. This was one of the first cases of automation gone awry. The crew was required by the airline to use the autothrottles for this approach, and they had placed too much trust in the automatic system.

10. A China Airlines B-747 flying at 41,000 feet on February 19, 1985, lost power to its number 4 engine. It was night and the crew was operating with the autoflight engaged. The autoflight system compensated for the power loss by a complicated combination of control inputs. When the captain finally became aware of the problem and disconnected the autoflight system, the airplane entered a spiral that could not be corrected until the aircraft had fallen over 30,000 feet. Again, overreliance on automation was cited as the cause of the incident.

11. A series of accidents involving Airbus Industrie aircraft have been at-

tributed to failures of automation or, more often, to the design of automation systems; that is, the full range of operations and constraints under unusual circumstances were poorly understood by the pilot (*Aviation Week and Space Technology*, January 31, 1995; February 6, 1995). As one example, during an approach to a runway at Nagoya, Japan, the copilot of a China Airlines A-300 inadvertently selected an automated go-around mode but attempted to continue flying the airplane to a landing. The autoflight system attempted to climb and increased engine power, but the copilot overrode the control in an attempt to land. The aircraft went out of control when the copilot could no longer overpower the autopilot, which had trimmed the aircraft for maximum climb. In 1992, another A-320 crashed on approach to the Strasbourg airport when the pilots inadvertently selected a high-angle descent rather than the standard 3-degree glide path. Redesign of the display was made subsequent to the accident. Collectively, these and other accidents described in greater detail in *Aviation Week and Space Technology* (January 31, 1995; February 6, 1995) have caused human factors researchers and aircraft designers to take a very hard look at the appropriate levels of flight deck automation and, in particular, important human factors lessons that should be learned in the introduction of any automation system. The relevance of these lessons to air traffic control is discussed in detail in the chapter on automation.

12. On January 25, 1990, an Avianca B-707 crashed during an approach to the John F. Kennedy Airport in New York. The weather had caused delays and holding. The crew had warned the air traffic controllers that they were low on fuel but failed to impress them with the seriousness of their dilemma. This accident, like that at Tenerife years before, demonstrated the weakness of voice communications that take place in other than one's native language. Many factors were present in this accident as in the Tenerife one, but basic was the inability of the crew to communicate effectively with the controller. Standardized, understandable voice phraseology was adopted by the FAA subsequent to this accident.

13. Finally, two accidents have called attention to the vulnerability of the national airspace system in ground operations. In 1990, two Northwest Airlines aircraft, a DC-9 and a B-727, collided on a fog-shrouded runway in Detroit. The crew of the DC-9 was cited for its taxiing onto an active runway without clearance. The airport was cited for the nonconformance of its signs and taxiway lights. The tower was cited for its lack of clear taxi instructions, knowledge of a problem intersection without adequate safeguards, and failing to broadcast a stop takeoff message when it was found that an aircraft had taxied onto the takeoff runway.

On February 1, 1991, a USAir B-737 collided with a Skywest Metro at the Los Angeles airport. The accident occurred at night while the Metro was awaiting takeoff clearance on a runway. The Metro had taxied into position at an intersection some distance down the runway. The B-737 had been cleared to land on the same runway. The tower controller could not see the Metro in the lights of the runway and, because a flight strip for the Skywest was not at the controller's position, forgot that it was awaiting takeoff. The National Transportation Safety Board cited the lack of management in the tower facility, from the perspective of both oversight and policy direction, and failure of appropriate coordination in following procedures in the tower. The controller was very busy and did not have adequate backup, nor was the surface radar available for monitoring the aircraft on the airport. Certain procedures of information exchange were violated. Both of these accidents have stimulated major efforts to improve navigation and surveillance on the ground.

14. In 1994, the FAA introduced the expanded national route plan, which enables commercial airlines to have greater flexibility in choosing their desired courses at high altitudes, thus initiating a trend toward greater authority by commercial pilots and airline dispatchers to manage their flight trajectories. This trend may be enhanced by the implementation of free flight, in which operators under instrument flight rules have the freedom to select their path and speed in real time (Planzer and Jenny, 1995).

The time line shown in Figure 1.2 constitutes but a partial list of key events and trends that have occurred and influenced the national airspace over the past 40 years. In particular, we have not highlighted many of the specific developments and technological evolution within the air traffic control facility as these pertain to current practices. These developments are thoroughly treated in the chapters that follow.

SCOPE AND ORGANIZATION OF THE REPORT

The panel's work is being reported in two separate reports. This report describes in some detail the human factors aspects of the baseline air traffic control system as it exists today. Although we consider automation issues to some degree, particularly those that already exist, we also focus on many more traditional human factors issues, such as current training procedures, display design, workload, and team communication and cooperation. This report sets the stage for an in-depth examination of proposed automation levels of the future air traffic control system, to be described in the panel's second report. Both volumes focus of course on air traffic control; however, we also address other complex systems, in which lessons learned from human factors research can be applied to the air traffic control system.

This report has 12 chapters. In Part I, we discuss the general system by which air traffic is controlled; the procedures used for selecting, training, and evaluating controllers; and the support provided by airway facilities staff. Chapter 2 considers the specific air traffic control systems and describes the different facilities and controller tasks. Chapter 3 addresses controller attributes and the tools used for personnel selection, training, and evaluation. Chapter 4 describes the operations and personnel of airway facilities.

In Part II we consider specific human factors issues within the current system: the cognitive tasks of the controller (Chapter 5), workload and vigilance considerations (Chapter 6), teamwork and communications (Chapter 7), system management (Chapter 8), and human factors issues in airway facilities (Chapter 9).

In Chapters 10 and 11 we consider two important methodological issues in human factors research: how research is tried out on air traffic control issues and how human factors knowledge is actually incorporated into design of the current air traffic control system.

In Chapter 12 we set the stage for our Phase 2 report by focusing on two aspects of automation. We first summarize and synthesize automation levels achieved in the current air traffic control system both in North America and in Europe, show how they have worked, and the lessons learned. In the second part of the chapter, we address several generic issues in the human factors of automation, drawing guidance from other domains such as process control, robotics, and the flight deck, where automation has evolved further than it has in air traffic control.

2

Tasks in Air Traffic Control

In this chapter, we describe the tasks of people who work in three types of air traffic control facilities: the air traffic control tower, the terminal radar approach control (TRACON), and the air route traffic control center or, as it is often called, the en route center. We also consider the tasks of people at two other locations: the Air Traffic Control System Command Center in Herndon, Virginia, and the flight service stations around the country.

In describing the duties of the controller in the tower, the TRACON, and the en route facilities, we point out their common as well as their distinguishing features. Although the descriptions are fairly generic, it should be emphasized that, within a type of facility, there are vast differences from region to region, dictated by the level of activity (compare the New York TRACON with that at Champaign, Illinois, for example) and by other unique features of the work culture.

AIR TRAFFIC CONTROL ORGANIZATION

The Federal Aviation Administration's current headquarters organizational structure is divided into six "functional lines of business," each the responsibility of an associate administrator reporting directly to the agency's administrator (FAA Headquarters *Intercom*, December 13, 1994):

1. The air traffic control organization, called Air Traffic Services, is responsible for operation of the 20 en route centers, almost 200 TRACON facilities,

hundreds of airport towers, the Air Traffic Control System Command Center, and flight service stations located throughout the United States and Puerto Rico. Air Traffic Services formulates plans and requirements for future air traffic control operations and evaluates and analyzes current operations. Encompassing both air traffic control and airway facilities activities, Air Traffic Services is responsible for requirements; system management; rules and procedures; national airspace system operations, transitions, and implementation; resource management; logistics; flight inspection programs; and airspace capacity planning.

2. The Research and Acquisition organization is responsible for the development of communications, navigation, and surveillance systems; system architecture; aviation research; research and development performed at the FAA Technical Center; and all technology acquisitions.

3. The Regulation and Certification organization is responsible for aircraft certification, flight standards, rule making, aviation medicine, and accident investigation.

4. The Airports organization is responsible for airport planning and airport safety.

5. The Civil Aviation Security organization is responsible for security operations and planning and for civil aviation security intelligence.

6. The Administration organization is responsible for agency human resources, budgeting, and accounting, as well as for the FAA's Aeronautical Center and for administrative functions at the nine FAA regions.

Regional administrators at the nine regions are responsible for the administrative functions of the multiple facilities in their regions. Regional administrators report to the Administration organization at FAA headquarters; they do not have line authority over the regional division managers for Air Traffic, Airway Facilities, Airports, or Civil Aviation Security, who report directly to their respective associate administrator or director at headquarters. Thus, the director of Air Traffic Services directly supervises the regional division manager for Air Traffic (FAA order 1100.148B).

This chain of responsibility and authority continues through the area level, at which air traffic managers, reporting to their respective regional division managers, are responsible for the day-to-day operations of an assigned group of air traffic control facilities. Air traffic managers are supported when necessary by assistant managers, area managers, and area supervisors, to whom air traffic controllers report (FAA order 1100.126F, April 13, 1990; FAA order 1100.5C, February 6, 1989).

Air traffic control services are provided at three types of facilities:

1. Terminals, including tower and TRACON controllers,
2. En route centers, and
3. Flight service stations.

Each type of facility provides a particular service to the users of the national airspace system and, in all three, automation is used to assist people in making decisions and in supplying up-to-the-minute information. In 1994, the three types of facilities carried out more than 164 million operations; by the year 2000, they are projected to carry out 207 million operations.

THE TOWER

The Task

Within the terminal area of air traffic control, there are two distinct functions provided by air traffic controllers. The tower controller, located in a glass structure "on top of the tower," controls aircraft on the ground, just after takeoff, and just before landing (Figure 2.1). Tower control tasks are usually divided between the ground controller and the local area controller. The TRACON controller, located in a windowless radar room either below the tower cab or somewhere else in the area, controls aircraft in the wider region of space around the airport. The key responsibilities of tower controllers are to:

1. Issue clearances for the aircraft to push back from the gate and then to leave the ground. These clearances generally involve confirmation of schedules

FIGURE 2.1 Control tower. Source: Federal Aviation Administration.

of flight plans that were filed previously through Flight Services and by airline dispatchers. For takeoffs and landings, they involve prior assurance of safe separation from other traffic.

2. Manage ground traffic to and from the gate. This involves lining aircraft up in a sequence for takeoff and coordinating the traffic on the runways so that it does not conflict with other ground traffic (aircraft or vehicles) or with aircraft that are taking off or landing.

3. Hand off the departing aircraft to and accept the arriving aircraft from the TRACON controller residing in the radar room.

The ground functions of taxi management are handled by the ground controller, and the takeoffs and landings are handled by a local controller. At smaller facilities, or at times of very low traffic density, the two functions may be carried out by a single individual.

Visual Resources

The most critical task of the tower controller (both ground and local) is to keep track of who is where. Because all aircraft are nominally within sight of the controllers in the tower, the most important resources at their disposal are their eyes, coupled with a voice communication link. The challenge is always to know how to communicate with an identified aircraft on the ground and in the air. Thus towers are constructed to provide as much full visibility of the entire airport surface as possible, although there are occasional deficiencies caused by airport structures (one such deficiency was identified as a causal factor in the runway incursion accident at the Los Angeles airport in 1991; see number 13 on the list in Chapter 1).

The task of knowing whom one is looking at is not a trivial one at a busy airport. Many aircraft look alike; vision is often degraded at night or when ground fog obscures parts of the runway; pilots can add to the tower controller's demands if they become confused and take a wrong turn on a taxiway or ramp; and the visual chaos is often enhanced by the diversity of ground vehicles, traveling this way and that, occasionally without communications to the tower.

Some assistance for the local and ground controllers is being provided by airport surface detection equipment (ASDE), a system that provides radar identification of ground vehicles and aircraft at the airport. It is being added to many airports that do not have this equipment and upgraded at airports that do have it; the installations and upgrading are behind schedule, however. Many towers are equipped with radar displays called DBRITE (digital brite, radar indicator tower equipment) to augment visual control of airborne traffic. The DBRITE provides the local controller with (1) a radar presentation of about 15 miles around the airport and (2) alphanumeric information on the aircraft that is received from the

DAL542 ¹	7HQ	30	330		FLL J14 ENO 00D212	
DC9/A	1827	18			OOD PHL	2575
T468 G555						
16 16						
486 09		PXT				*ZCN

FIGURE 2.2 Flight strip. Source: Federal Aviation Administration.

automated radar terminal system at a nearby TRACON (discussed at length below).

The information from the DBRITE provides the tower controller with a view of arrival aircraft that are not yet under his or her control, so the aircraft that are the controller's responsibility on the ground can be safely and efficiently coordinated with planned arrivals. DBRITE enhances the local controller's ability to control higher volumes of airport traffic by providing key information about the aircraft identity, type, altitude, first fix after departure, and speed while the aircraft is on final approach.

Flight Strips

The controller's task of maintaining location information is greatly facilitated by paper flight strips (Figure 2.2). These physical representations of each aircraft, which are computer generated at the time the flight plan is filed, represent a visible reminder of an aircraft's status in the sequence of taxi-takeoff (for departure) or landing-taxi (for arrival). As they are physically moved around the controller's workstation, they are a reminder of what each represented aircraft is doing on the terminal surface, thereby generally helping to maintain the big picture of who is where. Along with the visual obstruction cited above, a lost flight strip was also identified as one cause contributing to the 1991 Los Angeles runway incursion.

Communications

Using standardized phraseology, the controller talks with the pilots on radio. A particular pilot knows that he or she is the recipient of a communication by the header ID ("United two twenty-four: hold short at runway two four left"), and other pilots can also hear the message. Such a "party-line" feature creates added auditory input in all cockpits, allowing all pilots to build a better mental model of what surrounding traffic is doing (Pritchett and Hansman, 1993).

Tower controllers communicate with pilots, with each other (ground to local), and with the TRACON controller in the radar room to accept arrivals and hand off departures. This latter communication is handled in three ways: first, voice communication is used to coordinate (e.g., accept or decline handoffs). For example, a local controller may refuse to accept a landing aircraft until he or she knows that the runway will be clear for a certain period of time before the approaching aircraft touches down. Second, information on the flight strips is also relayed between tower and radar room, specifically by the FDIO (flight data input/output) computer system. Third, the communication is mediated by the pilot, who, when instructed, changes radio frequency and contacts the next appropriate control facility. Then the handoff is complete.

Traffic Management

Throughout this process, the tower controller at busy facilities feels continuing pressure from the clients being served: from outbound aircraft, to move each one as soon as possible to the desired "Number 1 for takeoff" position, and from inbound aircraft, to get them off the active runway as soon as possible and, once off, to get them taxied to the gate as soon as possible. When the taxiing aircraft must be cleared to cross an active runway that is accepting a steady stream of departing and arriving aircraft, the scheduling demands can be challenging indeed.

THE TRACON

The Task

The tasks of the terminal radar or TRACON controller, are (1) to manage the safe and expeditious flow of a departing aircraft accepted from the tower to a handoff to the en route controller, a job usually handled by the departure controller, and (2) to manage the arriving aircraft from a handoff from the en route controller to a handoff to the tower controller on a final approach for landing, a job usually handled by the approach controller. A key component of the TRACON controller's job, like that of the tower controller, is sequencing or "lining up" the aircraft in an orderly inbound or outbound flow, at regular spacing. Maintaining the safe separation between aircraft is as important as it is for the tower, but for the TRACON controller it is an even more challenging task because separation is now a three-dimensional problem and aircraft are constantly climbing and descending (in addition to their lateral movement). Thus, the TRACON controller must be ever sensitive to the critical separation criteria for all aircraft operating under instrument flight rules: 1,000 vertical feet and 3, 4, or 5 miles lateral separation, depending on the size of the aircraft. (Large aircraft require more lateral separation because of the wake turbulence they create with their wings.) The pressures for efficiency often dictate separations that are not much greater than this, and the countervailing pressures for safety dictate that these criteria shall never be violated, such that operational errors are recorded.

The skills necessary to balance these criteria and maintain the regular, evenly spaced flow are considerable, as we detail in subsequent chapters. Aircraft must be constantly adjusted in their speed, heading, and altitude to meet the dual criteria of safety and efficiency while they follow generally standardized approach and departure routes (a good description of this task is provided by Luffsey, 1990). Empty "slots" in the airspace are sometimes filled by trying to fit in an additional aircraft. The adjustment of flight paths is accomplished through voice communications with the pilot and his or her rapid and accurate compliance. At the same time, the TRACON controller tries to maintains sensitivity to the pilot's needs to avoid excessive and abrupt maneuvering.

Each controller will work his or her set of aircraft through a given sector of the airspace. These sectors are sometimes oddly shaped three-dimensional volumes, which include not only the standardized arrival and departure routes, but also features like terrain, structures, special-use (restricted) airspace, and missed approach paths. Sometimes a given controller will work only arrivals, sometimes departures, and sometimes both, an assignment that may vary throughout the day or night, as the TRACON becomes more or less busy.

The Information: The ARTS III System

The critical information available to the controller is collected by the ARTS (automated radar terminal system) computer system. The most sophisticated version is the ARTS III computer system which supports all high-activity (level 4 and 5) TRACON facilities (Figure 2.3a and 2.3b). The information that is integrated in the computer is provided by primary and secondary radar and by the flight data input/output computer. The primary radar (airport surveillance radar) receives returns from all aircraft in the air. The secondary radar is an active system that receives digital signals from all aircraft equipped with a transponder. The FDIO hosts computer-based flight plans. From these three sources, for each aircraft equipped for instrument flight rules, the ARTS III system has available a data block, which contains information on aircraft call sign, type of aircraft, destination airport, first navigational fix after departure, mode C altitude (if equipped with a mode C transponder), ground speed, and scratchpad information useful to the controller (Figure 2.4). Such information may be accessed at any time via the computer interface. Some of it is contained on the computer-generated flight strips, and some is presented on the radar display.

The sophisticated ARTS IIIA system, used at all level 4 and 5 TRACON facilities, contains automated monitoring systems that provide conflict alerts and minimum safe altitude warnings. Level 2 and some level 3 facilities are supported by the less sophisticated ARTS 2 system, which does not have the automated options and is served only by primary radar signals.

TASKS IN AIR TRAFFIC CONTROL 39

FIGURE 2.3a ARTS III radar display. Source: Federal Aviation Administration.

The Radar Display

For the controller, the most critical source of visual information is the radar display, which supports maintenance of the big picture. The primary radar receives a return from anything in the sky and paints this on the scope. The secondary radar receives additional information from aircraft equipped with transponders; this information includes aircraft identity, derived from a code transmitted by the aircraft transponder, and altitude information, if the aircraft is equipped with a mode C transponder. When the ARTS computer has a flight plan

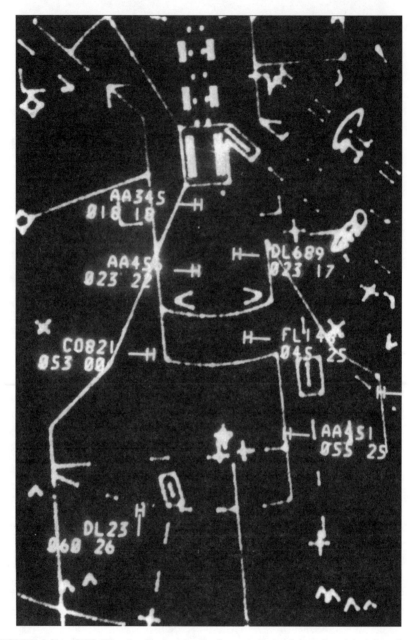

FIGURE 2.3b ARTS III radar display detail. Source: Federal Aviation Administration.

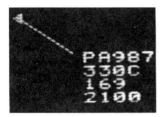

FIGURE 2.4 Data block. Source: Federal Aviation Administration.

associated with the transponder code transmitted by the aircraft, a data tag containing the aircraft's call sign (identity), mode C altitude (if the aircraft is equipped), and ground speed will be displayed on the controller's radar display.

The ARTS III computer works to integrate the primary and secondary radar information to provide the most accurate estimate of current aircraft location. The raw blip of the primary radar is always on the screen. In addition, a line at the end of the data tag represents the computer estimation of the position based on both primary and secondary radar. This upgraded system yields more accurate position by time data. Its first priority is to track aircraft using the secondary radar. If for some reason it cannot get a good return on the secondary radar, it will track on the primary radar return. The ARTS tracker is a predict-and-confirm system. Once a target is identified, it establishes a history of direction and speed in the computer memory. Then the computer places the alphanumeric information on a predicted path and waits for the 4-second radar sweep to confirm the position. To the controller, the information always appears to be associated with the actual target. When the system fails to confirm its prediction, it begins a search around the predicted path and alerts the controller with the letters CST (meaning "track lost") displayed in the altitude field of the tag.

The controller can communicate with the computer generating the display via a trackball and keyboard system. The trackball can be used to position a cursor on top of a given aircraft symbol, and the keyboard can then be used to enter information into the host ARTS computer pertaining to that aircraft.

The particular example of the radar screen, shown in Figure 2.3a and 2.3b, indicates the orderly flow of aircraft across the TRACON area, entering or departing at "gates" in the four corners adjacent to the en route area, and being merged and lined up just short of the two active runways. This is a situation that the skilled controller can handle by an adjustment in the flight path of one or the other.

The screen represents a compromise between information and clutter. Naturally the controller would like to have maximum information about each flight, at a location immediately adjacent to the aircraft's accurately depicted position. In the same display the controller also may have information regarding other spatial features, like the location of ground hazards, approach and departure routes, navigational fixes, and even, ideally, severe weather patterns, which are available to most controllers with the new ASR-9 radar. At the same time, this plethora of information, particularly during busy traffic periods, can present a very cluttered display. Data tags may overlap or overlie other display features. Some intelligent features within the ARTS computer will adjust the location of tags to prevent excessive overlap, and the controller does have some options for decluttering the display.

Automation also provides conflict alerts for projected and current loss of separation and alerts for loss of separation from the terrain (a minimum safe altitude warning). Data blocks blink and aural alarms are sounded. The controller has some options to declutter the display including removing fixed objects, decreasing sensitivity to weather, and simplifying or removing data blocks.

Flight Strips

Flight strips are issued by the flight data input/output computer. The existence of the data blocks lessens the controller's degree of dependence on the flight strips, in comparison to the era before the ARTS computer was implemented. But these strips still remain an important augmentation to the controller's memory of what each aircraft is doing or is about to do. Controllers may write on the slips, indicating instructions just issued to aircraft, or they may cock the strips of certain aircraft at odd angles, to remind them of certain unusual circumstances that may need to be addressed in the future, information that will not be known by the ARTS computer (and hence cannot be portrayed in the radar data block).

Traffic Management

Although controllers strive to preserve an orderly flow of traffic, several forces exist to counteract this goal. The following is a partial list:

- The sector may be filled by a very heterogeneous mixture of aircraft, including slow-flying but fast-maneuvering general aviation aircraft and fast-flying but slow-maneuvering transport aircraft, particularly the wide-bodied "heavies" like the 747 and the DC10. Aircraft differ in the extent to which they leave dangerous but invisible wake vortices in trail, which require different separation standards between light and heavy aircraft, both in terms of the vortices they leave and the susceptibility to the vortices they encounter.
- The sector may be pressured to accept an excessive number of aircraft for approach and departure at heavy traffic periods in the morning and the late afternoon. In a high-workload sector, there may be as many as 10-15 aircraft at one time.
- Weather can severely disrupt the traffic management plan that a controller is trying to execute. A reported wind shear along one of the arrival strings shown in Figure 2.3a can wreak havoc on the orderly flow. So can a runway that is

suddenly closed or a sudden change in wind direction that will force a reversal of the runway direction for arrivals and departures.
• Controllers may have to deal with aircraft flying under visual flight rules (VFR) that are not communicating with them and that appear to be in a conflicting path with controlled aircraft. Some of these aircraft may not be equipped with a transponder, making it harder for the controllers to see them, since no data tag will be generated.
• Pilots may not always carry out the instructions issued by the controller or may carry them out differently from what was intended. They may fail to slow or climb through an assigned flight level, thereby not only destroying the controller's careful tactical plan for flow, but also potentially creating a conflict situation with other aircraft, drawing the controller's attention away from other regions of the sector.

Communications

The problems resulting from pilots' occasional failures to comply reinforce the importance of communications, which, along with the radar display and the flight strips, represent the third critical element of controller interactions with the environment. As with the tower controller, TRACON communications are highly standardized and controllers are trained to deliver these in a clear, coherent fashion, as well as to monitor pilot "readback" of the communications string to ensure that it was correctly heard. However, such readback is not always accurate, and controllers may sometimes fail to detect the inaccuracies (Monan, 1986:Chapter 5). Furthermore, a heavy workload may force the controller to speak more rapidly than is optimal for pilot comprehension, particularly when longer strings are required. Finally, communications back from the pilot to the controller may be considerably less standardized than in the other direction, because there are far more pilots than controllers, and their level of skills and fluency in the English language are far more diverse.

As in the tower, communication between controllers in the TRACON is as important as it is between controllers and pilots. Communications and coordination between controllers in adjacent sectors is critical when aircraft cross the sector boundaries. This communications link is just as critical when aircraft make the transition between the TRACON and the adjacent tower or en route airspace.

The handoffs from one controller to another are typically accomplished by the automated handoff, in which case a quick sequence of keyboard interaction sends a message to the receiving controller, which gives the latter individual the opportunity to accept, also with a keypress, once the controller feels that the sector is ready to absorb the additional traffic. As in the tower, the handoff process between facilities is also mediated by voice communications with the pilot, as the appropriate frequency to contact the receiving controller is announced. If for some reason an automated handoff cannot be made, then voice communications are used.

The Environment

The TRACON environment is dark, an appearance dictated by the low contrast of the greenish-yellow blips of the primary radar on the dark screens. Its level of activity (and number of people and amount of chatter) varies radically as a function of the time of day. At busier times, more controllers handle progressively smaller regions, as multiple sectors (which are combined in low-workload periods) are pulled apart as traffic increases. Supervisors within the facility coordinate the staffing in this dynamic fashion and may themselves step in to assist a particularly busy sector as required. It is in part this flexibility of personnel assignment that allows the air traffic control system to respond adaptively to such changing situations as weather and aircraft emergencies. Furthermore, at some facilities, personnel may at times alternate between the TRACON and the tower upstairs.

Equipment and Other Failures

The TRACON controller's efforts to manage the orderly flow may occasionally be disrupted by equipment failures. Sometimes these disruptions may produce only minor annoyances, as when a data tag for an individual aircraft is temporarily lost. In such cases there is backup information on the flight strips, which can replace the temporarily hidden data tags, and communications can be initiated with the pilot to receive whatever information is necessary. More serious are the more severe breakdowns in the ARTS system, which may result when extremely high traffic density exceeds the computer's capacity. In such cases, the system is designed to fail somewhat gracefully, so that the more powerful automation options (e.g., the predictor algorithms) are lost before the data tags are eliminated. And even in the absence of any information from the ARTS computer, primary radar and flight strips will still allow some representation of aircraft position. Equally serious, if not more so, are the rare losses of power and communications. Although all of these catastrophic failures are rare, they are nevertheless real possibilities that controllers must be prepared to handle.

Finally, there are nonequipment failures, such as a crash on a runway or a lost primary radar contact, that have an equally serious need for sudden crisis management, typically by following prepared procedures. The most generic response of the TRACON system to failures of all kinds is to temporarily sacrifice efficiency and preserve safety at all costs—a goal that may well be met by increasing the minimum separation boundaries, if radar resolution or communications ability is degraded, or by diverting aircraft to adjacent sectors or facilities.

THE EN ROUTE CENTER

At the air route traffic control centers, also called en route centers, the controllers primarily use radar information to provide guidance to aircraft flying across the country. However, there are certain areas for which there is no radar coverage; for these areas, nonradar procedures are used. The implications of this are that the minimum spacing between aircraft must be quite large, sometimes causing a loss of efficiency in traffic flow. Two major nonradar areas are the oceanic areas of the Atlantic and the Pacific oceans, for whose control the New York and the Oakland en route centers, respectively, are responsible. Oceanic controllers rely on aircraft position and intent data provided by pilots. The discussion of en route centers in this chapter focuses on operations for nonoceanic flights.

Flights are guided along what is a generally orderly series of linear routes across the sky at different flight levels. The linear paths are defined by navigational aids called VORs, each with a given name. Two crossing radials from a VOR may define an intersection, designated by a pronounceable 5-letter code. Generally eastbound flights travel at odd flight levels, and westbound flights at even levels. In the United States, there are 21 centers that in 1994 handled a total of 39,000,000 operations.

Like the TRACON controller, the en route controller must balance concerns of expeditious flow against those of safety. However, the safety separation standards are greater in en route: 5 miles or, depending on the aircraft's altitude, 1,000 or 2,000 feet, rather than 3 miles, 1,000 feet. This increase in separation standards is dictated jointly by the greater difficulty that the more distant radar coverage has in establishing precise location, as well as by the faster speed of travel. The effect of this greater separation on traffic flow is minimal, because the density of traffic is considerably less than in the TRACON area.

Each en route center is also divided into a series of irregularly shaped sectors that have both horizontal (lateral) and vertical boundaries. Adjacent controllers may be working aircraft above or below one another. Also, a given sector may overlay a TRACON space beneath. Each sector may be worked by a team of two controllers: a radar position (R-side), whose primary responsibility is to monitor the radar display and ensure separation, and a data position (D-side), whose primary responsibility is to handle data and coordinate. During periods of low traffic, a single controller may handle both responsibilities. When manned by two controllers, however, communication between them is vital.

The Information: The En Route HOST Computer

Information for the en route controller is gathered from air route surveillance radar and integrated with other information from the FDIO computer by the en route automated system called the HOST. This is a software system, developed in the 1960s and currently "hosted" on hardware introduced in the 1980s. The HOST provides flight data processing and radar data processing. Flight data processing is the system that develops flight plans from information received from automated flight service stations, controller input, and air carriers (dispatchers) that have a direct link to the HOST for filing flight plans. Flight data processing is interfaced with towers, TRACONs, sectors within the en route centers, and with other en route centers so that flight plans can be automatically sent.

As for the TRACON controller, the primary tool for the en route controller is the radar display, called the plan view display, or PVD. The PVD is somewhat different from the TRACON display in that it is a fully digitized display of primary and secondary radar targets that are presented to the controller in symbolic format from the IBM HOST computer radar data processor. Hence, what is depicted is an intelligent estimate of the current location of each aircraft, based on computer aggregation of returns from the air route surveillance radars located within the en route center's area. These returns are processed by the HOST computer.

The Interface

Examining a typical PVD display in Figure 2.5, we see a pattern that bears many similarities to the TRACON display. However, since the information depicted is entirely digital (i.e., no raw radar returns), the operations can be carried out in a more brightly illuminated environment; that is, there is no need to present a light stroke on a dark background. More information is presented and a set of equal-time cross-hatched lines behind the aircraft indicates its past trajectory, providing a good representation of current heading and recent air speed. The levels of automation here are similar to the TRACON, in terms of flight path predictors, minimum safe altitude warnings, and conflict alerts.

Here, too, the flight strips remain an important part of the controller's task. However, because of the generally higher levels of automation and the digital information available at all centers regarding flight activity of all aircraft, there are now options to allow many of the manual operations of flight strip updating and manipulation to also be carried out by computer. Communication also is managed in very much the same way between TRACON and en route centers.

Traffic Management

The primary objective of the en route center is to maintain the expeditious but regular delivery of an aircraft stream to the receiving TRACONs, providing them as rapidly as they can be received (but ideally no faster, since this will produce bottlenecks in the sky). To assist in this process, each center is equipped with a traffic management unit that attempts to coordinate flow across the entire

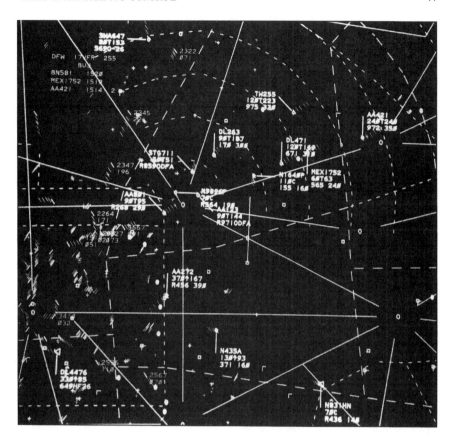

FIGURE 2.5 Plan view display: En route center. Source: Federal Aviation Administration.

area and may also coordinate with the air traffic control system command center in Herndon, Virginia (discussed below), to accomplish this.

Individual controllers at their workstations monitor the ongoing flights of a multitude of aircraft flying at various speeds and respond to pilot requests and adjust to weather conditions by issuing instructions to alter air speeds, flight levels, and (if necessary) headings, in order to maintain maximum but regular flow. At the same time, the ever-vigilant monitoring for predicted conflicts is ongoing. A well-organized, conflict-free flow through a sector can be suddenly compromised by an aircraft that wishes to climb or descend to seek more favorable tail winds or to avoid turbulence; the winds themselves can have different and unpredictable effects on the speed attainable by different types of aircraft. Some flight path adjustments are therefore second-order ones, issued in response to projected conflicts that may themselves have developed following the granting of a pilot request.

TRACON AND EN ROUTE: SIMILARITIES AND DIFFERENCES

In addition to the separate areas of airspace for which they are responsible, there are certain other generic differences in the kind of tasks and traffic dealt with by the TRACON and the en route controllers. These differences are best characterized by the more continual changes in aircraft altitude experienced in the TRACON, the greater frequency of flying by visual flight rules by aircraft in the TRACON region (and greater variety of aircraft types), and the greater extent, in the en route centers, to which control can be carried out strategically, through the execution of longer-range plans, than tactically, in terms of resolving more suddenly arising conflicts.

At the same time, as noted above, there are far more similarities than differences between the two operations, particularly as these translate into the human factors issues that are discussed in this book. For example, the strategies and tactics for failure management that are described for the TRACON are similar for the two kinds of controllers. In the rest of the book, we often speak generically about controllers, referring to those that occupy both kinds of facilities.

One final common feature that characterizes the tasks of controllers at all levels is the nature of changes in work shifts. The impact of the specific shifts (day or night) during which controllers work is addressed in Chapter 6. We note here the critical importance of shift or station changes, when one controller assumes the duties of another at a station. In these circumstances, the traffic situation must be accurately understood by the replacing controller: Who is where in the sky or on the ground? What actions are pending? What conflicts may be forecast in the distant future but are not yet sufficiently imminent to warrant corrective action? Data show that this period, the first minutes following a new time on position, is one in which errors are more likely to occur (Cheaney and Billings, 1981).

CENTRAL FLOW CONTROL

Metering the number of aircraft in the national airspace on a daily basis is an important task. Flow control is designed to meet user needs to the best ability of the system—that is, to ensure that the national airspace system accepts the maximum number of aircraft yet maintains high levels of safety. Factors that affect flow control are the physical structures of airports, including runway and taxiway availability; the number of arrivals and departures that can be operated safely in a given hour; controller equipment status, including what equipment has failed that reduces their capability to handle workload; the status of the national airspace system equipment; emergency situations; and the main factor, weather.

The Air Traffic Control System Command Center, located in Virginia near Dulles Airport, is responsible for the management of traffic throughout the air traffic control system. This facility, in conjunction with traffic management units at each of the en route centers and some designated TRACONs and towers, establishes the daily traffic flows into and out of 28 major airports based on all the factors listed above. En route centers also provide flow control within their airspace to ensure that sectors do not become saturated.

Flow control is dynamic, and the flows may change on a minute by minute basis. The primary method for ensuring that the traffic is metered is to hold aircraft on the ground and release them into the system at intervals so that they can be sequenced into the approach without further delay when they reach their destination. Airborne holding is another method and is being used on a limited basis (fuel costs being the limiting factor). Both methods have pros and cons for the commercial airline industry. Holding on the ground saves fuel costs and usually ensures no delay at the destination but has an effect on the customer and airline competition. Airborne holding costs more in fuel, presents greater safety hazards, and creates more workload for the controllers, but it ensures that when an approach slot is available there is an aircraft there to take it, thus making the best use of airspace for meeting demand.

Flow controllers at the local facilities, called traffic management coordinators, primarily utilize the HOST system to tell them where the traffic is located and at what time it is expected to affect the airport. The HOST is supplemented by a system called aircraft situation display, which uses a computer to present a picture of all the airborne aircraft in the country at any time. The display gets its information from the HOST by data link through a computer located at Cambridge, Massachusetts. This system is very helpful to the en route centers' traffic management units in determining when a sector will become busy and where to reroute traffic to keep workload efficiencies at a maximum.

The working day for flow control starts very early in the morning, when the daily plans of operations are formulated by two organizations. First, dispatchers at the airline operations centers of the major airlines arrange the daily plan of flights to and from the major hubs. The other organization is at the central flow control center at Herndon, Virginia, where each morning a national plan for traffic flow is developed, taking into consideration issues like weather, which may restrict flow to certain regions, and critical events like the Super Bowl, which may create bottlenecks in certain areas.

As the national airspace increases its activities and flights begin departing in the East, then the Midwest, and then the West, air traffic control is distributed to the facilities (TRACONs and en route centers). Minor traffic bottlenecks and buildups are addressed by distributed local negotiations between adjacent facilities via their traffic management coordinators. If a TRACON is temporarily saturated, controllers there will coordinate with one or more of the feeding en route centers to slow down the delivery of aircraft. Similar local negotiations

FIGURE 2.6 Central flow control aircraft situation display. Source: Federal Aviation Administration.

may be carried out between adjacent centers, just as they also take place between adjacent sectors within a facility. Central flow control at the command center continues to actively monitor the biggest picture via the aircraft situation display (ASD) (Figure 2.6). Occasionally it becomes actively involved in implementing ground holds or managing the flow of traffic to and from international destinations, but by and large the philosophy is a fairly hands-off one, to allow local solutions to be achieved within the facilities, unless problems develop that they cannot handle.

Such problems may be of two sorts: first, there may be anticipated problems such as the gradual buildup of traffic in a region, a buildup that may need to be addressed by three or more facilities (TRACONs and en route centers), making achievement of a solution difficult with a single phone call. Second, there may be truly abrupt or catastrophic failures in the system, as when severe weather closes an airport or a power outage at some major facilities drastically degrades the ability to monitor traffic position.

In these infrequent instances, central flow central must "jump into the loop" as an active participant in control (Huey and Wickens, 1993), suddenly utilizing the full situation awareness that has been maintained during the previous routine period to rapidly implement strategic adjustments to traffic plans. (The analogy with situation awareness of the individual operator is apparent.) Such crisis negotiation is accomplished by extensive vocal communications over phone lines to facility traffic management units and to airline dispatchers. Such verbal negotiations may well be supported by the spatial display of national flow available to flow control managers at the facilities as well as at central flow control.

The response of the national flow system has in the past worked well. No accidents, for example, have ever resulted from the catastrophic impacts on the flow control system of severe events like power loss or runway closure. The system typically responds adaptively to minimize the consequences of out-of-the-ordinary situations such as power outages, severe weather, and communication failures by increasing the amount of voice communications and separation margins whenever possible. However, it is important to ask: (1) how the response might vary if human-human communications are replaced by human-automation communications and (2) how such a response would be made more vulnerable in an airspace that is far more densely populated than at present, a density that is intended to be the direct result of the increased flow capacity made available by the same automation.

FLIGHT SERVICE STATIONS

The air traffic controller specialists in the flight service stations provide a myriad of services, primarily to general aviation pilots. The services provided are flight plan filing, preflight and en route weather briefings that include the status of navigational aids, airport conditions reports, search and rescue operations, assistance to lost or disoriented aircraft pilots, provision of instrumental flight rule and special visual flight rule clearances, soliciting pilot reports on flying conditions, and providing special services such as customs and immigration notification.

Pilots can receive these services by visiting a flight service station, by telephone, or through air-to-ground communications. In 1994 there were 131 flight service stations, of which 60 are automated. Current congressional plans call for reducing the total number of facilities providing flight services to 61.

The automated system, called Model I Full Capacity, is a 1970s-era weather and flight notification distribution system. In the early 1980s it replaced a leased weather display and teletype system. The system interfaces with the national airspace data interchange network communication system and the en route centers' HOST system. It has reduced the workload of flight service station controllers and provides for a much quicker briefing to pilots, but it leaves much to be desired in terms of functionality and basic human factors engineering.

The typical automated flight service station contains the following operational positions: preflight weather briefing, inflight, flight data/notice to airmen, weather observer, and area supervisor. At designated stations, specially trained controllers provide en route flight advisory services—that is, timely and pertinent weather data tailored to specific altitudes and routes using the most current available sources of aviation meteorological information. These specialists are in constant communication with the National Weather Service's meteorologists at its field offices and center weather service units.

A modernization program began in the late 1970s. Its purpose was to achieve equal or improved service to the user, while reducing personnel and maintenance requirements through the consolidation of 317 manual stations into 61 stations with modern automation tools. The program has been successful to some degree, but it has created many issues at locations where an old facility has been closed or is projected to close. Users have been concerned that they would not receive the same level of service, especially at remote locations—Alaska is a good example. Walk-in services for many pilots were out of the question; some stations were not even accessible from the airport. As a result, business is sometimes done entirely by telephone. The primary concern has been that, with fewer stations, the automated flight service station air traffic controllers would be busy on the telephone and users would be delayed in getting service. To offload some of this unmet demand, the FAA implemented several broadcast programs provided by private contractors to distribute weather information. For example, the DUATS program provided contract awards to two companies to provide free access to on-line computer services for weather information and flight plan filing.

The consolidation process has been delayed for over a decade and has been subject to political pressure. The key issue is the downsizing and relocation of controllers from the closing stations to the now-centralized ones. Communities were solicited to bid for station sites, and the selections were driven by their cost to the government. Subsequently, some automated flight service stations were located in areas that are difficult to staff.

The National Association of Air Traffic Specialists is the bargaining unit representing the nonmanagement flight service station specialists. They have accepted the new concept but have been very concerned about the relocation of positions and the loss of jobs.

SUMMARY

Controllers work in three types of air traffic control facility: the tower, the terminal radar approach control (TRACON), and the en route center. The air traffic control organization, called Air Traffic Services, manages all of these facilities. This organization is responsible for formulating plans and requirements for future operations as well as evaluating and analyzing current operations. Division managers at the nine regions manage the air traffic control activities in their region. These regional administrators are supervised by the director of Air Traffic Services.

Controllers in all types of air traffic control facility develop strategic plans for traffic flow, monitor these plans with visual inputs to update the big picture of the traffic flow, and communicate with pilots and other controllers to ensure continued safety and efficiency. Controllers in the towers depend heavily on direct visual sightings of traffic at the airport, while those in the TRACON and en route environments are supported by computer-based, partially automated radar displays. The level of automation varies from facility to facility. Controllers depend on paper flight strips to represent the progress and special status of individual aircraft as they pass through the controller's sector of the airspace. All controllers must be prepared to deal with unanticipated events—for example, equipment failure, weather emergency, or pilot noncompliance with instructions—in a flexible manner that preserves safety even if it temporarily disrupts efficiency.

Although controllers in any of the three basic positions—tower, TRACON, or en route center—share many competencies, there are important differences among their tasks. Furthermore, there are differences between sites that perform the same functions and even within a site from sector to sector. Anecdotal evidence suggests that each site is likely to have its own culture, composed of shared beliefs among a particular set of operating personnel.

Ideally, the introduction of new technology into a large organization would be uniform throughout all its branches. Such uniformity of implementation is particularly difficult in the air traffic control environment because of the facility-specific culture and task environment. Furthermore, it has not been possible to create a common training or job performance evaluation program that covers all air traffic control specialists because of the local variation in job requirements.

3
Performance Assessment, Selection, and Training

The loss of half of the controller workforce from the 1981 strike placed significant pressure on the FAA to maintain continued high levels of performance with the influx of large numbers of new trainees who had little or no experience in performing air traffic control tasks. During the 1980s, in order to meet staffing demands, the FAA hired between 1,800 and 3,400 applicants a year as air traffic control specialist trainees. Although personnel researchers and managers had been working on performance appraisal, selection, and training programs over the years, the need for focused and efficient efforts in these areas became critical as applications from inexperienced individuals flooded into the government's Office of Personnel Management. For example, after the strike, approximately 67 percent of those hired as air traffic controllers had no prior experience in aviation, compared with 30 percent prior to the strike. Other differences are that the post-strike groups had slightly more formal education beyond high school and included slightly fewer minorities. As stated by Manning et al. (1988:1):

> The continued safety of the NAS [national airspace] requires that ATCSs [air traffic control specialists] be carefully selected and trained. Each candidate for the occupation is continually evaluated, from an initial aptitude selection test battery through grueling performance-based screening at the FAA Academy, and finally in on-the-job training, conducted at the assigned facility. Because of the safety-related, critical aspects of the job, identifying and screening for characteristics in individuals that will predict success in air traffic control is especially important. In fact, research has demonstrated that not all individuals have aptitudes required to perform the duties required of an ATCS.

All entry-level employees attend the FAA Air Traffic Control Academy in Oklahoma City for initial training in their respective type of facility (tower, TRACON, en route, field service station).[1] Air traffic control specialists are certified at one of two different levels of training. *Developmental controllers* are in apprenticeship positions and are learning skills both on the job, under the observation of more skilled controllers, and in periodic training sessions within the facility. *Full-performance-level controllers* are fully certified and well trained for their tasks. Many of them engage in the training of developmental controllers.

The objective of personnel selection and training is to produce successful controllers in the most cost-effective manner, cost being the expenditure of both money and time. Several methodological questions bear on reaching this objective. The key questions include:

- What criteria is the selection system designed to predict—performance on the job? Performance in training?
- What measures are used to define performance and how are they obtained?
- What cognitive and perceptual factors are related to effective performance on the job?
- How well do selection instruments measure the appropriate cognitive, perceptual, and personality factors? How well do selection instruments predict performance in training and on the job?
- How effective are various training programs both at the Air Traffic Control Academy and in the field in producing full-performance-level controllers? What are the attrition rates at various phases?

Performance criteria are used by researchers to define the level of effectiveness to be achieved by personnel. Selection is based on the definition of the abilities that candidates need to achieve good job performance, and training is the activity by which individuals are taught specific aspects of the job. Without good measures of job performance, it is difficult to develop effective selection and training programs. Although it appears that the basic abilities needed by controllers have not changed over the years, they have had to undergo new training in response to the introduction of new equipment and procedures. With additional automation on the way, it is possible that changes will take place in both the array of capabilities and the attitudes needed by candidate controllers, in the ways in

---

[1] Exceptions to this policy are made when an employee has had other training or experience that satisfies the requirements, such as attending certain private schools that specialize in air traffic control training or certain military air traffic control training, or working as a controller prior to the 1981 strike.

which these capabilities and attributes are refined by training, and perhaps in the ways in which controller performance is assessed.

This chapter discusses the historical development and current status of performance assessment, selection, and training of individual air traffic controllers and how these programs might change as more control functions are automated. Team training is discussed in Chapter 7.

PERFORMANCE ASSESSMENT

Our discussion of performance assessment is divided into two sections. The first section deals with the practices and issues surrounding the evaluation of controllers for the purposes of determining success in on-the-job training, effectiveness in job task performance, and eligibility for salary increases and promotion. The second section concerns the development of job performance measures for use as criteria in personnel selection research studies. Each section presents measurement issues and complexities.

Performance Assessment for Management Decision Making

Full-performance-level air traffic control specialists are responsible for the safe and efficient flow of aircraft through the airspace and the ground space they control. A variety of methods are used to evaluate the performance of these specialists, including real-time monitoring on the job, specially designed simulation exercises, checklists, and annual written performance appraisals. Real-time monitoring on the job provides immediate detection of violations in standards for aircraft separation and other errors. These data can be used to make immediate decisions about the controller's work status and the need for additional training. Checklists and written reviews are used once or twice a year to provide general technical assessments. In en route centers, dynamic simulations are employed extensively for training and skill assessment of developmental controllers.

The most widely used method of technical performance assessment has historically involved immediate supervisors observing controllers at work and completing a checklist (Table 3.1 is an example) indicating the level of performance in each general task area (e.g., maintaining separation, communication). Until 1993, supervisors conducted 40-minute, over-the-shoulder evaluations of this type twice a year. In 1993 the over-the-shoulder program was deemed inadequate and, in 1994, work began on a new assessment program called the operational assessment program. The proposed new program, still in draft form, is planned to assess each controller's technical performance on an ongoing basis in the areas of separation standards application, communications, position/sector management, equipment operation, and customer service delivery. The proposed program calls for quarterly assessments and includes a summary of the supervisor's evaluation of the controller's strengths and weaknesses. Stein and

TABLE 3.1 Performance Checklist

Job Function Category		Job Function
Separation	1.	Separation is ensured.
	2.	Safety alerts are provided.
Control Judgment	3.	Awareness is maintained
	4.	Good control judgment is applied.
	5.	Control actions are correctly planned.
	6.	Positive control is provided
Methods and Procedures	7.	Prompt action to correct errors is taken.
	8.	Effective traffic flow is maintained.
	9.	Aircraft identity is maintained.
	10.	Strip posting is complete/correct.
	11.	Clearance delivery is complete/correct/timely.
	12.	LOA's/Directives are adhered to.
	13.	Provides general control information.
	14.	Rapidly recovers from equipment failures and emergencies.
	15.	Visual scanning is accomplished.
	16.	Effective working speed is maintained.
	17.	Traffic advisories are provided.
Equipment	18.	Equipment status information is maintained.
	19.	Computer entries are complete/correct.
	20.	Equipment capabilities utilized/understood.
Communication/Coordination	21.	Required coordinations are performed.
	22.	Cooperative, professional manner is maintained.
	23.	Communication is clear and concise.
	24.	Uses prescribed phraseology.
	25.	Makes only necessary transmissions.
	26.	Uses appropriate communications method.
	27.	Relief briefings are complete and accurate.

NOTE: A 3-point scale is used to rate the job functions: (1) satisfactory, (2) needs improvement, and (3) unsatisfactory.

SOURCE: Wing and Manning (1991).

Sollenberger (1996) have been working on a performance measurement checklist based on psychometric measurement principles.

Checklists are also used to evaluate controller candidates who are undergoing on-the-job training to become full-performance-level controllers. These evaluations are completed daily by the assigned instructors and monthly by their supervisor. The results are used to diagnose areas in which additional training is needed and to determine when the developmental controller is ready for certification. Further detail on training and the assessment of trainees as they move through the various stages of the program appears later in this chapter.

In addition to these checklist evaluations, supervisor-prepared annual written performance appraisals are given to both full-performance-level controllers and those still in training, the developmentals. According to an FAA memorandum of understanding on performance assessment (April 1994), each controller receives a mid-appraisal progress review as part of the process. Annual ratings have three levels: exceeds standards, fully successful, unacceptable; they are assigned to various performance elements, including operating methods and procedures, communications, and training. Although the ratings are not anchored with specific behavioral examples at the three category levels, some examples are provided for performance that exceeds the standard. Final ratings are based on weighting the elements (different weights are used for full-performance-level and developmental controllers). The results of the annual performance assessment are used in making decisions about promotions and salary increases.

Relying on performance appraisal systems based on supervisor judgment has been a long-time concern of many private and public organizations, particularly with regard to the validity and reliability of the assessment instruments. However, according to the Research Council's Committee on Performance Appraisal for Merit Pay (Milkovich and Wigdor, 1991), the extensive research literature in this area does not provide strong guidance in choosing a performance appraisal system. The committee found mixed results regarding the advantage of job-specific ratings over global ratings. Moreover, although some researchers believe that scales based on job analyses and behavioral examples are advantageous in providing employees with constructive feedback, the committee found no clear evidence that behaviorally anchored scales are superior to other scale formats in informing the decision-making process.

The use of objective measures to supplement supervisor judgment is most effective in assessing performance for jobs in which the tasks can be quantified. To some extent, the job of the air traffic control specialist provides such an opportunity. However, the problem has been that job analyses have shown wide variation in the specific content of the job depending on the type and level of facility to which the controller is assigned (Hedge et al., 1993). Not only is it true that controllers working in terminals perform different tasks from controllers working in en route centers, and that controllers in facilities with low-volume traffic have different job requirements from controllers in facilities with high-volume traffic, but also each controller's job is tied to a specific air or ground area or sector that has a unique set of features with which the controller must be extremely conversant in order to perform effectively. Indeed, there is evidence that knowledge of the specific airspace features around a facility is one of the most critical aspects of controller expertise (Redding et al., 1992). Because of these task variations and constraints, it has not been possible to develop a uniform performance test or set of tests that would fairly measure controllers' performance across the board. A more detailed discussion of the controllers' cognitive tasks and mental models can be found in Chapter 5.

Over the years, however, some attempts have been made to address these issues by creating simulated scenarios in a generic airspace. The use of simulation for assessment of air traffic controllers was first proposed in a study by Buckley and his colleagues in 1969. In this study, high-fidelity simulations of different traffic configurations in a generic airspace were used to measure individual differences is en route air traffic controller performance. In 1983, Buckley et al. extended this work and identified four important and independent categories for scoring controller performance across different sector geometries and traffic densities:

- Confliction: number and duration of conflicts,
- Occupancy: time and distance flown under control,
- Communication: number and duration of ground-to-air communications, and
- Delay: number and duration of delays.

Although there is a great deal of merit in the approach proposed by Buckley et al. (1983), the argument against using generic exercises has been twofold (Borman et al., 1992; Hedge et al., 1993). First, becoming proficient in a new sector takes time, and, second, being tested has required going to a central location. Although it is possible that such an approach may be useful in developing performance criteria for selection, given the current technology, it does not appear workable as a performance measure to be used in making decisions about the future job responsibilities or salary levels of full-performance-level controllers.

A further complication in using generic simulated exercises for the evaluation of full-performance-level controllers is the difficulty in obtaining reliable measures unless traffic densities are higher than the busiest live traffic in any sector (Buckley et al., 1969). As a result, the generic simulation is not an accurate test of performance requirements in the workplace. In this regard, it is of interest to note that, with current live traffic loads, the base rate of operational errors reported in one year is extremely low—there are approximately 800 errors spread across more than 15,000 active controllers generating about 3 billion opportunities for error (Hedge et al., 1993). It should be noted that reported errors represent the lower bound, since many of the minor errors go unreported.

According to reports from the field, simulation exercises incorporating local features are currently used in en route centers and TRACONS to measure controller technical performance; however, most of these simulations are designed for training and testing developmental controllers or for upgrading the skills of full-performance-level controllers, using sector features with which the controllers are familiar, rather than for evaluation of job performance.

Performance Measures as Criteria for Selection

Performance in Training

Criterion measures used to date for the selection of air traffic control specialists have not been based on data collected on representative work samples of full-performance-level controllers; instead, they have been derived from the performance of developmental controllers during different stages of training (Hedge et al., 1993). Researchers at the Civil Aeromedical Institute have used a variety of measures that are based on weighted combinations of scores from written and simulated training exercises and from the time required in hours, days, and months for a developmental controller to move through each stage of on-the-job training and reach certification as a full-performance-level controller (Manning et al., 1988, 1989). The measures for training time for various stages of on-the-job training have been calculated separately for tower, TRACON, and en route facilities. Success rates at different training stages have also been calculated. For purposes of analysis, those who do not succeed are classified in the following categories: remaining in the same type of facility as a developmental but transferring to a lower-level facility, switching options (e.g., en route to tower), and separating from the air traffic control specialist occupation (Manning et al., 1989). All of these data are available through the training tracking system established by the researchers at the Civil Aeromedical Institute.

In an article discussing progress toward developing criterion measures for air traffic control specialist performance, Hedge et al. (1993) show the various criterion measures that have been used and their intercorrelations. These measures are classified in the three areas of field training performance that include training time and subjective performance ratings by instructors; experimental measures of job performance such as high- and low-fidelity task simulation studies (Buckley et al. 1969, 1983) and job performance ratings; and operational job performance ratings used for decisions about salary increases and promotions. Figure 3.1, taken from Hedge et al. (1993), summarizes the results of studies examining the relationships among these criterion measures. Among the experimental measures, the most highly correlated are supervisor general and specific ratings (.86), performance on low-fidelity videotape simulations and peer nominations (.70), performance on high-fidelity simulations and over-the-shoulder ratings (.60), and specific supervisor ratings and specific peer ratings (.59). The strongest relationship between experimental and operational performance measures was between peer nominations and supervisor annual ratings (.56).

Performance on Job Tasks

Because good job performance is the ultimate goal of selection, it is generally acknowledged (Wigdor and Green, 1991) that selection variables should be linked to measures of operational job performance rather than to measures of

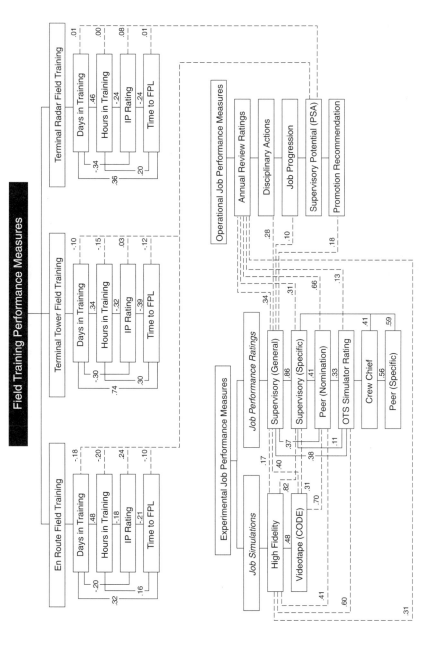

FIGURE 3.1 Summary of empirical evidence for interrelationships among criterion measures. Source: Hedge et al. (1993).

training performance. As noted above, in the domain of air traffic control, the most frequently used measures of job performance are supervisor ratings. What is recommended (Wing and Manning, 1991) is collecting data on full-performance-level controllers who are performing a representative set of work tasks. The precedent for such an effort is the Job Performance Measurement project, a Department of Defense project begun in 1980 to develop robust measures of performance in entry-level military jobs so that, for the first time, military enlistment standards could be linked to performance on the job (Wigdor and Green, 1991). The impetus for the project was the need to establish the credibility of military selection procedures after technical errors in computing test scores were discovered. The goal was to determine how well the Armed Services Vocational Aptitude Test Battery was able to predict performance on high-fidelity measures of job performance. Prior to this 10-year study, training performance had been used as the primary criterion measure.

Throughout its development, the Job Performance Measurement project addressed many important methodological issues that apply directly to the development of hands-on job performance measures for air traffic control. Among these are: (1) identifying and selecting representative tasks as work samples; (2) developing performance measures and establishing criteria for what is to be considered as effective performance (what the full-performance-level controller does on the job and what he or she can do as demonstrated by simulation exercises); and (3) creating a comprehensive data collection plan.

The separation and control hiring assessment (SACHA) program is an effort currently being undertaken at the FAA to develop a selection system that predicts performance of air traffic control specialists at work. It is anticipated that performance will be measured though work samples, behavioral ratings, and time required to achieve proficiency on different aspects of the controller's job.

A critical part of this work, which draws directly on the experience gained in the Job Performance Measurement project (Wigdor and Green, 1991), is the development of job performance criteria based on hands-on tests. As noted above and discussed in Chapter 5 on controller cognitive tasks, there are several complexities associated with developing a representative set of work samples for test purposes. Key among these are the variability in controller jobs and the differences in the sectors being controlled. However, based on a task analysis of controllers in towers, TRACONS, en route centers, and flight service stations, there appears, at the broad level of job duties and worker requirements, to be some commonality among the first three positions (Nickels et al., 1995). Job duties and responsibilities are grouped in the following categories:

- Perform situation monitoring,
- Resolve aircraft conflicts,
- Control aircraft or vehicle ground movement,
- Manage air traffic sequences,

- Route/plan flights—manage airspace,
- Assess weather impact,
- Respond to emergencies and conduct emergency communications,
- Manage sector or position resources,
- Respond to system or equipment degradation, and
- Multitasking.

At the present time, FAA researchers and contractors are working from the results of the task analysis to develop simulation scenarios that can be used to test the core technical skills of full-performance-level controllers. These scenarios will probably use generic sectors that must be learned by controllers in the criterion sample. It is anticipated that training on the generic sectors may require three or four days. Performance testing is expected to require two to four hours. The development of both high- and low-fidelity simulations is under consideration.

To supplement the data from work samples, FAA staff members working on the program are also developing a series of behaviorally anchored rating scales to be used as criteria in the selection system. Initial effort in preparing written examples of different levels of behavior in various categories was accomplished in workshops attended by subject matter experts in air traffic control. To date definitions and performance examples have been developed for the following categories: coordinating, communicating and informing, maintaining attention and vigilance, managing multiple tasks, prioritizing, technical knowledge, maintaining safe and efficient air traffic flow, reacting to stress, teamwork, and adaptability/flexibility. When complete, these scales will be used to assess the performance of a representative set of full-performance-level controllers.

SELECTION

The goal of any personnel selection system is to accurately identify applicants who will be successful in performing the job. For over 50 years, researchers have been working on developing effective selection tests for air traffic control specialists (Sells et al., 1984). As stated in the previous section, the criterion used in determining the validity of these tests for predicting success in the workplace has been performance in training (Sells et al., 1984; Manning et al., 1989). What follows is a brief history of selection research in the FAA. Underlying all this research is the use of task analysis techniques to identify the critical characteristics of the air traffic control specialist's job and the abilities needed to perform the job effectively. It is important to note that all selection tools and procedures have been tested to determine if they have an adverse impact on minorities, as defined by the Uniform Guidelines for Employee Selection Procedures established in 1978. In cases in which an adverse impact was found, the necessary adjustments were made (Manning et al., 1988).

Screening for Cognitive Abilities

Selection research in the 1950s involved a number of studies using commercial aptitude tests to predict performance in controller training programs. The results of this research led in 1962 to the first Civil Service Commission test battery for selecting air traffic control specialists. This battery, which contained a series of tests on arithmetic reasoning, spatial relations, abstract reasoning, and air traffic control problems, was used for 20 years. In 1981, approximately 2 months after the strike, a new selection battery developed by the Office of Personnel Management (OPM) was implemented as a first-stage selection screen. The OPM battery consists of three tests: the multiplex controller aptitude test (MCAT), the abstract reasoning test (ABSR), and the occupational knowledge test (OKT). The current version of the MCAT includes paper-and-pencil simulations of activities required for controlling air traffic; several of the items portray situations that may result in aircraft conflicts, whereas others require time distance computations and manipulations of spatial relationships. An air route map showing allowable flights paths is provided (see Figure 3.2 for example). The ABSR is a 50-item test assessing the ability of applicants to infer relationships between symbols. The OKT contains items in seven knowledge areas related to controlling air traffic. Based on early experimental administrations of the OKT, Lewis (1978) found that the test was a better predictor of success in second-stage screening than self-reports of prior experience.

A weighted average of the MCAT (80 percent) and the ABSR (20 percent) is used for the initial qualifying score; any applicant who receives a score of less than 70 is eliminated from the candidate pool. Those with scores of 70 and above can improve their total by the results of the OKT and by points assigned for veteran preference. The combined total score is referred to as the rating.

Because of the historically high percentage of candidates failing to complete training and become full-performance-level controllers (approximately 44 percent in en route centers), in 1976 Congress recommended that a standardized, centralized program be put in place at the Air Traffic Control Academy. The goal was to put in a second-stage screen that would weed out the candidates who were less likely to succeed in field training. As a result, two nine-week programs were developed: one to screen candidates initially selected for the en route option and the other for candidates in training for tower positions (including TRACONS). In 1985 the two programs were combined into a single screen, and assignment of candidates to options occurred after the screen was completed (i.e., all candidates were screened and then assigned to positions). This second-stage screen, which combines selection and training for candidates with no prior experience, contains a set of nonradar-based air traffic control principles and rules and presents a series of laboratory simulation exercises to test the application of the principles. The laboratory exercises are standardized, timed scenarios that are graded. These exercise grades are combined with written knowledge and skills tests to calculate

Aircraft	Altitude	Speed	Route
10	7,000	480	AGKHC
20	7,000	480	BGJE
30	7,000	240	AGJE
40	6,500	240	CHKJF
50	6,500	240	DIKGB
60	8,000	480	DIKJE
70	8,000	480	FJKID

Sample Question

Which aircraft will conflict?

A. 60 and 70
B. 40 and 70
C. 20 and 30
D. None of these

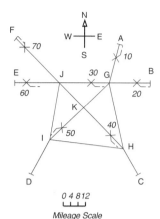

FIGURE 3.2 Sample information from the Multiplex Controller Aptitude Test. Source: Federal Aviation Administration (1995).

a composite performance score that determines whether the candidate passes or fails the screen.

In 1989, Manning et al. examined the degree to which performance on these first- and second-stage selection and screening tests predicted success in training. The data for this study included: (1) field training performance measures for 3,185 en route and 1,740 terminal developmentals and (2) baseline data on 125,000 applicants who took the OPM selection battery as well as over 9,000 entrants to Academy programs. All correlations reported in this study were corrected for restriction in range. The mean score of Academy entrants on the MCAT was 90, and the mean overall selection rating was 91.6. Candidate performance scores in the Academy nine-week screen were consistently higher for the academic portions of the course than for simulated laboratory exercises; scores on these two parts were combined to obtain a composite performance score.

The first analysis in this study was designed to determine how well the scores on the OPM tests predicted success in the nine-week Academy screen. The results show corrected correlations of .55 between the MCAT and the composite performance score of en route trainees in the nine-week Academy screen and .58 between overall selection rating (MCAT + ABSR + OKT + veterans points) and the composite performance of en route trainees in the Academy screen. For terminal trainees the correlations were slightly lower (.48 for the MCAT and composite performance score in the Academy screen and .52 for overall selection rating and composite performance).

The second analysis examined the strength of the relationship between each of the two screens (OPM and Academy) and the time required for a controller to reach full-performance level. Again, all correlations were corrected for restriction in range. Among the OPM tests, the MCAT is the best predictor of time to complete the phases of field training and reach full-performance level for en route controllers (–.26), whereas the OKT is the best predictor of time for VFR tower and TRACON controllers to reach full-performance level (–.21 and –.16, respectively). For the Academy screen, the best predictor of time to reach full-performance level for all three positions was the composite performance score (en route, –.36; tower, –.42; TRACON –.24). In general, the results suggest that both the OPM test and the nine-week Academy screen are useful predictors of training performance as measured by time required to reach the full-performance level.

In another study, Manning et al. (1988) compared the attrition rates in field training before and after the introduction of the common Academy screen. They reported that, prior to the introduction of the screen, 38 percent of those in field training left the agency; after introduction of the screen, the field loss rate was cut to 10 percent. Even with this level of effectiveness in refining the population of candidates entering field training, it was felt that the nine-week screening program was too expensive and time-consuming. The FAA's policy was to hire candidates and pay them as employees while they attended the nine-week screen regardless of whether they passed the screen and continued on to field training. Another consideration in reducing the screen time was a desire to minimize the disruption in the lives of those who failed to complete the screen, particularly with regard to pursuing other employment opportunities.

The approach to this problem was to ask researchers at the Civil Aeromedical Institute to develop a shorter, more efficient screen to replace the nine-week course. The result was a one-week computer-based pretraining, preemployment screen that includes tests designed around the abilities and aptitudes identified for effective performance of air traffic control specialist tasks (Weltin et al., 1992; Broach and Brecht-Clark, 1994). The major categories of aptitudes examined include sensory/perceptual, spatial working memory, verbal working memory, long-term memory, and attention allocation. The tests designed to assess the attributes are described in detail by Weltin et al. (1992). One test, the air traffic scenario test, provides a low-fidelity dynamic simulated work sample; the other two tests measure various cognitive abilities. Essentially, the candidates practice with the computer tests for 3.5 days and then are tested with a series of exercises.

In 1991 a concurrent validation study was conducted to compare the power of the one-week pretraining screen with the nine-week screen to predict training performance (Wetlin et al., 1992; Broach and Brecht-Clark, 1994). In this study, training performance was defined as a combination of field training times and scores in the radar course. Although the relationship appears weak, the results showed that the new one-week screen was slightly better in predicting training performance (according to the above definition) than the nine-week screen. The corrected correlations were .25 and .21, respectively. In 1992 the one-week screen replaced the nine-week screen. Currently, however, the one-week screen is not being used because it was not validated against performance on the job. As the new selection system proposed under the FAA SACHA program is developed, it is planned to incorporate new versions of these computer-based tests.

Biographical and Personality Characteristics

Other variables involved in selection research include (1) biographical data, such as high school performance, age, and prior air traffic experience (Collins et al., 1990) and (2) personality characteristics, such as openness to experience, extroversion, agreeableness, conscientiousness, anxiety, curiosity, and tendency to anger (Schroeder et al., 1993; Nye and Collins, 1991). Personality characteristics of air traffic control specialists have been studied using the following types of tools: the sixteen personality factor questionnaire developed by Cattell (1970), the state-trait personality inventory developed by Spielberger (1979), the occupational personality questionnaire, and the NEO personality inventory, which contains scales for five major personality constructs (Barrick and Mount, 1991), and a variety of self-report surveys.

The contributions of personality variables to the screening of air traffic control candidates were studied by Schroeder et al. (1993). They examined five personality factors and found that, collectively, they explained an additional 3 percent of the variance in training performance over that explained by the cognitive measures. Specifically, air traffic control trainees, when compared with normative samples on the NEO personality inventory, exhibited lower average scores in neuroticism and higher average scores on the dimensions of extroversion, openness to experience, and conscientiousness. Another study of personality variables conducted by Nye and Collins (1991) found that air traffic control trainees appeared to have slightly less anxiety and anger and more curiosity than individuals in the normative group. Overall these relationships appear to be weak at best.

Biographical characteristics of air traffic control trainees have been studied by VanDeventer et al. (1983, 1984), Manning et al. (1988), and Collins et al. (1990). Since the 1981 strike, the population of air traffic controllers has changed significantly with regard to prior experience—more than two-thirds of those hired after the strike have no prior experience in aviation compared with less than one-third before the strike (VanDeventer et al., 1983). Other differences are that the post-strike group is slightly more educated and contains slightly fewer minorities and slightly more women (Manning et al., 1988).

The demographic variables that appear to have the greatest relationship to success in training are high school math grades, age, self-expectation of performance as a controller, and prior military experience (VanDeventer et al., 1983; Collins et al., 1990). Among these, age at entrance is particularly interesting.

VanDeventer et al. (1983) report that the pass rates for the second-stage screen fall off significantly at age 29 to less than 50 percent. It appears from their analysis that the age group most likely to succeed is under 26. As a result, the FAA established a policy of not accepting applicants for air traffic controller positions who are over 30 years of age.

Potential Influences of Automation on Selection

Researchers at the Civil Aeromedical Institute have conducted two studies examining the potential influences of automation on the selection of air traffic control specialists (Della Rocca et al., 1990; Manning and Broach, 1992). Both studies were based on task analyses of advancements proposed in automated en route air traffic control and the advanced automation system sector suites. Della Rocca et al. (1990) report two analyses that identified task changes as a result of proposed automation; basically they suggest that the controller's task would become less tactical and more strategic; the controller would have more information and be provided with an array of computer aids for conflict avoidance. However, the underlying cognitive and sensory attributes would not change significantly over those required in the current system.

In the other study (Manning and Broach, 1992), nine air traffic controllers who had analyzed future requirements of proposed automation were asked to describe how they would expect controllers to perform a selected set of tasks with this automation and to identify the underlying abilities needed for effective performance. The expert controllers suggested that automation would lead to less verbal coordination and less need for the controller to process detailed information. However, they also believed that the underlying abilities needed to perform the job would not change from those currently required. As new automated solutions are proposed, it will be necessary to continue the task analysis process.

The conclusion that automation will not require a change in the underlying abilities to perform air traffic control tasks was derived by considering the specific forms of automation proposed for the advanced automation system. It should be noted that other forms of automation are possible, and the implications of automation for selection should be reconsidered, whenever new forms of automation are contemplated. For example, it is possible that different abilities—or different weightings of abilities—will be required depending on whether automation shifts controller tasks:

- Toward more decision making and away from calculation or spatial perception,
- Toward strategic or toward tactical emphasis,
- Toward supervisory monitoring and away from hands-on control,
- Toward more human-computer dialogue and away from human-human voice communication,

- Toward increasing emphasis on efficiency while preserving safety, or
- Toward team behavior and away from individual behavior.

Researchers with the FAA's SACHA program are continuing efforts to develop a new selection screen that will eliminate the need for the OPM screen. Since researchers have found that learning occurs by taking the MCAT test and that those who take it more than once achieve higher scores, it was decided to replace it with a more reliable measure of underlying cognitive ability.

The new battery will be built on the results of the recently completed job and worker requirements analysis (Nickels et al., 1995) and will include measures of cognitive, perceptual/spatial, and interpersonal characteristics. Plans are to incorporate the computer-based selection tests developed for the one-week screen into this test battery. Currently, it is anticipated that a concurrent validation study of this new battery will be conducted within two years using proficiency measures from high-fidelity simulations of air traffic test problems. Full-scale implementation is expected by 2000. It is also expected that changes in the program will occur as additional information is gained concerning plans for automation.

One useful source of information in developing the new selection system is the continuing research of Ackerman and his colleagues at the University of Minnesota. Most recently, Ackerman et al. (1995) examined the power of a broad set of ability and personality traits to predict skill acquisition during different stages of training in a TRACON simulator. Their results show that, whereas cognitive and perceptual ability scales provided the strongest predictions, overall predictive power could be enhanced by pooling ability measures with measures of personality and self-concept. One of the most powerful predictors was perceptual speed. Work in this area should be conducted on a regular basis to reflect the potential changes in the relationships among the air traffic controller tasks and the power of the predictor variables.

TRAINING

Air traffic control specialist training is accomplished in several phases. Air traffic control specialists are employed by the federal government under the general service (GS) pay system.[2] As noted in the previous section, a nine-week course for applicants with no prior experience in air traffic control was intro-

---

[2]The grade for an individual position is determined by several factors, including the level of service demand at a particular air traffic facility, which can change depending on the traffic count. Facilities range from level 1 (lowest activity) to level 5 (highest activity). Flight field stations have three levels, and the highest grade for those controllers is GS-12. The terminal facilities have five levels, and highest grade is GS-14. En route centers have three levels, and the highest grade is GS-14. Supervisor grades for all three types of facilities range between GS-12 and GS-15. Entry-level positions are generally GS-7; however, individuals sometimes qualify for entry at the GS-9 level.

duced at the Air Traffic Control Academy in 1976. This course had two functions—to provide initial skill training in nonradar tasks and to act as a screen for the next stage of training. In the past, those who passed the course (approximately 60 percent) entered field training as developmentals. Those who completed the nonradar portion of field training returned to the Academy for radar training and then went back to the field for additional on-the-job training and eventual certification. At each stage, pass/fail points were built in.

In recent years, several changes have been made. One change is that the philosophy throughout the program has become "train for success." Instead of imposing several pass/fail screens in the training process, the idea is to provide a supportive learning environment for the developmental—one that incorporates deficiency diagnosis and skill enhancement training as part of the overall program. This philosophy has led to a redesign of training in the facilities to ensure continuity of on-the-job instruction throughout the process. The underlying rationale was to create a training system that is supportive and reinforcing rather than one that is punitive.

A second change is that more training is conducted at the facilities. Specifically, for those in the en route option, all radar training now takes place at the facility; however, those assigned to towers and TRACONs still take the Academy radar course. Thus, except for the introductory course, all en route training occurs at the assigned facility.

A third change is that new applicants are being accepted only if they have previous experience in air traffic control; as a result, the first introductory course at the Academy is not being taught at the present time. The current sources of these experienced trainees include former members of the Professional Air Traffic Controllers Organization (PATCO), the military, the collegiate training initiative (CTI) for air traffic controllers, those currently functioning as air traffic assistants, and those in special cooperative and predevelopmental programs in the FAA. At the present time, most controllers are former PATCO or CTI graduates. The CTI program plans to expand from its five original institutions to 10 or more, with the goal of producing approximately 700 graduates a year. It is anticipated that some of the new programs will be located at institutions that are in regions and areas in which staffing has been difficult.

As the requirement for new controllers increases and as new equipment is introduced, it is anticipated that selection techniques will become increasingly important for identifying those applicants with appropriate abilities for the job. In addition, modifications in training are also expected. As a result, researchers are actively working on new programs to be put in place by 2000. An important aspect of this effort is the significant strides being made with microcomputer-based simulations for training. One example is the work currently being conducted using the TRACON simulation developed by Wesson International (Ackerman et al., 1995).

According to the Air Traffic Control Technical Training Order 3120.4H

(1995), once a developmental reaches his or her assigned facility, a team is formed to manage the individual's training. The team includes the developmental, two assigned on-the-job training instructors, the developmental's supervisor, and perhaps the facility's training administrator. All technical training requirements and instructional program guides are provided by headquarters. Some training is standard for all developmentals; other training is facility-specific. As a result, training and its evaluation may vary substantially from one facility to another. For the most part, facility training begins with classroom instruction, followed by work on the simulators. The content of these activities is tied to the facility's areas of responsibility. Many of the simulation exercises are developed around live traffic scenarios that have occurred in the sectors covered by the facility. These DYSIM (dynamic simulation) exercises are used to assess the developmentals' performance on different aspects of the job. In the future, SATORI (situation assessment through re-creation of incidents), which provides a graphic display of data with synchronized tapes of verbal interaction, can be used to review the performance on DYSIM problems. That is, a trainee's performance can be replayed as a means of providing immediate, detailed, corrective feedback (Rodgers and Duke, 1994).

In the en route centers, the first training is on the D-side (nonradar side) followed by the R-side (radar side) training. For both sides, the cycle begins with classroom and simulator training, followed by on-the-job training. The average and maximum time for each training activity is established by the facility. As developmentals move through the on-the-job training phases, they are observed and rated on a checklist each day by one of their on-the-job training instructors. These evaluations are used to determine the readiness for certification or the need to return to the classroom or simulator for skill enhancement training. Each month a supervisor also observes the developmental and completes the checklist.

The purpose of continuous evaluation and the support of a training team is to provide the developmental with every opportunity to gain the necessary competence to achieve certification and move forward. If a developmental exceeds the maximum time allocated for training in a phase, then he or she may be given remedial instruction or be assigned to another option or lower-level facility. There are 13 training phases to be completed before a trainee reaches the full-performance level. At each phase there is a certification examination. Certification on equipment is provided by the Academy; position certification is provided by the facility. All examinations are developed by the Academy. The time required for a developmental to reach full-performance level may be in excess of three years.

When new hardware and software are introduced, new training is required for all controllers who will be using the equipment. At the time of purchase, the FAA determines who will be responsible for the development of a training package to support the equipment. One choice is to have one contractor develop the entire package (hardware, software, and training); other options include selecting a different contractor or developing the training package in-house.

Training on new equipment may be conducted in three different ways. In some cases, the FAA will choose to train a special cadre of specialists, who in turn train the appropriate personnel on site. In other cases, a contractor is given the responsible for on-site training. A third approach is to give the training package directly to the facilities where it will be used and have the facilities conduct their own training. Larger facilities have training departments that manage and administer training for the facility, whereas smaller facilities may have only one person responsible for training.

It is of interest to both the FAA and those enrolled in training that training program effectiveness be maximized. Academy and field training programs have attempted to reach this goal through the use of classroom instruction, specifically designed and controlled simulated exercises, and on-the-job apprenticeships. Although there have been no systematic comparative evaluations of these procedures as employed by the FAA, there is, however, some anecdotal evidence that training has a positive impact on controller performance. For example, the approach controller at Sioux City Gateway Airport who guided the flight crew of a severely crippled United DC-10 to landing on July 19, 1989, had recently benefitted from completing an excellent facility-based training program. This controller was praised by the FAA administrator for his professionalism, skill, training, and personal dedication. A review of the transcript of his radio communication with the flight crew reflected exceptional performance under extremely difficult and stressful conditions (personal communication, National Transportation Safety Board, 1996). Positive impact is, of course, demonstrated by the fact that full-performance-level controllers are produced by the training system.

A critical question in designing effective training programs concerns how well training in a simulator or on the job transfers to actual performance on the job. On-the-job training as practiced in air traffic control facilities is essentially an apprenticeship program designed to systematically move an individual from the status of developmental to full-performance level. Thus, by the time the developmental reaches full performance, he or she has been performing the job for some time under the guidance of an on-the-job training instructor, thus making transition essentially seamless.

The concept of apprenticeship learning emphasizes the idea that, if individuals are trained on elements that are identical to those in the job, the degree of transfer will be maximized. Recently this idea has been highlighted in the theory of situated learning (Lave and Wenger, 1991; Greeno et al., 1993), which further states that learning is a social activity that is facilitated by the context in which it occurs. Some general principles of situated learning are: (1) an individual's knowledge about an action is dependent on the situation, (2) learning occurs by doing, and (3) to understand learning and performance, it is important to understand the social situation in which the learning and performance occur. Although the concepts of situated learning on the job have been well established, there may be concerns about safety, job stress, and the ability to provide the trainee with the full range of experiences in the real-world setting. Furthermore, in some settings, such as the flight deck, the stress of live performance and the difficulty of providing immediate and accurate feedback makes on the job training a less-than-ideal learning environment (O'Hare and Roscoe, 1990). As a result, it may be useful to use simulation techniques as a supplement (Druckman and Bjork, 1994).

A key issue in the design of simulators is the degree of required fidelity and realism. According to Miller (1954), there are two types of fidelity—engineering and psychological. Engineering fidelity refers to the degree to which the physical features of the simulator represent the real-world equipment. It has been suggested by Patrick (1992) that psychological fidelity is far more important than engineering fidelity, particularly for training of cognitive and procedural tasks. The focus of this view is to determine the factors that are required to produce psychological fidelity in the simulation. A more complete discussion of situated learning, training transfer, and the features of effective simulations for training can be found in Druckman and Bjork (1994).

Another consideration in designing simulated exercises is the potential effectiveness of decomposing the task and providing separate training for the subtasks. Naylor (1962) proposed that such part-task training would be most effective when the task is complex and the components are not highly integrated. Evidence for this approach is provided in a study of training for airplane flight skills (Knerr et al., 1987). By analogy, it is reasonable to assume that part-task training may be effective in helping air traffic controllers acquire skills for those tasks that are complex and not structurally integrated.

The FAA is currently using the operational computers in the air traffic control facilities (HOST and ARTS) to provide simulation training. In the en route centers' simulation, it is called DYSIM, and in the terminal facilities, it is called ETG (enhanced target generator). Both systems generate simulated targets that can be maneuvered by a pseudo-pilot operating from a remote radar display with a keyboard. In these simulations, the radar display and keyboards are the same as those used in the actual control room, thereby creating what is called full-fidelity simulation. Scenario scripting is done by outlining the path of each aircraft in a specific simulation syntax that is difficult to master. This development process is extremely time-consuming. When the simulation is operating, it cannot be stopped and replayed for lesson reinforcement. In the ARTS ETG, in some instances the computer capacity of live traffic will limit the number of simulated targets available or even drop the simulated targets in the middle of a training session.

To overcome the drawbacks of the DYSIM and ETG, the FAA is currently in the process of studying methods for providing simulations through the use of personal computers. The personal computer-based radar simulation will not be full fidelity, i.e., the simulators will not be the exact duplicate of what the controller will use in the live control environment, but they will emulate the radar display and the keyboard entries used in the live environment. In addition, the FAA is conducting research and development tests at some terminal facilities using these simulators. State-of-the-art personal computers that can be purchased off the shelf are being used; they are user-friendly and scenarios can be generated within minutes, compared with the days required for ETG and DYSIM. Training scenarios can be stopped and rewound for lesson reinforcement and played back at a later time for review. Aircraft can be piloted by use of a pseudo-pilot as with the DYSIM and ETG, or they can be piloted through the use of voice recognition. In the initial stages of a controller's development, this can be a very useful tool, as it saves on the use of additional personnel needed to play the pseudo-pilot role. These systems are also less expensive than using the actual radar displays for simulation. Four personal computer-based control positions cost approximately $250,000—the equivalent cost of one radar display.

In 1994, the FAA formally requested bids to provide personal computer-based radar simulators for the terminal environment. Contract award was expected sometime in 1996.

SUMMARY

Performance assessment for full-performance-level controllers has primarily involved checklist ratings by supervisors. Although some work has been put into designing simulated exercises, there appear to be too many complexities to consider standardized simulated exercises for purposes of performance assessment. According to the FAA, improvements are being planned for the over-the-shoulder method—an approach that offers direct assessment in the actual job environment, which is accomplished readily and can be repeated as often as needed. As automation is increasingly applied and system-level performance goals are established, additional work on controller performance evaluation will be required to parse out those aspects of system performance that can be attributed to the controller. This task becomes increasingly difficult with additional automation of functions, because human performance may be masked by machine/computer performance. The challenge is to develop precise definitions of system performance that permit the identification of the contributions of both controllers and machines/computers—and that allow for their assessment, both individually and collectively.

There has been a significant amount of research on personnel selection within the FAA, and researchers have conducted numerous studies examining the relationships between predictors and criteria. The principal drawback in this work has been the lack of good performance criteria. Now, with the SACHA program, researchers are working toward the development of job samples that can be used to collect hands-on performance. The panel encourages development of these job-related criteria.

A comprehensive, integrated selection battery is needed that can be given to potential candidates in a short period of time. The content of such a battery should include tests of the skills and knowledge relevant to current and proposed job tasks as well as assessments of personality and demographic variables and their relationship to performance. As automation is introduced, it will be important to reevaluate the elements of the selection battery.

A program is needed to provide formal evaluation of operational air traffic control training; most facility-based on-the-job training is idiosyncratic, with each facility making its own decision. There are currently evaluations of on-the-job training and simulator training for other jobs, but not for those of the air traffic control specialist.

The panel encourages the use of simulation for training at each facility, particularly in light of the need for full-performance-level controllers to efficiently receive refresher training as well as training in the operation of new equipment. Reduction in staffing levels puts additional pressure on the development of such a capability.

4

Airway Facilities

Airway Facilities is the FAA organizational element responsible for ensuring that systems and equipment are available to users of the national airspace. Such users include: air traffic controllers at en route centers and terminals (TRACON and tower controllers), traffic management personnel at the national Air Traffic Control System Command Center and at local traffic management units, flight service station specialists, and pilots, both as direct users of such equipment as navigational aids and communications systems and as indirect users receiving assistance from controllers. This chapter summarizes the breadth and suggests the depth of the activities performed by Airway Facilities specialists that bear on issues relating to automation of functions that support air traffic control.

The division of responsibility between Air Traffic and Airway Facilities personnel is more complicated than a simple distinction between attending to aircraft and attending to equipment. Air traffic controllers always consider the status and performance of the equipment on which they rely and develop strategies to maintain flight safety and efficiency despite equipment limitations, and Airway Facilities staff always consider the safety and efficiency of air traffic in the scheduling and prioritization of their tasks.

To fulfill its current responsibilities, Airway Facilities monitors, controls, and maintains the equipment on whose reliability, availability, and performance controllers and pilots rely. In addition, Airway Facilities shares with the Air Traffic organization the responsibility for installing and evaluating new, increasingly automated equipment as well as software and hardware upgrades to existing equipment. A critical procedural and legal responsibility is the certification of equipment, systems, and services. This certification responsibility involves the validation by Airway Facilities specialists that the equipment, systems, and services are performing within specified tolerances, as well as the legal attestation of certification with accompanying accountability.

SCOPE OF RESPONSIBILITIES

Airway Facilities monitors, controls, maintains, and certifies equipment, systems, and facilities that support air traffic control. Although a comprehensive list of items under its purview includes equipment at flight service stations (and their interfaces) and field sites of various types (e.g., remotely located navigation aids and communications links), our emphasis in this report is on the equipment and systems that support the activities of controllers at the en route centers and the terminals (including TRACONs, towers, and equipment local to airports). Taken together, such items include:

1. Equipment internal to facilities: the HOST computer that processes the radar data and flight data presented to controllers (and the backup direct access radar channel used at en route centers to provide relatively unprocessed radar data to controllers when the HOST fails); the display channel processors that further process and provide data for the controllers' displays; the radar data displays, alphanumeric readout displays, keyboards, and other elements of the controllers' workstations; flight data entry equipment and associated printers used to prepare the controllers' flight strips; intrafacility communications equipment; power supplies for air traffic control systems; and building systems (e.g., heating, air conditioning, electricity).

2. Equipment that interfaces with the facilities and the interfaces themselves: radars (long-range, airport surveillance, and weather radars); communications equipment (air to ground and interfacility communications); and airport local equipment with associated display/control devices used by tower controllers (runway lighting, low-level windshear alert equipment, runway visual range equipment, weather instrumentation, airport surface detection equipment, and microwave and instrument landing systems).

Although this list is not exhaustive, it suffices to suggest that the critical task of supervisory monitoring and control of equipment and systems that ultimately support the activities of air traffic controllers merits careful attention with respect to human factors and automation issues.

EQUIPMENT SUPPORTING SUPERVISORY CONTROL OPERATIONS

The Trend Toward Centralized Monitoring and Control

The equipment and systems that support air traffic control are widely distributed. Equipment sites may be generally categorized as: (1) remote from an en route center or a terminal (e.g., remote radar sites), (2) on the air traffic control room floor (e.g., the controllers' plan view displays and other workstation equipment), and (3) in the equipment "back room" at an en route center or terminal (e.g., HOST and other computers).

Airway Facilities specialists monitor and control equipment both at the site of the equipment itself and through centralized maintenance control centers. At the site of the equipment, they inspect any display and status indicator panels located on the equipment and can usually perform control actions (including such actions as shutdown, start-up, and adjustments) and initiate built-in diagnostic routines (often with the assistance of plug-in diagnostic tools). Air traffic controllers also monitor the status and performance of their workstation devices; their shout for Airway Facilities assistance is a common form of alarm for such equipment.

In addition (Federal Aviation Administration, 1994c), critical equipment and systems are increasingly required to include in their design an automated remote monitoring subsystem (RMS) that performs the following functions and reports the results to a centralized maintenance processor subsystem (MPS): acquiring necessary data from the monitored system; determining the status (e.g., failed, degraded, normal), the state (e.g., the availability of a redundant component), and the performance characteristics of the monitored equipment's hardware and software functions; determining whether it is appropriate to activate status alarms; and monitoring parameters of equipment and systems to support the certification process.

The MPS stores the data for later retrieval and further processes the data to permit its display at a centralized maintenance control center, at which Airway Facilities specialists monitor the equipment's and the systems' status, state, performance, and associated alarms and indicators; request data for diagnostics and certification purposes; and perform many control actions.

Maintenance Control Centers

Airway Facilities activities are organized by sectors (the Airway Facilities' use of the term *sector* does not correspond to its use by air traffic controllers). Each sector consists of an organization that monitors, controls, and maintains the set of equipment assigned to it. There are two fundamental types of sector (Blanchard and Vardaman, 1994): air route traffic control center (ARTCC) sectors are responsible for equipment and systems associated with en route centers; general national airspace system (GNAS) sectors are typically responsible for equipment and systems associated with terminal facilities and with remote facilities. The GNAS and the ARTCC sectors must coordinate their activities, because each monitors and controls systems on which the other relies. In addition, sectors of either type may have to coordinate their activities with several other sectors, which may share equipment resources. Each sector is supported by a maintenance control center (MCC).

Although the correspondence between the monitoring and control capabilities at the MCC and those provided directly at the site of the equipment varies considerably with the specific equipment and systems, and although some equipment does not interface with the MCC, the general principle is that Airway Facilities specialists are supported by a centralized monitoring and control workstation suite, which functions as the command center for activities supporting en route centers and terminal facilities. Airway Facilities activities throughout these facilities are generally directed by their operations managers, for whom the MCCs represent command centers. The MCC workstations merit significant attention because they are the hub of the supervisory monitoring and control activities that support air traffic control at en route centers, at TRACONs, and at airport towers and because they are focal points for the impacts of automation on such Airway Facilities activities.

The MCC for an ARTCC sector is termed a system maintenance control center (SMCC). An archetypal SMCC is useful for investigating the monitoring, control, and associated automation aids currently available to Airway Facilities staff. The SMCC provides a focal point for identifying both current system monitoring and control capabilities and issues bearing on the applications of automation.

The SMCC typically consists of an extensive set of separate indicator and alarm panels, control panels, keyboards, video displays, and printers that—taken together but hardly integrated—provide the capability to monitor and control limited aspects of: processing and distribution of radar and flight data, configuration of computers and peripheral equipment, communications equipment, audio tape recording of air traffic controllers' communications, environmental systems such as heating and air conditioning, and facility power subsystems.

Each MCC is, in effect, a concatenation of separate workstations designed by separate vendors and developed under separate acquisition programs. MCCs exhibit neither a consistent approach to automation across the systems monitored nor a consistent human-computer interface.

OPERATIONS

In general, Airway Facilities is responsible for the following activities pertaining to equipment, systems, services, and facilities (FAA Order 6000.15B):

monitoring of status, configuration, and performance; control (including adjustment and configuration); diagnosis of hardware and software problems; restoration of systems and services experiencing outages; certification; removal, maintenance, and replacement of items for periodic and corrective maintenance; logging of maintenance events and related data; and supporting aircraft accident and other incident investigations. Automation has been applied to maintenance activities through built-in equipment-level diagnostic tests and off-line diagnostic tools. A logging system that prompts the manual entry of maintenance and incident data supports both maintenance and incident/accident investigations (the system is described in FAA Order 6000.48). These applications of automation are widespread within Airway Facilities and therefore merit consideration.

Certification

Certification activities represent a high-visibility, critical responsibility of Airway Facilities that has long been considered a candidate for increased automation support. Because of safety and associated legal liability concerns, the FAA has established procedures whereby equipment, systems, and the services they provide (e.g., radar data) can be accepted for use by air traffic controllers only if they have undergone a process of verification followed by formal, written certification.

The verification and certification are performed by technicians who must be "certified to certify" and accept legal accountability by signing the certification log. Certification is performed when the equipment or systems are first accepted for use, when they are restored to use after interruption or maintenance, and periodically as scheduled (Federal Aviation Administration, 1991c, 1991d).

FAA Order 6000.39 (Federal Aviation Administration, 1991c:3) defines the two general types of certification as:

> Service Certification. The verification that the appropriate combination of services, systems, and equipment advertised to the user have been certified and that they are providing or capable of providing the functions necessary to the user, and followed by the prescribed entry into the log. The certifying official uses personal knowledge, technical determination, observations, and inputs from other certified personnel to accomplish certification.
>
> System/Subsystem/Equipment Certification. The technical verification performed prior to commissioning and/or service restoration after a scheduled/ unscheduled interruption affecting certification parameters, and periodically thereafter inclusive of the insertion of the prescribed entry in the facility maintenance log. The certification validates that the system/subsystem/equipment is capable of providing that advertised service. It includes independent determination as to when a system/subsystem/equipment should be continued in, restored to, or removed from service.

Restoration to Service

FAA order 6000.15B (Federal Aviation Administration, 1991d) identifies as a critical responsibility the restoration of equipment, systems, and facilities after service interruptions (unanticipated shutdowns) and other outages. General examples of outages include: internal failure of equipment hardware; power outages due, for example, to cable cuts or lightning that affect some or all facility systems; communications outages occurring within or between facilities or between ground and air; facility fires creating outages or the requirement to vacate the facility; major software failures that lock out input/output devices or produce faulty data; radar failures that temporarily eliminate this essential data source; and equipment, systems, and facilities shut down intentionally for maintenance purposes and for the installation of new or modified hardware and software.

Restoration to service of failed equipment, systems, or entire facilities requires close cooperation between Airway Facilities and air traffic controllers, as well as staff on site and across sector, regional, and national levels, because outages at one facility can affect systems and services at other facilities, and because the responses to outages may require the support and approval of staff at other locations. FAA Order 1100.124 and FAA Order 1100.139 (Federal Aviation Administration, 1970, 1974), addressing the respective responsibilities of Air Traffic and Airway Facilities staff at computer-equipped en route centers and terminal facilities, directs that "No individual or organization shall be permitted to take unilateral action which may have a detrimental effect on the scheduling, testing, maintenance, and utilization of the air traffic control system."

Although Airway Facilities is exclusively responsible for monitoring and maintaining equipment, systems, and facilities, FAA Order 6000.15B (Federal Aviation Administration, 1991d) specifies that Airway Facilities must keep Air Traffic advised of the operational status of all systems, subsystems, facilities, and equipment and that it is the responsibility of Air Traffic to determine the priority of restoration when more than one item of equipment, system, or facility has become inoperative. All intentional shutdowns (e.g., to install new equipment or to perform system certification) must be requested by Airway Facilities and are approved at the discretion of Air Traffic.

In addition to coordinating with local Air Traffic staff, each MCC reports all equipment and system outages and restoration activities to the National Maintenance Coordination Center (NMCC), located in Herndon, Virginia. The NMCC monitors the following situations and coordinates resolutions with the national centralized Air Traffic Control System Command Center in Herndon: facility and service outages (equipment failures, software failures, power failures, and telecommunications failures); natural disasters (severe weather alerts, earthquakes, hurricanes, tornadoes, etc.); and other disasters (criminal acts, acts of terrorism, and air traffic accidents). The NMCC staff coordinate the responses of cooperating sectors and facilities and, when appropriate, notify or mobilize engineering and logistics support, civil emergency preparedness organizations, federal criminal justice personnel, and military organizations.

In all cases involving interruption and restoration of items affecting air traffic control, Airway Facilities functions in a supportive capacity. Air Traffic must decide the priorities by which Airway Facilities applies its resources. However, in so doing, Air Traffic must consider recommendations from Airway Facilities that take into account the likelihood (considering logistics, staffing, and the course of problem determination) of restoring the affected item(s) within desired time frames, the levels of functioning available with degraded equipment, and potential temporary work-around strategies. That is, Air Traffic can establish desired priorities but cannot restore the affected items alone. Therefore, the outage of automated systems or functions becomes a problem that must be jointly solved. Such cooperative problem solving is currently addressed by experience and procedures rather than by automated supports, at both the local and national levels.

STAFFING

The FAA's 1993 demographic profiles of the Airway Facilities workforce identifies a total population of over 11,000, including over 9,000 engineering and technical staff. Of the technical staff, approximately 7,300 are classified as electronics technicians, each of whom typically specializes in one of the following: radar, communications, navigation, automation, and technical management. The electronics technicians are of primary interest with respect to the direct support of air traffic control operations and are those most significantly affected by proposed automation.

The primary staffing unit for technical activities in support of en route and terminal operations is the Airway Facilities sector. There are currently 20 ARTCC sectors and 57 GNAS sectors, although the FAA is in the process of consolidating the 77 sectors into 33 system management offices (SMOs). Including engineering, technical, and administrative staff, approximately 8,500 members of the overall workforce are assigned to sector organizations: 6,500 to GNAS sectors and 2,000 to ARTCC sectors.

Each sector is staffed as a "self-contained and self-sufficient" work unit. Each sector is organized into system operations and maintenance engineering groups reporting to the sector manager. The system operations group is responsible for the monitoring and management of the systems-level operations. It includes the operations managers and the technical specialists, whose activities are focused on the MCC described above. The operations manager has traditionally been an automation specialist knowledgeable in both hardware and software aspects of several systems. Each technical specialist is typically expert in computer operations, radar, communications, navigation, *or* environmental systems and equipment, although some may be certified in more than one area. The operations managers and technical specialists of the systems operations group are supported by computer operators and by maintenance engineering technical support staff, who perform detailed diagnostics, data analysis, and maintenance tasks (Federal Aviation Administration, 1991a). Actual assignments of numbers and types of technicians are determined by specific sector requirements according to guidelines prescribed by FAA Order 1380.40C (Federal Aviation Administration, 1991a). These requirements can vary considerably across sectors. In addition, the regional level includes a staff member whose function is to perform technical inspection of the region's sectors, both periodically and on special occasions (e.g., after an aircraft accident or incident or when sector performance has fallen below an acceptable level) (Federal Aviation Administration, 1991b).

A New Job Classification

Until 1994, the focus of electronics specialists was on specific subsystems or items of equipment to which they were assigned. Assigned specialties included radar, navigation, communication, and automation (computers). The automation specialist job classification was created to recognize the trend toward computerization and the need to develop technical skills that apply across computer-based systems. The operations manager, selected for knowledge spanning multiple systems, was often selected from the ranks of the automation specialist. However, in practice, the proliferation of computer subsystems has often drawn the automation specialist toward specific systems, such as the HOST computer.

In recognition of the need to develop generalists who focus on system-level functions and the delivery of services across interacting systems, the FAA has recently created the GS-2101 job classification, whose knowledge and skill areas emphasize systems engineering skills (Booz, Allen and Hamilton, 1993, 1994; Federal Aviation Administration, 1993c, 1993d). The GS-2101 job classification was established in order to address two perceived needs: (1) a response to rapid changes in information technology (e.g., the trend toward networks) and (2) an emphasis on the management of systems and services rather than on maintenance of equipment components. The knowledge, skills, and task emphases of the GS-2101 specialist include: ability to work with automation tools for diagnostics and maintenance, ability to perform centralized monitoring and control, ability to perform system- and service-level certification, breadth of knowledge across systems rather than depth of knowledge of specific items of equipment, knowledge of how information flows between systems, ability to work with information management systems, maintaining end-product services for users, performance of independent actions, and ability to work well in interaction with others (users who are treated as customers and colleagues).

Airway Facilities staff currently view the national airspace system as a hierarchical structure of elements that contribute to systems that provide services. The art of Airway Facilities consists of maintaining uninterrupted services despite the degradation or failure of individual elements. Airway Facilities specialists therefore become creative developers of "work-arounds, band-aids, and patches" that keep systems operating. Before the advent of the GS-2101 job classification, Airway Facilities personnel increasingly considered themselves systems engineers as well as electronics specialists. Therefore, it is unclear whether the new job classification provides a title that recognizes an existing approach to the job or whether it represents a distinct, new approach. The formal position descriptions, qualification standards, and job classification descriptions do not determine the answer to this question by themselves; the answer will be determined more by the selection process, training procedures, task assignments, and the characteristics of the new equipment that the GS-2101 specialists will be required to maintain. Currently, the GS-2101 incumbents are the same specialists who have been maintaining the existing equipment and systems. Therefore, the GS-2101 represents more an approach to the future than a response to the current task demands.

Selection and Demographics

The selection of Airway Facilities technicians is neither centralized nor standardized. Each region hires new technicians by evaluating the experience and education reported in candidates' applications against knowledge and skill criteria for the specializations that the regional office requires. Guidance applicable to each specialization is available in formal qualifications standards and position descriptions. There is no prehire selection test for Airway Facilities personnel. Those hired typically have backgrounds in electronics, usually developed in military service or through technical education.

The introduction of the GS-2101 job classification changes the knowledge and skill qualifications used to guide hiring (Federal Aviation Administration, 1995). Knowledge of computer systems, computer programming, networks, telecommunications, and systems analysis methods are included in the GS-2101 qualification standard, in addition to the traditional knowledge of electronic principles. GS-2101 specialists must also possess good interpersonal skills, because their jobs include significant interaction with other staff.

In terms of demographics, the average age of the Airways Facilities workforce is 44.6 years, and 64 percent are older than 40. The average length of service is 18.5 years, and 47 percent have 20 years or more of service. Currently 13 percent of the workforce is eligible for retirement; within 10 years, 35 percent of the ARTCC sector workforce and 27 percent of the GNAS sector workforce will be eligible for retirement (Federal Aviation Administration, 1993a).

These data combine to suggest that the workforce will see, within 10 years, a simultaneous retirement of significant percentages of its experienced technicians and the equipment on which they have developed their experience.

Training

A salient feature of the technician hiring, training, and placement process is that new hires are placed at the time of hire rather than on the basis of performance during training. With respect to electronics technicians, each region determines its staffing needs, reviews candidates' work and education histories, and on that basis assigns them at the time of hire in one of the following areas: radar, navigation, communications, or automation. Technicians may also be assigned at the time of hire to specific systems, subsystems, or equipment items within an assigned specialty. The implication for training is that trainees do not pursue a generalist course covering all specialties (in which case their training performance might suggest appropriate job assignments or on-the-job training assignments). Rather, since the assignments are made a priori, trainees pursue at the FAA Academy strings of courses that are tailored to the assigned specialty and, within that specialty, to the assigned systems, subsystems, or equipment.

A brief set of general electronics foundation courses are prescribed for all electronics technician trainees and administered mostly at the FAA Academy. These courses include instruction in general electronics, semiconductor and digital techniques, introductions to computers and microprocessors, and fundamentals of engineering mathematics. At completion of these few foundation courses, each trainee moves along a personalized track the course contents of which are tailored to a specialty area and assigned systems or equipment (Federal Aviation Administration, 1995; FAA Catalog of Training Courses).

The training process has two goals: (1) certification of the technician's abilities with respect to given systems and equipment, so that the technician may be authorized to certify them for use in air traffic control and (2) career progression, so that, by demonstrating proficiency, the technician can progress to journeyman status. In principle, these goals are met by providing theory through course material and application through subsequent on-the-job training.

In order to meet the specific needs of different regions and their sites, the Airway Facilities training program is extremely flexible. Training may be administered at the FAA Academy or at the site. Training methods include classroom instruction, correspondence study courses, computer-based instruction, and on-the-job training. A curriculum modernization study is under way to determine the most effective combination of media and content for each current course objective (Federal Aviation Administration, 1994a). A given trainee may be assigned to receive instruction in a specific item of equipment or across several systems. After initial and on-the-job training, technicians may receive additional training when systems are modified, when new systems are introduced, or if refresher training is needed. New hires undergo a post-hire assessment whereby, through evaluation of experience or through testing, he or she may be permitted to bypass appropriate initial training courses. The goal is to train technicians as quickly and inexpensively as possible while maintaining performance standards

(Federal Aviation Administration, 1976, 1985). Local (regional and sector level) supervisors play the major role in determining the training program for each technician and in evaluating the proficiency and progress of trainees.

Performance Appraisal

Three types of performance evaluation are used for technicians: tests of mastery of training material, certification tests, and yearly performance appraisals. Formal tests, often standardized, are used to evaluate the technicians' mastery of training material. The "train for success" philosophy is applied: at the discretion of their supervisors, trainees are permitted to retake failed examinations. A formal personnel certification program, prescribed by FAA Order 3400.3F, establishes the examinations that technicians must pass to achieve certification. Certification is defined as "confirmation that the individual possesses the necessary minimum knowledge and skills to determine the operational status of a service/system/subsystem/equipment" (Federal Aviation Administration, 1992a:3). Personnel certification is a critical goal for electronics technicians, because it permits them to exercise, when so assigned, the duty of certifying systems, services, and equipment for use in air traffic control. Personnel certification is therefore fundamental to the technician's viability within the organization; it also contributes to career progression. Certification examinations are conducted by the technicians' local supervisors. The examinations typically consist of observation by the supervisor of walk-through exercises performed by the technician. The supervisor also uses personal knowledge of the technician's proficiency, based on observations during on-the-job training, as well as inputs from other certified personnel.

Technicians undergo performance appraisal yearly. The appraisal is conducted by their direct supervisors, who apply a five-step rating scale (unsatisfactory, partially satisfactory, satisfactory, exceptional, and outstanding). Monetary rewards are calibrated to the rating scale. Three general appraisal factors are used: organizational effectiveness (includes the performance of assigned technical tasks and duties), customer focus (the satisfaction of Airway Facilities and other users of the services provided), and teamwork (emphasizes effectiveness of working relationships with other Airway Facilities staff and with airspace users).

Formal discussion of performance is also undertaken semiannually, without written performance appraisal or rating. Airway Facilities is currently seeking approval by the Office of Personnel Management of a condensation of the performance rating scale to pass/fail and the elimination of associated monetary rewards in favor of biennial step increases.

SUMMARY

Airway Facilities monitors, controls, and maintains the systems and equipment on whose reliability, availability, and performance air traffic controllers and pilots rely. In addition, it shares with the Air Traffic organization the responsibility for installing and evaluating new, increasingly automated equipment as well as software and hardware upgrades to existing equipment.

A critical procedural and legal responsibility is the certification of equipment, systems, and services. Restoration to service of failed equipment, systems, or entire facilities is another critical task and requires close cooperation between Air Facilities and Air Traffic, between Airway Facilities staff on site, and across sector, regional, and national levels.

Airway Facilities specialists monitor, control, and maintain equipment and systems that support air traffic control both at its site and at increasingly centralized maintenance control centers. Each MCC is, in effect, a concatenation of separate workstations designed by separate vendors and developed under separate acquisition programs. MCCs exhibit neither a consistent approach to automation across the systems monitored nor a consistent human-computer interface.

Until 1994, the focus of electronics specialists was on specific subsystems or items of equipment to which they were assigned. The automation specialist job classification was created to recognize the trend toward computerization and the need to develop technical skills that applied across computer-based systems. In recognition of the need to develop generalists who focus on system-level functions and the delivery of services across interacting systems, the FAA has recently created the GS-2101 job classification, for which the knowledge and skill areas emphasize systems engineering.

It is unclear whether the GS-2101 job classification provides a title that recognizes an existing approach to the job or whether it represents a distinct, new approach. The formal position descriptions, qualification standards, and job classification descriptions for the GS-2101 do not determine the answer to this question by themselves; the answer will be determined more by the selection process, training procedures, task assignments, and the characteristics of the new equipment that the GS-2101 specialists will be required to maintain. Currently, the GS-2101 specialists are the same people who have been maintaining the existing equipment and systems. Therefore, the GS-2101 represents more an approach to the future than a response to the current task demands.

The training process for Airway Facilities specialists has two goals: (1) certification of the technician's abilities with respect to given systems and equipment, so that the technician may be authorized to certify the systems and equipment for use in air traffic control, and (2) career progression, so that, by demonstrating proficiency, the technician can progress to journeyman status. In principle, these goals are met by providing theory through course material and application through subsequent on-the-job training. Local (regional and sector level) supervisors play the major role in determining the training program for each technician and in evaluating the proficiency and progress of trainees.

Currently 13 percent of the workforce is eligible for retirement. Within 10 years, 35 percent of the ARTCC sector workforce and 27 percent of the GNAS sector workforce will be eligible for retirement. These data combine to suggest that the Airway Facilities work force will see, within 10 years, a simultaneous retirement of significant percentages of its experienced technicians and the equipment on which they have developed their experience.

Part II

HUMAN FACTORS AND AUTOMATION ISSUES

5

Cognitive Task Analysis of Air Traffic Control

So far we have focused on describing the tasks that air traffic controllers must perform in managing traffic, the physical facilities in which they do so, and the means by which controllers are selected and trained for those tasks. In this chapter, we describe the controller's task from a somewhat different, more psychological perspective, identifying the cognitive and information-processing steps demanded of the controller and, by extension, the sources of vulnerability in the controller's performance.

Several task analyses of air traffic control tasks have been carried out (e.g., Hopkin, 1988a; Ammerman et al., 1987; Murphy, 1989; Stager and Hameluck, 1990; Harwood et al., 1991; Seamster et al., 1993; Endsley, 1994). Our analysis draws heavily on the work completed by these investigators, particularly by Ammerman and colleagues. It also attempts to place their analyses in a more cognitive framework, by emphasizing the relationship between the tasks performed and the different cognitive or information-processing mechanisms employed by the controller (Wickens, 1992).

We begin by presenting a general cognitive model of the controller's task. We then describe the ways in which human cognitive processes both are an asset in air traffic control and are vulnerable to environmental and system variables, discussing factors that moderate these vulnerabilities. Such an analysis has equal relevance for training as it does for design. Our treatment in this chapter is closely related to the discussion in Chapter 6, which specifically links these cognitive elements to the concept of mental workload.

COGNITIVE MODEL OF THE CONTROLLER'S TASK

Controllers are generally successful and skilled in the performance of their tasks. Our proposed model of the cognitive processes by which these tasks are accomplished is shown in Figure 5.1. It is meant to be generic enough in form that it can accommodate equally the characteristics of tower, TRACON, and en route controllers. At a very global level of detail, we see the controller's task as one in which actions (at the right) are driven by events (at the left). The figure depicts five cognitive stages that intervene between events and actions: selective attention, perception, situation awareness, planning and decision making, and action execution. To elaborate more, the actions performed by the controller (such as communications and manual manipulations) are the result of following well-learned procedures and strategic plans, which are continuously formulated and updated on the basis of current awareness of the situation in the airspace and, in particular, the projection of that situation into the future. This awareness, referred to as the big picture (Hopkin, 1995), is based in turn on external events involving aircraft, weather, and equipment, as these events are selected for processing and then perceived by the controller via radar displays, radio messages, paper printouts, and (occasionally) telephone calls. The five stages do not constitute a rigid sequence. Steps may be skipped; for example, planning and decision making may be unnecessary if the appropriate action is known on the basis of past experience. The processes can be iterative; for example, perception is the basis for situation awareness, but situation awareness can guide selective attention and influence subsequent perception. Finally, each of the five processes draws on knowledge stored in long-term memory, and each of them may modify or add to that knowledge.

External Events

External events that call for controller actions occur primarily in the airspace outside the tower or en route center. These include the filing of flight plans, pilot requests for clearance, changes in aircraft trajectories, handoffs from other controllers, and changes in weather. Other important events, however, may occasionally occur at an airport or in an air traffic control facility itself, such as blocked runways and instrument or power failures. The immediate manifestation of most of these events is the presentation of new information to the controller. The information presentations that are spawned by the events are easily identified and categorized through task analysis. These categories include visual changes on the primary radar display, information contained on the flight strips, auditory input from voice communications by pilots and other controllers, visual and auditory alerts provided by automated handoffs or by projected or real loss of separation, as well as other input regarding weather conditions and runway status.

Information delivered on all of these channels must first be selected. The

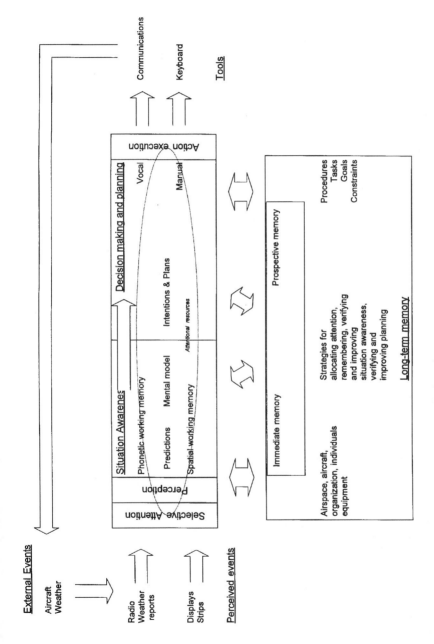

FIGURE 5.1 Cognitive model of the controller's task.

skilled controller knows where to look (or what to listen to) in order to gain critical information at the time that the information is both needed and available (Moray, 1986). Breakdowns in this selective attention process may occur, however, if an event occurs in a visual channel where it is not expected or if the display space is so cluttered that the event cannot easily be seen.

Not only the location but also the very nature of the events may be characterized by the extent to which they are expected and anticipated. Expectations may be based on specific past events (a plane is expected to continue on a given heading, a pilot is expected to read back the clearance provided and to change the aircraft's speed, altitude, or heading accordingly; Monan, 1986), or they can be based on general scripts of how the air traffic control process operates (Schank and Abelson, 1977) (e.g., a pilot newly arriving in the sector is expected to initiate communications and exchange information according to a well-established protocol). In either case, perceptual processes are influenced by long-term knowledge. Psychologists speak of top-down processing in describing the influence that expectations and knowledge have on perception. The extent to which controllers, like experts in all fields, easily perceive what is expected cannot be overstated. Conversely, however, the vulnerability of the controller's perception of the unexpected is a fact of life.

Working Memory

Once information is perceived, it may be retained in working memory. The human working memory system represents the "workbench" at which most of the conscious cognitive activity takes place (Baddeley, 1986). Working memory may temporarily retain information that is either verbal or spatial. Verbal working memory is the "rehearsable" memory for sounds, typically digits and words, and is the memory system that the controller uses when receiving a request or readback from the pilot (Morrow et al., 1993). Hence, working memory represents a critical component of communications. It is also the mechanism used when the controller, after reading a data block or flight strip, must temporarily retain the written information prior to translating it into a spatial representation. Spatial working memory is used to maintain analog, the representation of the airspace (Logie, 1995). The contents of spatial working memory replicate to some extent the controller's radar display, but they also incorporate, in three-dimensional spatial form, the critical altitude component that is represented only digitally on the radar display.

Information in working memory is further interpreted on the basis of knowledge stored in long-term memory. Information that matches stored "schemas" may result in the identification of familiar situations, predictions of future events, and retrieval of associated responses (e.g., weather problems at the airport require delays, which can be achieved by setting up a holding pattern). More effortful reasoning or computation may be required to identify other significant relationships and predictions (e.g., if two aircraft continue on their present trajectories, they will conflict). These processes of comprehension and prediction provide a mental picture of the situation confronting the controller (Seamster et al., 1993) and underlie the controller's situation awareness.

Situation awareness has been defined by Endsley as "the perception of the elements in the environment within a volume of time and space, the comprehension of their meaning, and the projection of their status in the near future" (Endsley, 1995:36). Situation awareness includes, ideally, an understanding of the current and future trajectories of all aircraft within the sector, some representation of traffic about to flow into the sector, awareness of other relevant but possibly changing conditions, such as weather and equipment status, and an understanding of how all the factors affect the achievement of air traffic control goals and constraints (such as permissible separations between aircraft, avoidance of terrain and restricted airspace, etc.). In future systems, it may also include awareness of the current operating modes of automated equipment (Sarter and Woods, 1995) and possibly of the momentary distribution of responsibility for traffic separation between ground and air in a free-flight regime.

Long-Term Memory

The processes involved in comprehending the perceived information draw heavily on knowledge structures in long-term memory. Characterizing less dynamic aspects of the controller's environment, such structures include knowledge of the airspace, including geography, terrain, air routes, fixes, and air traffic control sector shapes around a particular facility (Redding et al., 1992), knowledge of radar and equipment characteristics and capabilities, knowledge of weather configurations, and knowledge of different aircraft performance and maneuvering capabilities. Experience in a domain often leads to long-term memory structures that permit more efficient and/or insightful encodings or "chunking" of multiple events (Chi et al., 1981). In the air traffic control domain, for example, experienced controllers may directly identify important types of events involving multiple aircraft (such as conflict) rather than focusing on individual aircraft (Seamster et al., 1993).

On the basis of situation awareness, the controller must select an action. Typical actions include maintaining separation and coordinating traffic flow by requesting changes in the heading, altitude, or speed of one or more aircraft. In most familiar situations, the appropriate action may be immediately retrieved from long-term memory of an extensive repertoire of well-learned and well-documented procedures. In other cases, determining the appropriate action may require greater cognitive effort (such as ordering heading changes to avoid a potential conflict situation). In an unfamiliar situation, the controller may verify the adequacy of a potential action by mentally simulating its consequences (Klein and Crandall, 1995), for example, trying to visualize them in spatial working memory.

The decision-making and planning processes also draw heavily on knowledge in long-term memory. Relevant long-term knowledge includes formalized procedures (e.g., for creating the appropriate separation) acquired through training and documented in manuals and texts; goals and constraints, such as the required degrees of separation under different conditions of weather and aircraft equipment; informal strategies or heuristics picked up by experience and the observation of others (Hopkin, 1988b); and knowledge of risks and future uncertainties associated with particular situations or types of actions (Wickens and Flach, 1988).

The product of the decision-making process may be an immediate action, a strategic plan for action at a later time, or a series of actions that must occur over a period of time (for example, as aircraft are scheduled to arrive at a particular point in the airspace in a particular sequence). Generally, the more strategic plans take place within the en route centers, particularly in the traffic management units therein. The successful execution of planned actions at a later time depends on the reliability of prospective memory (Harris and Wilkins, 1982), that is, the ability of the controller to remember to take a particular action at a point in time in the future. Prospective memory is also required to confirm the effects of controller actions, like checking the altitude of an aircraft that had previously been directed to a new flight level to determine that the goal has been attained. Such memory is aided by many reminders in the environment, in particular the annotation of flight strips.

Another kind of long-term knowledge that is involved in virtually every stage of cognition is one that involves the strategies and heuristics that are developed over time for efficiently managing cognitive processes (Gopher, 1993; Huey and Wickens, 1993; Seamster et al., 1993). These strategies may improve the efficiency and validity of perception, situation awareness, planning, and action execution. For example, strategies for allocating perceptual attention among external events may help controllers handle situations with a high event rate (Stein, 1993; Gopher, 1993). Strategies for prioritizing tasks may also help when workload is high (Gopher et al., 1994). Strategies for remembering important information (for example, by rehearsing or using efficient encodings for aircraft identifications and trajectories, flight plans, and clearances) may help controllers construct a coherent picture of the situation and carry out planned actions. Other strategies apply to novel or complex situations and include judging the time available before some action must be taken (for example, to avoid a conflict) and using the available time effectively to verify the adequacy of situation understanding and the proposed plan (Cohen et al., 1996b; Raby and Wickens, 1994; Orasanu, 1993; Fischer and Orasanu, 1993). For example, controllers may quickly review their mental model or plan for completeness (have all aircraft been considered?), reliability (can the aircraft make the requested maneuver in time?), and consistency (are all the aircraft carrying out the requested procedures?).

Attentional Resources

In addition to inputs to the controller's task defined by external events and by long-term knowledge, another input is that of the controller's own cognitive effort or attentional resources, which are allocated to sustain task performance. As discussed in more detail in the following chapter on controller workload, events may vary on four related dimensions that influence the resources that the controller must allocate to deal with them. The first two dimensions are (1) the frequency with which the events occur in time and (2) the extent to which the events are complex. Both a high event rate and high complexity of the individual events can increase cognitive workload. The third and fourth dimensions can mitigate the demand on workload; events thus vary in the extent to which they are (3) expected and in the extent to which they are (4) familiar or routine. Extensive experience with high-workload situations, or in handling a particular type of complex event, reduces the cognitive resources required to deal with the event. Even in the absence of such familiarity, if high density or highly complex events are anticipated, then advance preparation can mitigate the demands on resources if the events actually occur.

These dimensions interact in a variety of ways. The pilot who reads back a clearance incorrectly generates a simple but unexpected event. To notice such unexpected events, the controller must continuously and carefully monitor all channels of information to assess if there is any change or conflict with expectations. This monitoring itself requires some cognitive effort. However, once the problem is noticed, it can be handled with little or no cognitive effort, as long as the controller is reasonably experienced and the event rate is reasonably low; the formulation of intentions and actions remains fairly routine. Unexpected events that are somewhat more complex (e.g., the announcement of an unanticipated newly arriving aircraft, a request for diversion) require somewhat more cognitive effort to be incorporated into the controller's mental picture of the airspace, but if the events are familiar, this effort is not prolonged. As Rasmussen (1986; Rasmussen et al., 1995) describes it, these events call for rule-based behavior. That is, with minimal problem-solving requirements, the controller may call up internally memorized rules to deal with the routine situations.

If events are relatively unfamiliar and complex, then major cognitive effort is required. These events often trigger what Rasmussen describes as knowledge-based behavior, the need for creative problem solving. When such events are also unexpected (e.g., an aircraft is unable to taxi off an active runway because of a malfunction, or an aircraft mistakenly executes an inappropriate maneuver in a crowded airspace), then knowledge-based behavior must be initiated on the spot, often under severe time pressure. In contrast, when the complex events are expected (e.g., a pilot in a crowded airspace requests a rerouting around weather), the required knowledge-based processing may be carried out at least in part in advance of the event itself, by preparing possible solutions in advance. As a result, less cognitive effort is then required to handle the event when it does occur.

It is important to recognize that the extent to which events of varying levels of complexity generate rule-based versus knowledge-based behavior is greatly dependent on the skill level and experience of the controller. As in any profession, a relatively complex event may be handled by rule-based behavior by the expert, but it may require knowledge-based problem solving for the novice. It is also important to keep in mind that many aspects of expertise are facility and even sector specific (Redding et al., 1992): the same event that triggers rule-based behavior in a facility (or sector) at which a controller has worked for years may trigger knowledge-based behavior in a facility or sector new to that controller, where local procedures may differ as well as the nature of equipment, the sector structure, the terrain, the traffic mix, and the air routes. The heavy impact of facility-specific learning in air traffic control has important implications for the difficulty of generic training (see Chapter 3).

As we discuss in the following chapter, it is assumed that controllers attempt to manage the task demands at a reasonably constant level of cognitive effort (Hart and Wickens, 1990), by drawing on strategies stored in long-term memory. When task demands become excessive because of combinations of high event rates and complexity, controllers attempt to maintain adequate performance without an excessive expenditure of effort by amending the strategies by which they deal with aircraft (Sperandio, 1976), as well as by changing their criteria for dealing individually with pilot requests. They may also shed tasks of lower priority or offload tasks either to the pilot or to other controllers who may be less busy. Hence, task and workload management (dealt with in Chapter 6) is clearly linked to team issues (dealt with in Chapter 7).

COGNITIVE VULNERABILITIES IN THE CONTROLLER'S TASK

The cognitive task analysis has revealed a diverse array of cognitive skills that the controller must marshall to handle the complex dynamic problems of managing multiple aircraft in an uncertain environment. For some of these skills, the human expert is uniquely qualified and so far has well exceeded the capabilities of even the most sophisticated forms of artificial intelligence. In the most general terms, we can characterize these strengths in terms of the controller's adaptability and flexibility in carrying out knowledge-based behavior. Although most control involves fairly routine following of procedures, the skilled controller is keenly attuned to subtle cues that may predict future *unusual* events and will possess in long-term memory a wide variety of adaptive strategies and plans to address these events if they do occur. And as the closed loop and iterative nature of Figure 5.1 make clear, the skilled controller is also able to monitor the implementation of the plan and to flexibly modify it in creative but adaptive ways should initially formulated plans appear to be unsuccessful. It is apparent that such adaptive flexibility is (a) not easily taught by formal training procedures but must be learned on the job and (b) becomes progressively more important with traffic that is more complex and less predictable or routinized in its behavior.

Although skilled controllers may thus be characterized by their cognitive strengths, it is also the case that human information processing is subject to several forms of vulnerability, all of which have implications for controller performance. In this section we outline seven major categories of vulnerabilities, each inviting performance degradation. Each of these vulnerabilities in turn may be exacerbated or attenuated by certain design or environmental factors, which are outlined in the following section on moderating factors. Each vulnerability in the following discussion relates to a particular aspect of the controller model shown in Figure 5.1.

Visual Sampling and Selective Attention

Because much of human visual search and pattern recognition is serial, with event-filled displays the controller is vulnerable to missing critical events through breakdowns in the serial visual scanning process (Stein, 1993). This is particularly true to the extent that many of these events must be inferred from signals that are not particularly salient to the untrained eye (e.g., a future conflict, a change in the altitude field in the data, a pilot's failure to implement a requested course alteration) rather than perceived from salient ones (e.g., a blinking data tag or automated alert for loss of separation). The quality of visual sampling is further inhibited by the amount of information or clutter in the visual environment, whether this is the view of a radar scope or the view from the tower of a busy taxi and ramp area. More visual elements to be scanned increase the likelihood that critical ones will not be attended to (Moray, 1986). And if these elements are similar, the likelihood is increased that elements may be confused. Yet in the case of the radar display, what is unwanted clutter at one instant may be valued information at the next, a dilemma that is not easily resolved.

Expectation-Driven Processing

Expectations influence perceptions. We see (or hear) what we expect to perceive, and this tendency allows the perception of expected and routine events to proceed rapidly and with minimal effort. Yet such expectation can be a source of vulnerability when events occur that are not expected, especially when these events are not perceptually salient or occur under conditions of high workload. Such perceptual errors, for example, lie at the root of the confirmatory "hear back" problem (Hawkins, 1987; Monan, 1986), when the controller incorrectly perceives that the pilot has correctly read back the clearance just provided; indeed, they underlie many other potential errors in voice communications. Expectation-driven perception may also underlie a controller's failure to perceive that an aircraft has subtly deviated from the expected flight path. Such errors can also be expected to form a major source of breakdown in communications within the air traffic control facility (see Chapter 7).

Working Memory

Working memory is very susceptible to interference, both from other items competing for the same processes and from other information-processing activities. Speaking, for example, will disrupt verbal working memory, and visual scanning and search will disrupt spatial working memory (Liu and Wickens, 1992). Working memory also suffers to the extent that it must retain items that are similar to each other and therefore confusable—like a fleet of aircraft with very similar call signs (Fowler, 1980). It can also be degraded by the need for storage of rapidly presented long strings of information (Morrow et al., 1993; Burke-Cohen, 1995).

Situation Awareness

Recent research in aviation, both civilian (Sarter and Woods, 1995; Adams et al., 1995; Wickens, 1996) and military air combat (Endsley, 1995; Waag and Houck, 1994), has revealed the wide differences in pilots' apparent ability to maintain situation awareness of the aircraft and its automation systems and of the surrounding airspace. Such differences undoubtedly relate to differences in the vulnerability of several processing components, related to the fundamental processes of selective attention, perception, comprehension, and prediction. Perception is influenced, as we have seen, by expectations. A good mental model of where events are likely to occur guides selective attention (usually reflected in visual scanning) to sample relevant parts of the display (Stein, 1992). But effective scanning and selective attention can be compromised by heavy workload demands.

The predictive component of situation awareness is heavily dependent on spatial working memory to "compute" likely trajectories based on current aircraft state, intended plans, and individual aircraft dynamics. Hence, this predictive component is highly vulnerable to competing demands for attention. Usually, aircraft move routinely and predictably through the airspace, and so prediction is not demanding. However, when multiple aircraft move in three dimensions and vary in air speed such that their predicted position at a future time is not a constant distance separation on the radar display, then such prediction on multiple aircraft taxes the controller's processing capabilities to the utmost and limits the resolution with which the future state of traffic in the airspace can be visualized. Such circumstances may be envisioned with possible implementation of free flight.

Communications

Voice communication involves a two-way process of sending (speaking) and receiving information (listening), and its vulnerabilities in the national airspace system have been well documented by Nagel (1988), who notes that the largest single cause of air traffic control incidents relates to breakdowns in information transfer (see also Kanki and Prinzo, 1995). Its successes and failures both depend on factors described previously: expectation-driven processing and working memory whose limitations may hinder the understanding of long communications strings (Burke-Cohen, 1995) that must be retained for even a few seconds before being translated into action, or that contain unfamiliar material (e.g., strange names, nonnative language).

Communications effectiveness also depends on shared assumptions, a shared mental model or shared situation awareness between speaker and listener (Salas et al., 1995). For example, a pilot unaware of other traffic that influences a controller's decision to issue inconvenient instructions may be more resistant to following them in a timely fashion. If a controller is unaware of a pilot's momentary high level of workload or the aircraft's current situation with regard to nearby weather, the controller may issue instructions with which compliance is more difficult. If one controller is unaware of the high workload (or low skill level) of a controller in an adjacent sector, the former may be more likely to take an action that can directly raise the workload of the latter.

It is clear that these communications issues directly affect the ability of controllers and pilots to function effectively as teams, and we discuss this further in Chapter 7.

Long-Term Memory

As we have seen, long-term memory is also relevant for maintaining situation awareness. Vulnerabilities in long-term memory are manifest in four kinds of activities. First, there may be breakdowns in what we call "transient knowledge" in long-term memory, which consists of immediate memory for events in the current situation and prospective memory for actions that the controller plans to perform within the next few minutes. These lapses or breakdowns occur when controllers fail to recall developing aspects of the current situation of which they were at one time aware. Self-generated activities, like writing an amendment on a flight strip, are less likely to lead to such forgetting than activities initiated by another (human or computer) agent (e.g., when an automated control system updates the electronic strip (Hopkin, 1988a; Slamecka and Graf, 1978; Vortac and Gettys, 1990). High levels of workload may also lead to the breakdown of prospective memory, causing controllers to forget to check on the status of certain aircraft. A drastic breakdown of this sort was partially responsible for the collision between two aircraft on the runway of the Los Angeles International Airport in 1991 (National Transportation Safety Board, 1992) and again at St. Louis' Lambert Field in 1994 (Steenblik, 1996).

Second, breakdowns in enduring knowledge in long-term memory are in part a result of shortcomings in training—forgetting of procedures, regulations, etc. But these may also reflect an inadequate mental model of the fixed features of the immediate airspace (flight routes, fixes, terrain, etc.). Given that the mental model of the expert controller is quite dependent on precise knowledge of these spatial features, when the controller transfers to work in a different region (i.e., different sector within a facility or different facility), considerable time will be required to attain proficiency in the new area. Third, breakdowns in procedural knowledge may result when different operating procedures or equipment are introduced. Negative transfer of old habits to the new situation may cause the old habits to persist (Singley and Anderson, 1989; Holding, 1987).

Fourth, breakdowns may occur because of inadequate knowledge or understanding of aircraft performance limitations and capabilities, inadequate strategies for dealing with future conflicts, and for optimizing deployment of the controller's attentional resources (i.e., knowing which aircraft need most attention now and which can be deferred).

Judgment and Decision Making

Decision making may become difficult in novel or unusual situations in ways that have little to do with workload demands (that is, even when the number and complexity of events is limited). For example, on occasion in situation assessment, the available data (e.g., regarding expected future traffic flow into the sector or expected changes in weather) may be incomplete, conflict with other evidence, or be unreliable and ambiguous. In planning and decision making, there may appear to be no feasible option that reliably achieves all of the controller's goals (e.g., maintaining separation while avoiding prolonged holding).

Still, the nature of most air traffic control decision making is relatively routine and enables controllers to select appropriate procedures to apply once they correctly identify and classify the existing situation. These types of decisions have been studied in many real-world situations by Klein et al. (1993). This research suggests that experienced decision makers learn a large set of patterns and associated responses in a domain. Rather than comparing options in terms of their predicted outcomes, proficient decision makers are more likely to recognize familiar types of situations and retrieve an appropriate response. Pattern recognition by itself, however, does not account for how decision makers handle uncertain or unfamiliar situations. Recent research (Cohen et al., in press; Pennington and Hastie, 1993) suggests that, in these situations, decision makers adopt strategies in unfamiliar or uncertain situations that build on but go beyond recognitional abilities. Such strategies attempt to identify and correct the shortcomings in recognitional responses to the situation. For example, decision makers identify and try to fill gaps in a situation model; in addition, they may test the model by identifying predictions and collecting additional data. Decision makers may elaborate the model by means of assumptions to fill gaps when data are not available or to explain data that appear to conflict with the model. Finally, they evaluate the plausibility of the assumptions required by the elaborated model, and, if the assumptions seem implausible, they may explore alternative elaborations. Strategies of this kind enable decision makers to handle uncertainty and competing goals without the formal apparatus of probabilities and utilities based on normative theory. Instead of manipulating abstract symbols, they focus on concrete, visualizable representations. Decision errors may sometimes result, however. For example, decision makers can forget or fail to evaluate the assumptions that are embedded within the picture of their current situation.

Many decisions in air traffic control are collaborative. For example, controllers in adjacent sectors may need to develop a joint strategy for avoiding a future conflict that affects aircraft in both. A controller may issue an instruction to a pilot that the latter finds difficult to accept, or the pilot may make an urgent emergency request that the controller finds difficult or unsafe to grant. Hence, much decision making may be viewed as collaborative, and some of it as negotiated. As we noted before, the success or failure of such collaborative decision making may depend substantially on the extent to which common situation awareness is shared by the participants (Salas et al., 1995).

Errors

The concept of controller error has two somewhat different meanings. *Operational errors* have a formally defined meaning in terms of loss of separation, and their occurrence has serious safety and personal implications for the controller. In contrast, we refer to *controller errors* here as any of a much wider range of inappropriate behaviors that result from breakdowns in information processing. Many of these may have only minor safety implications (e.g., pressing the wrong key for accepting an automated handoff). For others, the safety implications may be severe, even if they do not contribute to a formally defined operational error (e.g., issuing an inappropriate instruction that creates a difficult and complex traffic situation for other controllers or pilots).

In discussing controller error, it is important to emphasize the fact that humans make errors in working with complex systems (Reason, 1990). This fact is the inevitable down side of the highly advantageous quality of human flexibility and adaptability, which we discussed as a great cognitive strength for air traffic control. Indeed, it is a strength that the human operator brings to any complex system (Rasmussen et al., 1995). Thus, aspects of design should focus less on the complete elimination of all human error (an unattainable goal) and more on error-tolerant design, either through incorporating (and preserving) redundancies or through implementing (or preserving) error recovery mechanisms.

Norman (1981) and Reason (1990) have defined similar error taxonomies within the framework of an information-processing model such as that shown in Figure 5.1, and other investigators have applied similar sorts of categorizations to the identification of controller errors (Stager and Hameluck, 1990; Rodgers, 1993; Redding et al., 1992). The models developed by Norman and Reason (see also Wickens, 1992) identify five categories of human error, outlined below.

Knowledge-based mistakes are errors in understanding the situation. For example, a controller may not realize that a conflict exists or is pending. Such errors result generally from a lack of knowledge or information regarding the situation—perhaps from impoverished displays, from poor information sampling (scanning), or from a controller's inability to extract the appropriate information from the display or to interpret that information correctly.

Ruled-based mistakes involve selecting an inappropriate rule of action to address a correctly diagnosed situation. For example, the controller may have correctly perceived the pending conflict but chooses to implement a solution that is inappropriate, perhaps requesting an aircraft to maneuver in a fashion that imposes limitations on its performance or that violates some other aspect of the airspace.

Lapses are a form of error that is relevant in the context of long-term memory. A lapse involves forgetting to take a planned action (a lapse of prospective memory).

The *mode error* happens when the controller performs an action that might be appropriate in one mode without realizing that the system is in a different mode, so that the same action is no longer appropriate. For example, the controller may forget that certain separation standards have temporarily changed (e.g., because of weather conditions). Mode errors are increasingly prominent in more advanced automation systems that themselves have multimode functions. The crash of Airbus A320 near Strasbourg, France, was apparently the partial result of a mode error, when the pilot apparently believed that the autopilot was in a 3.3-degree flight path angle descent mode, when in fact the same "3" setting triggered a 3,300 ft/minute descent mode.

Slips of action occur when the correct intention is formulated, but the incorrect action slips out of the controller's fingers (in the case of keyboard entry) or mouth (as when the controller delivers an instruction intended for one aircraft to a different one) (Norman, 1981). The cause of these slips remains poorly understood, although it is appreciated that they are as likely to occur with experts as with novices (Reason, 1990). One reason is that slips are more likely to occur when the operator is not fully paying attention to error-producing components of the task. For the novice, who *must* pay attention in order to accomplish the tasks at all, such a state of inattention is not really achievable. However, the expert, for whom several tasks can be performed at an automatic (i.e., inattentive) level, it is easy to understand how slips can occur.

A major cause of slips is when procedures or actions to carry out an intention differ from one case to another, but the physical environments (and physical signals triggering the action) are quite similar. These circumstances produce negative transfer from the old to the new, and, as we saw in the section on long-term memory, can also be considered as breakdowns in long-term memory. If feedback from the action is made readily visible or audible (as is the case when one hears one's own voice), then the slips can often be self-detected and corrected before they lead to undesirable consequences. As a result, the system becomes more tolerant of errors—"error tolerant."

MODERATING FACTORS

Environmental design and system factors may either exacerbate or attenuate the vulnerabilities of the human information-processing system. Some of these factors are summarized in the following section, organized as above in terms of the major categories of vulnerability.

Visual Sampling

Difficulties with visual sampling can be exacerbated by the low arousal resulting from sleep loss, fatigue, and circadian rhythms, by cluttered displays or a cluttered visual environment, by display environments that have many similar appearing elements; and by the distraction of high workload.

Many of these problems can be attenuated by automated assists that recognize critical events and translate them into salient abrupt onset signals (e.g., conflict alerts, minimum safe altitude warnings), by decluttering options, by display technology that integrates (or brings close together) related items (Wickens and Carswell, 1995) and that distinguishes confusable items by physical properties (e.g., color coding), and by concern for fatigue and workload issues.

Expectation-Driven Processing

The problems resulting from the bias to perceive the expected event (and therefore misperceive or fail to perceive the unexpected) is exacerbated for the perception of rare or atypical events. High workload often leads to less complete perceptual processing of all events and hence disproportionately enhances the likelihood that the unexpected will be perceived inappropriately. Nonredundant channels of communication, poor data quality, and rapid communications via speech channels all tend to make this misperception more likely to occur.

The vulnerability of processing of the unexpected can be attenuated by incorporating redundancy into any critical message that may be unexpected, and by training the importance of clarity of communications, giving emphasis to key words and phrases that may be unexpected in the circumstances.

Working Memory

The limitations of verbal working memory are exacerbated to the extent that long messages are communicated solely by the voice channel, since working memory is heavily involved in the processing of speech (Burke-Cohen, 1995). The vulnerability of working memory is exacerbated still further by the presence of any concurrent verbal activity, whether this activity is in the environment (the controller is trying to listen when there is related verbal activity heard in close proximity) or is carried out by the listener (trying to remember a communication while concurrently speaking or listening). In general, high workload and stress make working memory more vulnerable to information loss, as does the existence of confusable material (similar-sounding words, names or acronyms, similar aircraft call signs).

The limitations of working memory can be partially addressed by redundancy. This may be accomplished by restating or repeating critical elements (Burke-Cohen, 1995) or possibly by designing redundant communications channels that will back up (but not replace) auditory communications with a visual "echo" of the spoken message, to be referred to if necessary. The data link system (Kerns, 1991) discussed in Chapters 7 and 12 can accomplish this function, although elimination of auditory channels via datalink would destroy redundancy. Attention to task analysis, minimizing unnecessary auditory stimulation, and minimizing the existence of potentially confusing (similar) auditory utterances also address problems of working memory.

Situation Awareness

Situation awareness is more vulnerable (and more difficult to achieve) in a crowded, complex, and heterogeneous airspace; when operating procedures are inconsistent; when the controller is handling a less familiar sector; when information must be translated from symbolic (verbal) formats into the spatial mental picture; and under conditions of high workload or distraction. All of these contributing factors to the loss of situation awareness exert even greater influences to the extent that the future state, rather than current one, is to be assessed. Situation awareness is also inhibited by the loss of data from poorly designed displays or from conditions that inhibit the communications from other aircraft or other controllers.

Situation awareness may be better preserved by display formats that are compatible with the controller's mental model of the airspace and by the integration and easy accessibility of all necessary information. Predictive displays help the controller anticipate future situations; tools that guide controller's attention to the right place at the right time support maintenance of improved situation awareness (Sarter and Woods, 1995). It is not clear, however, that three-dimensional displays offer similar improvements (May et al., 1995).

Communications

We have already noted aspects of expectation-driven processing and of working memory that can exacerbate or reduce deficits in communications. There is also evidence that many aspects of the information that is exchanged are conveyed via nonlinguistic features: the tone of voice can convey urgency or uncertainty, and speakers can augment their voice message by pointing or gesturing (Segal, 1995). System design changes like the datalink system and those that physically isolate one controller from another remove some of these vital channels. Furthermore, replacement of voice communications received by the pilot from air traffic control by datalink may eliminate many important nonlinguistic cues available to the pilot, such as the degree of urgency of an instruction. In contrast, efforts to support shared situation awareness, perhaps through common displays or common training, can facilitate communications.

Long-Term Memory

Failures of transient long-term memory (prospective memory) may be induced by high workload and distraction and by the removal of the operator from the role of an active decision maker in choosing relevant actions, whose impact should be later remembered. Failures of remembering appropriate procedures are invited whenever the procedures are suddenly changed. A corresponding invitation to memory failure occurs when a controller must move to a new sector or facility. Failures of memory are also exacerbated by poor training and the absence of opportunities for recurrent training of infrequently used (but critically important) skills. These issues are relevant to the potential impact of automation, discussed in Chapter 12 and to be advanced further in the panel's Phase II report.

Many long-term memory problems are addressed by care given to training. Also, good displays, with reminders of pending actions, address certain problems of forgetting. The limitations of long-term memory reflected in negative transfer from one environment to another may be mitigated if care is given to the consistency of operating rules and procedures whenever possible.

Decision Making

Decision making is vulnerable when information is incomplete, conflicting, or unreliable or when goals conflict. Deviations from optimal decision making, as noted above, have been demonstrated in experimental tasks in which workload, stress, and task complexity were minimal. Vulnerability may be increased significantly, however, in real-world tasks with high workload (e.g., large numbers of aircraft) and time constraints (e.g., time until separation minimums will be violated).

Training in normative decision-making methods (e.g., decomposing decisions into options, outcomes, and goals, assessing probabilities and utilities, and mathematically combining the assessments) is unlikely to result in improved performance by air traffic controllers (Means et al., 1993). However, decision making may be improved by training and displays that are sensitive to strategies that do work well in real-world environments. Training, for example, may sensitize controllers to trade-offs among speed, accuracy, and task prioritization (Means et al., 1993; Cohen et al., 1996). In addition, it may foster techniques for identifying and correcting problems in situation understanding and plans, to the extent that time is available prior to taking action. For example, controllers may learn to recognize gaps in their knowledge of relevant information, conflicts in the data, or unreliable assumptions underlying their understanding of the data (Cohen et al., in press a, b). Similarly, displays may make explicit the time available before action must be taken and alert decision makers to other high-priority tasks. Displays may also highlight conflicts or sources of unreliability that deserves controller attention.

Errors

Many of the exacerbating factors discussed above are likely to induce errors of different kinds. Poor displays and inadequate training lead to mistakes; high workload leads to lapses, etc. In particular, however, changes in procedures and poor design attention given to compatibility and confusability are invitations to both mode errors and slips. Mode errors are also induced by multimode automated systems, in which similar actions can have very different consequences.

Practically all of the attenuating factors discussed in the section on visual sampling will help to remediate controller errors. However, we place particular emphasis here on good interface design (adhering to basic human factors principles; Norman, 1988); on building adequate feedback into a system so that the controller has a clear visible or auditory display of actions taken (and implemented within the system) and their progress in completion; and on incorporating an error-tolerant philosophy into system design, such that redundant elements can catch errors with greater reliability and such that there are recovery paths from errors that may be made but noticed later (Norman, 1988; Rouse and Morris, 1987).

Trade-Offs in Human Factors Solutions

It should be noted in the above list that certain proposed or considered system modifications may affect one or more of the vulnerabilities. Sometimes these may coincide in attenuating the influence of more than one factor. For example, proposed datalink interfaces (Corwin, 1991; Kerns, 1991), by virtue of their visual displays of communications, which may be coupled with redundant voice communications, should simultaneously reduce expectation-driven processing and phonetic errors in working memory (Corwin, 1991; Kerns, 1991). Attention to consistency of design and procedures will attenuate the undesirable effects of both long-term memory forgetting (negative transfer) and slips.

At the same time, such technology may exert detrimental effects: for example, keyboard entries via datalink will increase the vulnerability to slips, the use of the keyboard may slow the transmission of information, and the removal of voice may eliminate important sources of nonverbal information. In a more general sense, our task analysis reveals that there are often trade-offs between human factors solutions that benefit one aspect of processing even as they inhibit another. Decluttering of a display to facilitate visual selective attention may hide information necessary to sustain situation awareness. Silencing auditory chatter to avoid interference with working memory may also have a corresponding negative effect on situation awareness, by removing useful communications channels (Pritchett and Hansman, 1993). Making procedures consistent to avoid slips and negative transfer may inhibit controllers in their requirements to be flexible problem solvers in unusual circumstances.

Finally, we note the existence of such trade-offs between workload and situation awareness that are achieved by the introduction of high levels of automation. Automated features, if they are restricted to alerts and display features based on computer computation, will thereby reduce the demands for cognitive spatial activity (i.e., will reduce workload) and will also be useful because they attenuate the vulnerabilities of visual sampling and spatial working memory. If, however, automation is extended to decision-making and action-taking activities with the desire to reduce workload still further, the advantage of reduced workload will be counteracted by a reduced situation awareness, resulting from poorer transient knowledge (Vortac et al., 1993), as well as a potential loss of skill.

CONCLUSIONS

Cognitive tasks analysis has provided a framework for understanding the implications of design and procedural changes and of training technologies on performance of the individual controller. Our analysis reveals the strengths of the skilled controller in the ability to flexibly adapt to novel or unusual situations, drawing on long-term memory to find solutions and to monitor the success of their implementation. Some vulnerabilities in the controller's information processing can be addressed by careful consideration of design factors, such as ensuring that salient signals (abrupt onsets) characterize important and unexpected events, minimizing the confusability of information, using computer technology to provide visual display of material to be retained in working memory, and providing reminders that will augment the controller's prospective memory for tasks to be performed. However, it is important to note that design or procedural solutions implemented to address one vulnerability may exacerbate the influence of another.

The panel identified a number of limitations in human perception that could be addressed by research and design:

• The less efficient processing of unexpected (and therefore rare) events is an inherent aspect of the human perceptual system, and design implementations and procedural changes must acknowledge this fact.
• Human perceptual processing is inhibited by display clutter. But what is clutter at one time (or for one controller) may be relevant information at another. Furthermore, loss of situation awareness may result from the absence of information, an absence that could be created by decluttering schemes. Research on the trade-offs between clutter and information in air traffic control would be invaluable.
• Controllers are limited in their ability to predict future traffic states with multiple aircraft of heterogeneous performance capabilities. Intelligent displays that can automatically accomplish this prediction and explicitly display it are valuable tools to assist controller performance. Research should address the efficacy of analog predictors of aircraft altitude.

The panel identified a number of opportunities for improvements in the air traffic control environment:

• Communication is facilitated by shared knowledge or situation awareness between speaker and listener, and it is important to preserve this wherever possible, perhaps enhancing it through display technology or training.
• A large component of the permanent knowledge structures that a controller brings to the job is spatial and procedural knowledge about the particular characteristics of the facility and its sectors. This fact limits the effectiveness of generic (i.e., not sector-specific) controller training.
• Efforts at improving controller decision making should focus on strategies that are effective in time-stressed environments: training in task and goal management strategies; sensitization to gaps in knowledge, conflicting evidence, or goals and to unreliability in situation understanding and plans; and information displays that promote appropriate trade-offs among speed, accuracy, and task allocation and alert controllers to significant uncertainties and goal conflicts.

The effect of high workload on cognitive vulnerabilities, such as error propensity and situation awareness, is complex and requires further research in an air traffic control context to illuminate. Great care and caution must be given prior to implementing procedural and equipment changes and current procedures must be carefully understood, because of the likelihood that new procedures can lead to negative transfer and slips. Although air traffic control errors are inevitable to some extent, they can be minimized by providing attention to good human factors design. Their impact on system performance can be minimized by adopting a design philosophy that preserves some redundancy of human information transmission (redundant displays, multiple operators) and by an error-tolerant philosophy that allows recovery from human errors before they are propagated to major system errors.

6

Workload and Vigilance

In the previous chapter, the underlying cognitive characteristics of the tasks that air traffic controllers carry out were described in some detail. Many of these cognitive operations impose demands on the controller's mental workload. It has already been noted that the projected increase in air traffic over the next decade threatens to overwhelm the capacity of the air transportation system. If safety is not to be compromised, it is vital that individual controllers are not subjected to overload due to high traffic density and complexity. Accordingly, in this chapter we examine the characteristics of the mental workload of air traffic controllers and its relationship to overall system performance.

It should be noted at the outset that the entire range of controller workload, from low to high, needs to be considered in air traffic control operations. It is most natural to think of high levels, or overload, when considering workload. Considerable evidence exists to indicate that human operators who experience high levels of workload can be susceptible to errors or performance breakdown. A study by Endsley and Rodgers (1996), for example, appeared to demonstrate a positive correlation between workload and operational errors, at least for high levels of workload. Indeed, high workload was identified as one of the contributing causes to the accident at the Los Angeles airport in 1991, in which a departing commuter aircraft had been positioned on the runway in the path of a landing USAir 737 (National Transportation Safety Board, 1991). However, underload can be equally pernicious. Hopkin (1995) suggested that the extensive research on overload in air traffic control has led to a relative neglect of underload; as we discuss later, operational errors have also been reported under conditions of low to moderate traffic complexity (Stager, 1991; Stager and Hameluck, 1990). Thus, it is important to understand both underload and overload, including the ways in which situation awareness mediates the relationship between workload and errors. Accordingly, in this chapter we treat the load on the controller as falling along a continuum from low to high, examining both workload and vigilance.

Another point worth noting is that there is no typical controller workload profile or characteristic style of vigilance that is representative of air traffic control in general. Workload patterns and the quality of vigilant monitoring are likely to differ between en route, TRACON, and tower controllers, between control centers of different levels, between radar and nonradar control, between different sectors, and so on. Ultimately any comprehensive examination of the workload of air traffic control must be stratified by these and other job- and system-related factors. We provide here a general analysis of mental workload and vigilance for en route and TRACON controllers, recognizing that there is likely to be considerable diversity within these categories.

MENTAL WORKLOAD

History and Definitions

The study of mental workload has occupied a prominent position in human factors research and practice over the past four decades. Workload assessment studies have been conducted since the 1950s and early 1960s (Brown and Poulton, 1961), and air traffic control was an early area of application (Kalsbeek, 1965; Leplat and Sperandio, 1967). However, much of the theoretical development of the field can be traced to a 1977 conference of the North Atlantic Treaty Organization and subsequently published book, *Mental Workload* (Moray, 1979).

Since that seminal volume, several thousand studies have been conducted on the theoretical underpinnings, assessment techniques, and practical implications of mental workload in a variety of domains. A partial bibliography created a decade ago had over 500 listings (Hancock et al., 1988). Even reviews of this work number in the dozens, and only a few are mentioned here: Damos (1991), Hancock and Meshkati (1988), Huey and Wickens (1993), Kantowitz and Campbell (1996), Lysaght et al. (1989), Moray (1988), O'Donnell and Eggermeier (1986), Warm et al. (1996), and Wickens (1992a). The number of publications attests to the importance accorded the concept of mental workload in the human factors research community. The pace of research has slowed somewhat in recent years, having been replaced by studies of situation awareness (Flach, 1994; Gilson et al., 1994; Wickens, 1992b). Nevertheless, an understanding of the factors influencing human mental workload is likely to be crucial to the design of air traffic control and other systems (Andre and Hancock, 1995). This need will persist in the future, as systems become more automated.

What is workload? First, the term generally refers to mental workload, that is, the load associated with the mental (including cognitive and affective) processes of the human operator, rather than (or in addition to) physical workload. This emphasis on mental workload is appropriate because the job of air traffic control, in common with most other modern work settings, is primarily cognitive and information-intensive, rather than physical and labor-intensive. Accordingly, most of what is discussed in this chapter deals with mental workload.[1]

The term *mental workload* has an immediate intuitive meaning, yet it has resisted precise definition. Various authors have conceived of workload as the objective task demands imposed on the human operator, the mental effort exerted by the operator to meet these demands, the performance of the operator, the psychophysiological state of the operator, and the operator's subjective perception of expended effort. Many definitions assume that mental workload is an intervening construct that reflects the relationship between the environmental demands imposed on the human operator and the capabilities of the operator to meet those demands. Workload may be driven by the objective load imposed on the controller from external environmental sources (airspace factors, displays, tasks, procedures, other controllers, and supervisors), but not inevitably, because workload is also mediated by the controller's response to the load and by his or her skill level, task management strategies, and other individual characteristics.

Although there is no agreed-on definition of workload, theories of workload based on the concept of attentional capacity or resources have been proposed (Kahneman, 1973; Kantowitz and Casper, 1988; Wickens, 1984). Each of these theories assumes that tasks (except those that can be performed automatically) require the allocation of the operator's attentional resources for efficient execution and that operator workload reflects the overall level of demand for resources. The theories differ in assuming either a single pool of resources that can be flexibly allocated to different activities (Moray, 1967) or multiple resources that differ qualitatively according to such features as input and output modalities, stages of information processing, and response requirements (Wickens, 1984).

Workload and Air Traffic Control Performance

What are the relationships between traffic factors, workload, and performance? At a global level, the controller's workload is related to the capacity to manage traffic. The more aircraft that have to be handled, the greater is the

---

[1] It is worthwhile to remember that consideration of physical workload may still be necessary on occasion. Even in the most information-intensive job, the human operator must interact physically with devices to exchange information. The placement and control features of these input and output devices, if poorly designed, may not only lead to injury (e.g., carpal tunnel syndrome) but also induce discomfort and fatigue. Furthermore, to the extent that the physical demands imposed on controllers (e.g., keyboard entry, movement of flight strips, reaching, and other manual behaviors) interact with cognitive activities and therefore contribute indirectly to mental workload, consideration of the physical workload is important.

workload until, at some point, "things give." The job of the air traffic controller is often seen as the quintessential example of a high-workload, stress-inducing occupation. However, although the workload can be high, there is little support for the view that the job is uniquely stressful (Costa, 1993; Hopkin, 1992). Studies show that controllers experience peaks and troughs of workload during operations. High workload can be a problem, not solely because it may impact negatively on controller performance (and hence potentially on safety), but also because it can set an upper limit to traffic-handling capacity. Decreasing sector size or increasing the number of controllers does not necessarily solve this problem because of the consequent increase in intersector and intercontroller coordination and communication. Moreover, decreasing sector size reduces the amount of time spent with each aircraft, so that the controller has less time to build up the picture of the traffic; as discussed further below, this may increase workload. Low traffic load may result in boredom and reduced alertness, with consequent implications for handling emergencies.

Although factors conducive to high and low workload are prevalent in air traffic control, this does not necessarily mean that all controllers experience extremes of workload. All successful controllers use various adaptive strategies to manage their performance and subjective perceptions of task involvement. Sperandio (1971) first showed that controllers handled an unexpected increase in traffic load adaptively by decreasing the amount of time they spent processing each aircraft. A controller may also stop doing less important, peripheral tasks, thus leaving more time for active control; or increase spacing, stack aircraft, or prevent them from entering the sector (hence reducing airspace capacity). Because air traffic control is a team activity, another possibility is that controllers may ask a colleague to take over a particular task. In general, controllers may use a variety of strategies to manage workload and regulate their performance: if they do not use any of these adaptive strategies, further increases in traffic load may result in errors.

These considerations suggest that one needs to distinguish between workload drivers (i.e., factors in the environment external to the controller), controller workload, controller strategies, and performance consequences. Figure 6.1 presents a schematic view of the interrelationships of these factors. The influence of environmental drivers can be modeled, but it must be supplemented by assessment of the controller to measure the actual workload experienced. The controller uses various strategies to cope with the external drivers. Controller performance represents the joint consequences of the effects of task drivers on workload and the mediating influence of controller strategies.

The remainder of this chapter examines each of these workload factors in turn. In the following sections, we: (1) model the effects of workload drivers, (2) discuss the effects of workload drivers on controller workload, (3) examine the relationship of workload to performance, (4) evaluate the role of controller vigilance, and (5) consider the influence of low traffic load and sleep loss or sleep

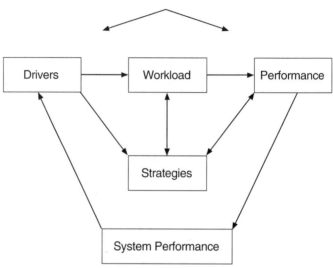

FIGURE 6.1 Interrelationships between workload drivers, workload, and performance.

disruption on performance. Note that, from a systems view, workload assessment and modeling are not necessarily of interest in themselves. The goal is to predict system performance under varying conditions of workload. To achieve this goal, one needs first to measure workload, predict the effects of different drivers on workload, and then attempt to predict controller performance in response to very high or very low workload. A more detailed discussion of workload assessment appears in Chapter 10, on research methods.

MODELING WORKLOAD

There is often a need for predictive workload assessment, not only for new systems but also for systems in which new hardware or software capabilities are to be introduced. Another factor in the thrust to develop predictive models is that federal agencies such as the FAA and the Air Force require workload certification of new aircraft before they can be acquired (Federal Aviation Administration, 1994). A number of such predictive workload models have been proposed, and a few are briefly discussed here.[2]

---

[2] Only models that assume that controllers are capable of concurrent time sharing of tasks are described. Several models that assume only serial processing of tasks and view multiple-task workload as a problem of scheduling (e.g., the SAINT model of Wortman et al., 1978) are not discussed here.

Some predictive models are based on the results of empirical studies, whereas others have been developed from first principles. One of the most commonly used models is time-line analysis, in which workload is modeled as a function of the proportion of time spent in performing a task relative to the total time available (Parks and Boucek, 1989). Levels of workload during a specified time interval can then be determined by summing the time lines of each task performed during that period and dividing by the time interval. Conventional time-line analysis makes the rather overarching assumption that workload during a given interval is 100 percent if the controller is fully occupied with a particular task during that interval, so that the introduction of other tasks during that same interval would lead to workload greater than 100 percent, or to overload. In contrast, as discussed previously, resource theories of multitask performance predict, and studies have shown, that overload will result only if each task competes for the same resources (or the same input and output channels) and if the total resource demand exceeds the operator's capacity. For example, a controller can easily talk to a pilot while scanning the radar display, but cannot easily talk while listening to another controller. To accommodate these findings, the W/Index predictive model includes metrics of task and resource conflict in estimating workload (North and Riley, 1989; Sarno and Wickens, 1995), and variants of this model have been employed to predict workload in a number of aviation settings, including at least one in an air traffic control setting (Burbank, 1994).

The effective time available for completing a task is also a feature of workload models proposed by Hancock and Chignell (1988) and by Laudeman and Palmer (1995). In the former, workload estimates are a function of time of task completion but are also modulated by predicted estimates of the skill level of and degree of mental effort expended by the human operator. Laudeman and Palmer (1995) extended conventional time-line analysis by assuming a linear function of increasing workload during the time available for completing a task. In their model, the function begins at zero before the task is attempted and returns to zero after task completion. By summing together the workload functions for each task, a predicted workload profile over time is obtained. The area under this overall workload function was proposed as an index of workload. Finally, Rouse et al. (1993) described a state-space, predictive model of subjective workload. In their model, subjective workload is a lagged function of the operator's actions, performance, the system state, and the operator's previous subjective experience. Subjective workload does not simply parallel operator performance, and workload is predicted to be dependent on adaptive changes that the operator initiates to moderate the impact of increases in imposed task load. Human operators often use such changing strategies and variations in operating procedures to maintain performance at some preferred level that is acceptable but not necessarily optimal or perfect (Hart and Wickens, 1990). However, whereas the model proposed by Rouse et al. (1993) implies that operators use adaptive strategies to keep subjective workload within acceptable bounds, including accepting lower levels of performance, the empirical evidence suggests that this is not the only adaptive strategy. Operators, including controllers (e.g., Sperandio, 1971) may change behaviors (e.g., operating procedures) not to minimize subjective workload, but to keep performance within acceptable limits.

Workload models have the unique advantage that they provide the only means of assessing workload ahead of time. The predictions can then be put to empirical test. Hence workload modeling is useful in design and in prototyping. A potential disadvantage is that this method is only as good as the underlying model. Furthermore, not all models have been experimentally validated. Nevertheless, on balance, workload modeling holds considerable promise for evaluating evolving systems and is likely to play an important role in the assessment of future automation concepts in air traffic control.

Workload Drivers

There are many workload drivers in the air traffic control environment. As stated earlier, however, it is important to note that task load should not be assumed to elicit a passive, fully predictable response from the human operator. Different controllers may respond differently to the same load factor (e.g., an increase in traffic density). The same controller may respond variably on two different occasions by using preplanning, task shedding, or other coping strategies to minimize mental workload on one occasion but not on the other. Skill and training also influence the response to workload drivers. We present here some of the more important sources of workload in air traffic control.

Airspace Load

A starting point is to examine those aspects of the air traffic control environment that contribute to task loading. In essence, this approach attempts to analyze the intrinsic load of air traffic control operations, as a prelude to assessing (or predicting) the load on the controller. At the simplest level of such an analysis, for example, the number of aircraft being handled by the controller could be defined as an important load factor. This variable is clearly insufficient on its own, however, because the demand imposed by a certain number of aircraft on the controller would also depend on other factors, such as traffic complexity, aircraft mix, weather, etc. Hence, one way to proceed would be to enumerate all potential variables and categorize them, deduce interrelationships between variables, and compute derived load factors from the raw airspace variables.

Using essentially this approach, Arad (1964) conducted a series of assessments of the objective job difficulty of different air traffic control operations. The goal of the studies was to use the results as a basis for sector design as well as other issues, such as planning of staff levels. Arad divided the drivers of load

Routine Control Load	Airspace Load
Number of controlled aircraft	Number of controlled aircraft
Sector flight time	Sector flow organization coefficient
Proportion of standard aircraft	Mean airspeed
Proportion of non-standard aircraft	Sector area
Proportion of terminal area hand-offs	Mean aircraft separation
Proportion transitioning	
Proportion of VFR to IFR "pop-ups"	

FIGURE 6.2 Some airspace load variables in air traffic control. Source: Adapted from Arad (1964).

into three general categories: background load, routine load, and airspace load. These load factors were defined by equations including such variables as the number of aircraft under control, sector flight time, sector area, mean aircraft separation, etc. Figure 6.2 lists the routine load and airspace load variables in Arad's (1964) scheme.

As mentioned at the outset of this chapter, controller workload is likely to vary among the different air traffic control positions. Airspace load factors also differ among the TRACON, en route, and oceanic control environments. Such differences must be taken into account in any comprehensive evaluation of controller workload.

Hurst and Rose (1978) followed up Arad's (1964) research with systematic observations of controllers on 47 radar sectors in the Boston and New York areas. Using modified versions of Arad's (1964) load factors, they examined the relationship between load and observer ratings of the activity level of controllers. These behavioral ratings of busyness were not related to the derived control load factors but were significantly correlated with peak traffic counts. Bruce et al. (1993) carried out a similar study on 65 sectors in 7 en route centers and found a significant relationship between traffic complexity (as assessed by traffic load factors) and the level of controller activity (e.g., verbalizations and manual activities). Unfortunately, the number of overt behaviors engaged in by controllers may or may not be accompanied by increased mental workload. As noted previously, behavioral ratings may be insensitive to the covert demands on the information processing of load factors. Hopkin (1971) pointed out that even more sophisticated behavioral measures (such as the use of keyboards and of communications equipment) may be largely insensitive to the load associated with problem solving and decision making by the controller. Hurst and Rose (1978) were aware of this limitation and suggested that physiological measures might provide additional validation of the impact of task load on controller mental workload. A more serious weakness of their approach, however, is the notion implicit in their study design and data analysis procedures that controller workload is a direct, open-loop function of task load, as in the stress-strain relationship of mechanical structures. Definition and quantification of airspace load factors are insufficient by themselves because of the multiple, closed-loop nature of the air traffic control operations. Air traffic control represents a dynamic system in which the controller's behaviors affect some of the same control load variables that are thought to impact on mental workload (e.g., airspeed).[3] Furthermore, as noted earlier, Sperandio (1971) has shown that controller workload does not necessarily increase proportionately with increases in airspace load, because controllers use strategies and vary operating procedures to achieve acceptable levels of performance.

The role of airspace factors in driving controller workload is being addressed by ongoing efforts to establish objective measures of sector complexity (Rodgers et al., 1995; Pawlak et al., 1996) and could also be informed by the recent development of the SATORI (situation assessment through re-creation of incidents) tool (Rodgers and Duke, 1994). SATORI allows for the graphical re-creation of all radar, weather, and communications data recorded at an en route center. SATORI contains within it several airspace variables in addition to those analyzed by Arad (1964), Hurst and Rose (1978), and Bruce et al. (1993). This software tool can be used to extract such variables as the number of way points, the volume of airspace, the number of navigation aids, military operations areas, as well as other sector and weather-related information that may be relevant as workload drivers.

Display Factors

The design of visual displays is a traditional area of concern in aviation human factors (Stokes and Wickens, 1988), and air traffic control is no exception. Early radar displays such as the plan position indicator (PPI) suffered from poor signal-to-noise ratios, glare, low contrast, and other factors that impeded quick and accurate detection of targets (Baker, 1962). With the development of signal preprocessing, alphanumeric displays, high-resolution graphics, and color-coded displays, the sensory detection problem was largely eliminated. But perceptual and cognitive processing remain important display design issues. The new display capabilities allow much information to be displayed. For example, the plan view display (PVD) can provide such information as the aircraft call sign, aircraft type, TCAS-equipped aircraft, reported altitude, assigned altitude, speed, time, target track, track history, and other information. Of course, the controller can

---

[3]Jorna (1991) makes the related point that the common view that air traffic control is a paced task over which the controller has little or no control is not entirely correct. Controllers may choose on occasion to divert aircraft into holding patterns if they have indications of high pacing of incoming aircraft, so as to not compromise safety (although delays occur) while managing their workload.

choose not to have all the information displayed, but nevertheless display clutter and potential increased workload are possible.

Other factors that influence display-related workload include type size, luminance, contrast, color, and visual coding of alphanumeric symbology. Standard human engineering guidelines and databases (Sanders and McCormick, 1992; Van Cott and Kinkade, 1972) can be consulted for appropriate design choices for each of these factors.

New displays for aiding the controller in prediction and extrapolation of flight paths have become available in recent years. As we noted in Chapter 5, the need to project aircraft paths into the future imposes a high demand on the controller and thus is likely to be a source of workload. Algorithms that allow for the accurate prediction of flight paths, with appropriate display of these predictions, will therefore considerably reduce the controller's workload in this phase. Efforts to examine three-dimensional display technology to facilitate controller visualization of the vertical dimension are also being initiated (May et al., 1995).

Controller-Pilot and Controller-Controller Communications

The primary means of communication between controller and pilots is verbal, through the use of radio telephony (RT). Commands and clearances from the controller allow the pilot to navigate through crowded terminal areas with the required amounts of separation. Controller-pilot communications are also vital for exchanging information about weather, traffic flying under visual flight rules, runway hazards, etc. Communications between controllers are also required for efficient handoffs between sectors, planning, scheduling, and other activities. Consequently, attention has focused on the nature of verbal communications and its role in the overall workload of the controller (Cardosi, 1993; Kanki and Prinzo, 1995; Prinzo and Britton, 1993). The analysis of the controller's verbal behavior as an embedded task index of workload has already been mentioned (Leplat and Browaeys, 1965). High levels of communications may not only increase controller workload but may also impact negatively on the controller's ability to get the big picture. Jorna (1991) stated that, when controllers spend more than half their time communicating with pilots, they report that their traffic awareness becomes disturbed. When this occurs, the effect of any task factor or workload driver (such as a visual flight rules to instrument flight rules pop-up) that normally has only a small impact on mental workload and performance may loom larger. Finally, different aspects of controller-pilot communications (e.g., message length and composition) also have an impact on pilot workload (Morrow, in press; Morrow et al., 1993) and, to the extent that this leads to communication delays or misunderstandings between pilot and controller, the workload of the controller can also be indirectly affected. Morrow et al. (1993) have outlined some principles for improving collaborative communication between controllers and pilots.

Workload and Current Automation

Although automation in air traffic control has been limited in scope to date (as is discussed more thoroughly in Chapter 12), implications of workload studies for current automation can be drawn. Automation has traditionally been introduced in many systems partially in an attempt to reduce or regulate the operator's required level of mental workload at times of high task load. This is the standard engineering solution to operator error (or substandard performance). In some instances, this may be the correct solution. The simplest view of the effect of introducing automation is that cognitive resources are freed for performance of other manual tasks. Workload research and theories of attention provide some basis for predicting the resulting impact on workload and performance. Clearly, automation of a previously manual task will have an effect only to the extent that the task is resource sensitive (Norman and Bobrow, 1975). Moreover, the multiple-resource theory of Wickens (1992a) predicts that the required resources must overlap with those required to perform the manual tasks. Alternatively, automation of a task will benefit performance and workload if it frees input or response channels that would otherwise be tied up (Navon, 1984).

Numerous dual-task and multitask studies have shown that removing a task from the operator's control can benefit performance and workload if these requirements are met (Damos, 1991; Wickens, 1992a). Tsang and Johnson (1989) found that lateral-hold automation in a flight control task reduced subjective workload, both when the flight task was performed alone and when it was combined with other cockpit tasks. In these and other multitask studies, the automated task was removed from manual control from the outset and remained so throughout the study. Given that more flexible automation is current in many systems (e.g., the cockpit flight management system, which has several modes; see Chapter 12), whether the workload benefits of automation also accrue when automation is invoked in a dynamic and flexible manner needs to be examined. In such cases, as opposed to when a task is permanently allocated to automation, the operator is likely to monitor the automation from time to time to ensure its proper functioning or to use its outputs, as in the case of decision-aiding automation. Thus automation of a task is not the same as removal of the task, so that the assumption that automation frees up cognitive resources and reduces workload may not hold (Wiener, 1988). Parasuraman (1993) found that periodic, dynamic automation of flight-related tasks enhanced performance on other flight tasks performed manually and reduced subjective workload. These effects were not simply the result of task subtraction (e.g., doing two tasks instead of three) as in multiple-task studies (e.g., Tsang and Johnson, 1989), because subjects were required to supervise the automated task and were able to do this satisfactorily (as assessed by post-session tests).

These studies suggest that, in principle, automation of tasks can be beneficial in air traffic control to the extent that the automated tasks are resource demanding. For example, the experience of controllers with the ARTS system has been that automatic handoffs between sectors have resulted in reduced workload for both the controller and the pilot. Automatic updating of aircraft expected times at all reporting points represents another example of automation that can reduce workload. Manual updating of times of arrival at subsequent locations, although a routine, well-practiced task for controllers, clearly takes up cognitive resources that could be freed by automation. This form of automation could mitigate the high workload of controllers at times of high traffic. As Hopkin (1991) pointed out, however, it is unclear to what extent this type of automation, although beneficial for workload, may simultaneously impair efficiency because of its negative impact on the controller's situation awareness. Although some routine actions can be resource demanding, they also strengthen memory, and automation can result in poorer memory for aircraft arrival times.

It should be noted that, although automation can reduce workload, depending on what is automated and when, in practice the workload benefits may be counteracted by other effects. Moreover, there is additional evidence that automation can increase rather than reduce workload (Wiener, 1988). Edwards (1976) first pointed out several years ago that automation does not necessarily reduce workload, but this early admonition was perhaps not widely heeded. On the presumption that reduced mental workload leads to safer operation, designers thought that automation would inevitably reduce human error and improve system safety. However, there is a potential fallacy in this line of reasoning that was recognized quite early, even by writers in the popular technical press (Bulloch, 1982). Thus automation may reduce, increase, or leave workload unchanged. These and other related characteristics of the effects of automation on mental workload are discussed further in Chapter 12.

Task Load, Mental Workload, and System Performance

We have seen that several factors make the analysis of mental workload in air traffic control operations a complex matter. Evaluating workload is not a simple matter of enumerating the task loading factors, such as airspace load variables. In addition, mental workload is modified by adaptive strategies used by controllers to regulate their performance. Finally, automation can increase or decrease mental workload.

Considering workload in relation to system performance adds another layer of complexity to the picture (Danaher, 1980; Stager, 1991). Intuitively, one might presume that an increase in an airspace loading variable (such as the number of aircraft being handled) beyond some threshold value would increase workload and reduce controller performance, possibly reducing overall system performance. One might also expect a greater likelihood of controller error in such instances. However, theoretical considerations and empirical data both indicate that the relationship between mental workload and performance is not so straightforward. Theory predicts that controllers may expand attentional resources in response to an increase in task load (Kahneman, 1973). The controller may experience increased mental workload, but performance is maintained. Although attentional resource theories predict that performance will decline if the upper limit of the controller's capacity is exceeded, demonstrating this in operational air traffic control settings has proved somewhat elusive.

Increases in task load may also increase controller mental workload but not change performance because of the use of compensatory or regulatory methods discussed previously (e.g., Sperandio, 1971). Nevertheless this would suggest that the controller has less spare capacity for dealing with unusual circumstances or emergencies, leading one to suspect that operational errors might increase. However, there is little direct evidence for such a scenario. In an analysis of operational errors in Canada, Stager and colleagues (1989) found that operating irregularities and incidents were not uniquely associated with high workload, but rather with low to moderate workload and moderate pace levels (intermediate values on the airspace load variables described earlier).

Another study of air traffic control errors by the Canadian Aviation Safety Board (1990) found that, of 217 incidents selected from 437 occurrences, 60 percent of the system errors were attributed to planning, judgment, or attention lapses on the part of controllers. However, most of the operational errors occurred during conditions of low traffic complexity. Similar results were obtained for U.S. air traffic controllers (Rodgers, 1993). Stager (1991) considers a number of human error taxonomies that might be examined to better understand this apparent paradox.

One way to resolve it is to distinguish between task load and controller mental workload. Low traffic load does not necessarily lead to low controller mental workload. As discussed in more detail in the next section, recent findings indicate that maintaining vigilance under low task load requires considerable mental workload. Hopkin (1988) noted that the emphasis in air traffic control research on high task load and stress has led to a comparative neglect of low task load and boredom. If maintaining vigilance is boring but demanding, and if, as has been argued, boredom is itself a stressor (Thackray, 1981), then the neglect of these factors becomes doubly serious. Hence, both very low and very high levels of task load (e.g., number of aircraft) can lead to substandard performance. This explanation does not account for the failure of existing studies to show a relationship between operational errors and high task load, unless one assumes that adaptive strategies are sufficient to limit the risk of error at high workload. However, it may be dangerous to assume that such a relationship does not exist merely because it has not been demonstrated empirically to date. A conservative conclusion would be that operational performance in air traffic control can be compromised by both very high and very low task load. Having discussed the upper levels of workload, we now turn to a discussion of the lower end of the continuum, or vigilance.

VIGILANCE

History and Definitions

Many activities require sustained attention for successful completion. When the activity needs to be continued for a long period of time without interruption, the ability to maintain vigilance for the occurrence of unpredictable but critical events may be compromised. Mackworth (1957) provided an early definition of vigilance: "a state of readiness to detect and respond to certain small changes occurring at random time intervals in the environment" (pp. 389-390). Although this definition is still used by most researchers today, the emphasis on detection may not be relevant to some modern systems, including the current air traffic control environment; as discussed earlier, the controller sensory detection problem has largely been eliminated because of improvements in sensor and display technology. Instead, vigilance for the discrimination or diagnosis of unusual conditions is required. Sometime such conditions may be missed altogether ("I didn't see it on the scope"), but more often they are not understood and responded to speedily. An expanded view of vigilance extends the concept beyond detection to discrimination, recognition, or diagnosis, and the measure of vigilant performance to include both accuracy as well as speed of response.

The decline in detection performance over time in vigilance tasks, or the vigilance decrement, has been confirmed in a large number of investigations (Davies and Parasuraman, 1982). The vigilance decrement refers equally to the decline in detection rate and the increase in response time over the duration of the watch. Several studies have shown that most of the decrement occurs within 30 minutes (Teichner, 1974), although for very perceptually demanding visual targets it can appear within the first five minutes (Nuechterlein et al., 1983). The cardinal features associated with vigilance decrement are the temporal uncertainty and low probability of occurrence of targets.

Since Mackworth's pioneering experiments, many studies of vigilance have been carried out. For reviews of this large corpus of work, see Craig (1985), Davies and Parasuraman (1982), Huey and Wickens (1993), Parasuraman (1986), and Warm (1984).

Task Factors Influencing Vigilance

Numerous task factors affect detection performance in vigilance tasks. These include psychophysical parameters, such as target intensity and duration, as well as temporal and spatial characteristics of the vigilance task, such as target frequency, regularity, number of stimulus sources, background event rate, and so on. For reviews of the effects of these factors on the vigilance decrement and on the overall level of vigilance performance, see Davies and Parasuraman (1982) and Warm and Jerison (1984). In general, vigilance is high for targets that are highly salient, temporally and spatially predictable, and occur frequently in the context of a low background event rate. Unfortunately, many real-world targets possess the opposite of some of these attributes: they occur very infrequently, and, although modern signal processing techniques can ensure that target intensity and duration are above threshold (although not always, as in the case of passive sonar targets; see Mackie et al., 1994), the temporal and spatial unpredictability of targets poses a considerable challenge to the controller.

The vigilance tasks that have been studied in laboratory experiments are quite varied in their characteristics. Despite this diversity, it is possible to describe vigilance tasks along some common dimensions. On the basis of a review of the literature and their own experiments, Parasuraman (1979; Parasuraman and Davies, 1977) suggested that many vigilance tasks can be classified according to a four-fold taxonomy: target discrimination type (successive or simultaneous), background event rate (low or high), sensory modality (visual or auditory), and target source complexity (single or multiple source of targets). Signal detection theory (Green and Swets, 1966) suggests two possible sources of the vigilance decrement: a decrement in perceptual sensitivity (d') and an increment in response bias (b) over time. The vigilance taxonomy was first applied to define the conditions under which each of these outcomes is likely. The increment in b over time indicates that operators become increasingly conservative over time in calling an event a target. This finding is ubiquitous in vigilance studies, suggesting that appropriate training to regulate the subject's response criterion can reduce the vigilance decrement.

Training studies have found the decrement in detection rate can be reduced, although not eliminated completely, and that the response criterion can be moved in the direction of optimality (Craig, 1985; Davies and Parasuraman, 1982). More generally, training the human operator's response criterion can enhance performance in many detection tasks, as shown by Parasuraman (1985) in studies of chest x-ray inspection by radiologists and by Bisseret (1981), who examined the detection of aircraft course conflicts by controllers. In this latter study, expert controllers had lower values of β than trainees—that is, they were more willing to call for a correction of a detected conflict. Bisseret (1981) suggested that trainees were less willing to respond because of their greater uncertainty, and therefore that training should emphasize appropriate adjustment of the response criterion.

The Workload of Vigilance

Vigilance tasks are boring and have traditionally been thought of as undemanding. This view follows from the traditional arousal theory of vigilance, which views the vigilance environment as an unstimulating one. In European scientific circles the words *vigilance* (French) or *vigilanz* (German) are often used synonymously with arousal or alertness, and reduced vigilance and lowered arousal are thought to be closely related. Considerable research exists to indicate, however, that although vigilance is influenced by arousal (as are many perceptual and cognitive functions), the vigilance decrement is not inevitably a consequence of reduced arousal (Parasuraman, 1984). Moreover, more recent research shows that, although maintaining vigilance can be boring, it imposes considerable mental workload on the operator. This finding is consistent with newer multidimensional conceptions of arousal that make reference to attentional resource theory (Matthews et al., 1990) or psychophysiological adaptation (Hancock and Warm, 1989).

Recent studies indicate that even superficially simple vigilance tasks can impose considerable mental workload, of the level associated with such tasks as problem solving and decision making (Warm et al., 1996). As noted earlier, the work of Warm and colleagues has established that subjective mental workload in vigilance is high and is sensitive to numerous task and environmental factors that influence task performance. The workload of vigilance does not simply arise from the operator's efforts to combat the tedium of having to perform a dull task (Sawin and Scerbo, 1994; Thackray, 1981). Using a simulated air traffic control display, Warm et al. (1996) showed that advanced notification of a conflict reduced rather than increased subjective workload, even though such decision aiding should increase boredom because it leaves the operator with little to do. These results support the view that the workload of vigilance is directly task-related, rather than a by-product of boredom.

Vigilance and Air Traffic Control

Maintaining vigilance for critical events such as loss of separation, altitude deviations, VFR pop-ups, incorrect pilot readbacks, and other infrequent events is an important component of the controller's task. However, despite the importance of controller vigilance to the safety of air traffic control operations, there are comparatively few studies of vigilance during simulated air traffic control. Studies in the operational setting are very rare.

Thackray and colleagues (Thackray et al., 1979; Thackray and Touchstone, 1989a, 1989b) have conducted a series of studies using task conditions that more closely simulate current radar displays in air traffic control. The results of a representative study are described here (Thackray and Touchstone, 1989b). In their task, which was presented on a console that closely resembled an actual air traffic control radar workstation, subjects (university students, not controllers) were presented with two diagonal, nonintersecting flights paths on a graphics display. Aircraft were identified by a data block giving the call sign, altitude, and ground speed. The aircraft could move in either direction along the paths, and the data blocks were updated every 6 seconds. The number of aircraft under control was 16. Subjects were required to detect one of three types of critical event: (1) a change in the altitude part of the data block to "XXX," simulating a transponder malfunction; and two aircraft at the same altitude and either moving (2) toward each other (conflict) or (3) away from each other (nonconflict) on the same flight path. Nine critical events, three of each type, were presented in random order during each 30-minute segment of a 2-hour vigil.

Subjects detected all the transponder malfunction targets and showed little change in speed of detection over time on task (mean detection time averaged about 9 seconds). For the same-altitude targets, however (conflict or nonconflict), about 4 percent of the targets were missed during the first hour and 13 percent during the second hour. Moreover, the latency of detected targets increased from about 19 seconds to about 28 seconds over the course of the watch.

The results of the study by Thackray and Touchstone (1989b) are consistent with the findings of classical vigilance studies using simpler, artificial stimuli and targets: a vigilance decrement over time was observed, both in the number of targets detected and in the speed of detection. However, subjects had to monitor a relatively large number of targets (16), and no decrement was found for the simpler targets (transponder malfunctions). In earlier studies with simpler critical events (e.g., altitude deviations) and lower numbers of aircraft, Thackray and colleagues found detection speed showed very little increase with time on task. Drawing on the vigilance taxonomy proposed by Parasuraman and Davies (1977), Byrne (1993) pointed out that the transponder malfunction target used by Thackray and Touchstone (1989b) was of the simultaneous type, whereas the altitude targets were of the high-event rate/successive type. He suggested that the greater demand for controlled processing imposed by the altitude targets was the reason why only these targets were associated with vigilance decrement. These findings suggest that the greater the information processing demands imposed by airspace load factors and target type, the greater the likelihood of a vigilance failure occurring during extended watches.

The studies by Thackray and colleagues could be criticized for their use of students as subjects. Furthermore, subjects were required only to monitor targets, without any of the other activities, such as communications and keyboard entry, that controllers engage in routinely. It is difficult to predict exactly what influence these factors would have on the pattern of results. One could argue, for example, that vigilance failures might be exacerbated with the additional demands imposed by these other activities. This would follow from theoretical and empirical vigilance studies indicating that the vigilance decrement is a function of the information processing demands of target detection, so that depletion of resources by other tasks would increase the decrement (Parasuraman et al., 1987). However, exactly the opposite might also be predicted, on the grounds that controllers are more vigilant when they have more to do than when they are bored (Sawin and Scerbo, 1994). Studies using more closely simulated air traffic control, or field studies may need to be conducted to resolve this issue.

Whatever the outcome of such studies, as discussed previously, it should not be concluded that vigilance problems cannot occur in real operations. The results of an important recent study by Pigeau and colleagues (1995) of North American

Aerospace Defense (NORAD) operators warn that such a conclusion would be premature. The subjects were 16 experienced surveillance operators who used normal operating procedures while they worked at actual NORAD consoles (which present fused, correlated data from several radar sites). Subjects had to identify either beacon tracks of aircraft with transponders or search tracks of aircraft without transponders (e.g., light general aviation aircraft) that were detected with search radar. Both simulated and live traffic were used, but detection performance was assessed only for the simulated tracks, which Pigeau and colleagues stated were indistinguishable from actual tracks. A number of task conditions were manipulated, including sector size, watch length, and shift time. A vigilance decrement in detection speed over time was obtained for the search tracks (which imposed a memory load associated with a successive discrimination type of target) but not for the beacon targets, which required simultaneous discrimination. However, the decrement was restricted to a particular sector size and occurred only during the night shift, so that the vigilance decrement was not as ubiquitous as found in laboratory studies.

Vigilance and Current Automation

Sheridan (1970) pointed out many years ago that automated systems change the role of the operator from a controller to a supervisor. Although various forms of automation have been implemented in current air traffic control systems, the controller still maintains fairly direct control over aircraft. Hence, controllers are very much in the loop in current air traffic control systems. To the extent that current automation has the aim of allocating certain routine data gathering and manipulation tasks to computers, leaving intact the controller's decision making and planning duties, automation may not harm controller vigilance. However, if automation does encroach on these higher-order task functions, there is the attendant danger that vigilant monitoring may be negatively impacted. Parasuraman and colleagues have carried out several studies indicating that the monitoring of failures in the automated control of a task is poorer than manual monitoring when operators are engaged simultaneously in other tasks (Molloy and Parasuraman, in press; Parasuraman et al., 1993, 1994, 1996).

Another factor relevant to automation concerns the workload of vigilance. Lowering the information-processing demands of the task environment can promote better vigilance. However, the danger in this approach is that it can be counterproductive if carried too far. The notion that an operator will have less to do, thereby allowing more time for vigilant monitoring, has often provided a rationale for implementing automation in systems. Vigilance itself has been seen as a low-workload task. As noted earlier, many studies have exploded these myths, but they still persist today in some quarters. In fact, in some cases, the human operator may be faced with greater monitoring workload levels with an automated system than existed prior to the automation, despite the fact that the automation was intended to reduce workload. The paradox (Bainbridge, 1983) is that implementing automation in an attempt to reduce workload may actually result in increased workload, because of the cognitive workload associated with monitoring the automation.

WORK-REST SCHEDULES, SHIFT WORK, AND SLEEP DISRUPTION

Consideration of performance issues at the lower end of the workload scale would not be complete without considering the related implications of work-rest schedules, shift work, sleep loss, sleep disruption, and fatigue on controller performance.

Current work-rest schedules for controllers in the United States call for an 8-hour shift (10-hour maximum with overtime), distributed into 7 hours on duty and 1 hour of breaks, 2-hour maximum time at position, and a minimum of 8 hours between shifts. Vigilance research suggests that performance decline can occur after about 30 minutes spent continuously at a task. However, as discussed previously, to date there is no evidence for any significant decrement in controller vigilance performance within the normal time-at-position limit of 2 hours. Given that breaks totaling a maximum of 1 hour are taken at periodic intervals throughout an 8-hour shift, performance is also unlikely to decline during the course of the shift (Swanson et al., 1989), although subjective feelings of fatigue may increase progressively with time (Rosa, in press). There is little evidence for any significant loss in the performance of tasks by controllers over the course of a normal 8-hour shift (Stager and Hameluck, 1988), although some studies find decreases in alertness and psychomotor ability on selected tests within performance assessment batteries. Rhodes et al. (1994), for example, reported a reduction in accuracy in the Wilkinson serial-reaction time test from the start to the end of a shift, especially during the midnight shift. However, Costa (1993) found no change in controller reaction time or in critical flicker fusion frequency after a 7-hour shift. Although Rhodes et al. (1994) stated that their test battery was "representative of some of the fundamental elements of the air-traffic controller's job," this needs to be verified, and in the absence of such validation the interpretation of such test changes remains unclear.

Although there is no systematic trend toward poorer controller performance toward the end of a shift, this is not say that there is no variation in performance. Stager and Hameluck (1988) analyzed operational errors at several Canadian centers as a function of both time on position and time on shift. Of 265 operating irregularities investigated, approximately 40 percent occurred within the first 30 minutes on position, 70 percent within the first hour, and 85 percent within the first 90 minutes. It is possible that the greater incidence of errors when they first assume a position could be because they are in the process of forming the picture of the traffic, although Stager and Hameluck (1988) found no direct support for this view. With respect to time on shift, there was no significant trend toward greater incidents toward the end of an 8-hour shift. Instead, many of the incidents occurred during the first 1 or 2 hours of the shift. In a cross-tabulation of time on shift and time on position for 101 controllers in 63 two-controller incidents, over 33 percent occurred within the first hour of work and the first hour on position. Of course, one source of difficulty in analyzing performance variations over time is the variation in traffic over the course of the shift.

These results indicate that current work-rest scheduling practices per se are not associated with increased operational errors. However, there is more to work-rest scheduling than time on position and time on shift. Although controllers must rest for a minimum of 8 hours between shifts, the type of rotation between shifts and its impact on circadian rhythms and sleep patterns can potentially impact negatively on performance. The effects of sleep loss on performance in many industrial and military systems have been amply documented (Wilkinson, 1992). Unlike some task environments having constraints that impose severe sleep deprivation (e.g., combat situations, medical practice; see Huey and Wickens, 1993), the air traffic control task imposes no such constraints. However, sleep-related disruptions in performance efficiency do remain a direct concern in air traffic control.

In considering the relevance of sleep disruption to controller performance, we have already noted Stager et al.'s (1989) report of substantial operational errors at modest and low task load periods, associated with late night and early morning hours. Such a finding is consistent with the well-documented fluctuations in performance efficiency associated with human circadian rhythms, which fall to a low point in the late night, early morning hours (Huey and Wickens, 1993; Horne, 1988). Furthermore, survey data from air traffic controllers has revealed that night work caused increases in the subjective rating of sleepiness and a reduction of total amount of sleep during the work week (Melton, 1985; Melton et al., 1973, 1975; Smith et al., 1971; McAdaragh, 1995).

Nevertheless, it is apparent that night performance in air traffic control is inevitable, so long as night operations continue in the national airspace. At issue is whether there are ways to ameliorate their negative effects. There is evidence that permanent assignments of some workers to a night shift never produces a full adaptation of circadian rhythms (and therefore restored performance efficiency) to the inverted day-night cycle (Huey and Wickens, 1993), and such a solution is discouraged in any case because of concerns about any disruption of family life for controllers assigned to a permanent night shift.

If then, shifts are to be rotated, two issues arise. How often should such rotations occur and, if they occur frequently, whether they should be phase advanced or phase delayed (see Figure 6.3). Regarding the first issue, the evidence is equivocal. Folkard (1980) has argued that for cognitive/memory tasks, such as those of the air traffic controller, relatively rapid rotation is more advantageous; whereas more recently, Wilkinson (1992) has argued that it is better to change

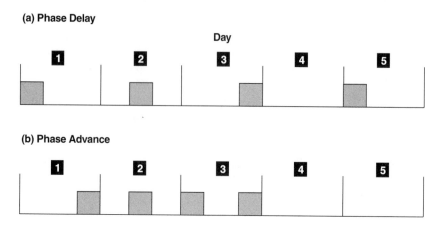

FIGURE 6.3 Extreme examples of phase delay and phase advanced schedules. The hashed box represents an 8-hour work period positioned within each 24-hour day.

shifts intermittently (i.e., once a week or longer), than continuously (i.e., working a different shift each 24-hour period).

In any case, no matter what the frequency of shift change may be, there is now fairly uniform evidence that phase-delayed shifts, such as that shown in Figure 6.3a, are less disruptive than phase-advanced shifts (Figure 6.3b) (Barton and Folkard, 1993), in the same manner that recovery from westbound transoceanic flights, which expand the day, is more rapid than from eastbound flights, which contract the day (Graeber, 1988). There are at least two reasons for the advantage of the phase-delayed schedule. First, it appears that human circadian rhythms have a "preferred" cycle that is slightly longer than 24 hours, and hence it is easier for those rhythms to adapt to a temporary lengthening of the cycle than to a temporary shortening. Second, the phase-delayed schedule distributes the work week over a longer period of time, with greater and more regular sleep opportunities between work time.

Unfortunately, the second reason is quite precisely the reason why air traffic controllers have generally preferred to opt for the phase-advanced schedule, as it is one that allows compression of 40 hours of duty time into a 4-day work week, allowing longer nonwork weekends (McAdaragh, 1995). In fact, a recent survey of 997 air traffic control specialists at 12 different facilities revealed that none had adopted the "performance preferred" phase-delayed schedule (McAdaragh, 1995). Thus, the current shift work preferences of controllers can degrade performance due to circadian rhythm disruption, sleep deficit, and accumulated fatigue.

In conclusion, the relation of shift work to air traffic control performance remains an important one, particularly as sleep-related disruptions appear to be most prevalent in low-load, vigilance-like monitoring tasks (Huey and Wickens,

1993). As we note in Chapter 12, it is precisely these kinds of tasks that may become more prevalent in the more automated controller workstation of the future.

CONCLUSIONS

Projected increases in future air traffic threaten to pose substantial demands on the capacity, and potentially the safety, of the air traffic system. Controllers may experience peaks and troughs of workload. If safety is not to be compromised, individual controllers should not be subjected to overload due to high traffic density. High workload can lower performance and set an upper limit to traffic-handling capacity. Decreasing sector size or increasing the number of controllers does not necessarily solve this problem, because of the consequent increase in intersector and intercontroller coordination and communication. Low workload may result in boredom and reduced alertness, with consequent implications for handling emergencies. However, although factors conducive to high and low workload are prevalent in the air traffic control environment, this does not necessarily mean that all controllers experience extremes of workload. Most controllers use various adaptive strategies to manage their performance and subjective perceptions of task involvement.

Various factors that influence mental workload in the air traffic control environment have been identified. These include airspace variables, display factors, and controller-pilot communications. Although studies examining each of these factors have been conducted, the precise relationships between these variables and workload, and the interrelationships of these variables, remain incompletely characterized. Moreover, the relationships between task load variables, controller mental workload, controller performance, and system performance are complex and not amenable to simple generalizations.

Evidence linking operational errors to performance and workload has found that errors occur under low task load conditions. Such conditions may increase demands on controller monitoring and vigilance. Vigilance declines as the information-processing demands of target identification increase (e.g., high memory load, high event rate, high spatial uncertainty). The mental workload of maintaining vigilance is also high, contrary to the belief that boring tasks are undemanding. Studies examining vigilance during simulated air traffic control have shown that performance can be good, but it declines under high task load conditions.

Current work-rest schedules do not appear to have a negative impact on controller performance, although subjective complaints of fatigue may occur. However, shift work and the consequent disruption of circadian rhythms and sleep loss continue to be a major source of concern. Current shift-work patterns (e.g., phase-advanced shifts and compressed work weeks) may result in degraded performance.

There are examples of current and past automation that have led to a reduction of controller workload. However, more generally, automation changes the pattern of the controller's workload. Automation can also increase workload due to demands on vigilance. These findings suggest that further implementation of automation in air traffic control systems must be preceded by systematic analysis of the impact of new technologies on controller workload and vigilance.

7

Teamwork and Communications

Teamwork among controllers and between controllers and pilots is critically important for safe and efficient air traffic control. The FAA, however, has generally considered the controller function to be an individual one and has therefore not focused on the teamwork aspects of controller tasks, selection, training, or performance appraisal. This chapter discusses teamwork and associated communications, supplementing the relatively meager literature on air traffic control with studies of cockpit teamwork, because pilots are part of the air traffic control team and because information pertaining to flight deck teamwork leads to promising hypotheses applicable to the study of air traffic control.

The definition of *team* we use in considering air traffic control functions is a broad one that includes individuals who are interacting face to face, by voice, or by written or graphic media to manage air traffic. The size of teams in air traffic control is variable. In addition to the primary actors (one or more pilots and a communicating controller), teams often use additional controllers sharing functions in a sector and may also include supervisors and instructors conducting on-the-job training. Controllers also interact with other controller teams to coordinate the management of flights. It is a characteristic of teams in this environment that they must interact with technology (i.e., radio, radar, computers) to do their jobs. Air traffic control teams are also faced with the responsibility of handling multiple tasks under time pressure at the group as well as the individual level (Waller, 1995).

Teams' functions in air traffic control are both transitory (the interaction between a controller and an aircraft) and relatively enduring (controllers sharing

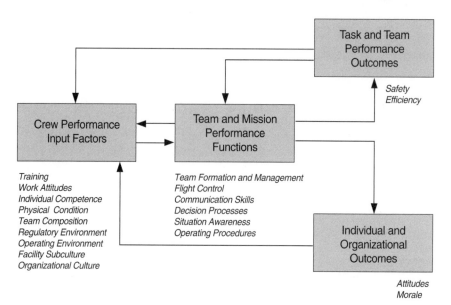

FIGURE 7.1 A model of team performance in air traffic control. Source: Adapted from Helmreich and Foushee (1993).

functions in the same sector on the same shift). Teams, however composed and however enduring, do not function in isolation but within an organizational and environmental system that profoundly influences their behaviors. It may be useful to consider controller teams in the context of a conceptual model of controller performance. Figure 7.1 outlines input, process, and output factors adapted from a model of flight crew performance developed by Helmreich and Foushee (1993). Input factors that precede team interactions range from individual to organizational to environmental. Individuals bring to the team task their physical condition (for example, fatigue level), technical competence (influenced by the nature and quality of training), and experience. Teams also vary in their composition (the mix of individuals) and compatibility, and all of these factors influence group processes and performance outcomes.

At a broader level, the behavior of teams is influenced by the regulatory environment, including doctrine and procedures, and the organizational culture, including norms and climate (see Chapter 8). The physical environment includes the reliability and usability of equipment as well as external factors such as weather and traffic levels. Because of the varying nature of the airspace to be managed and the organizational structure of the FAA, particular facilities tend to have distinctive subcultures. The existence of multiple subcultures makes it difficult to generalize about team behavior in air traffic control.

Group processes include both technical activities, such as navigation, aircraft separation, flow management, traffic situation monitoring, and response to user requests (Danaher, 1980). They also include interpersonal activities, such as the formation and psychological maintenance of a team concept and communications activities to maintain situation awareness and decision making at the group level. Group processes lead to multiple outcomes, including safety and the avoidance of operational error, the efficiency of traffic management, and work attitudes and morale. The model is recursive, in that process variables can influence input attitudes and outcomes and will affect both future input and process factors.

Research into air traffic team issues needs to take into account the multiple input and group process factors specified in the model to maximize the generality and validity of findings. Unfortunately, because of the variability of these factors across facilities, few, if any, investigations can control or measure all of them. These limitations need to be assessed when considering the implications of studies of controller team performance.

In this chapter, we describe team performance and interpersonal communications in the aviation system and illustrate them through their role in selected accidents and incidents. Relevant research into communications and teamwork in both air traffic control and on the flight deck is reviewed. Efforts to improve team performance through formal training programs in both air traffic and on the flight deck are described and evaluated. Finally, the implications of several types of automation for air traffic control teamwork and communications are discussed.

TEAM PERFORMANCE ISSUES

Under the definition of team we use here, any errors that involve interpersonal communication are classified as team-related. For example, failures in the transmission and receipt of clearances and separation errors associated with increased or decreased workload can be considered as team rather than individual cognitive or workload issues because they relate to the interface between humans. This is not just an academic distinction, because whether an error is classified as an individual rather than a team or system failure has implications for an organization's response strategy regarding sanctions, retraining, and work design.

Because the FAA has generally considered the controller function to be an individual one, in terms of skills and accountability, strategies for error reduction (aside from punishment of individuals found responsible) tend to focus on technological innovations such as automation (e.g., Helmreich and Schaefer, 1994). This is in common with practices in many technical endeavors, such as aviation and medicine. The logic implies that, if humans commit errors, their removal from the system should eliminate these errors. One result of this philosophy of error reduction has been minimal efforts to address team issues either in training or in work design. Recently, however, team aspects of error and superior performance have been recognized, and new team training efforts (discussed below) have been initiated in air traffic control.

Team Issues in Accidents and Incidents

The nature of team communication breakdowns can be illustrated by describing several accidents in which the controller-pilot interface was identified as a contributory element, either positive or negative. Research into input and group process factors associated with incidents also contributes to an understanding of the team role.

Eastern Airlines Flight 401

The 1972 crash of a wide-bodied jet transport in the Florida Everglades provided an early look at flawed communication between an air traffic controller and a flight crew as well as the impact of distractions on flight crew performance (National Transportation Safety Board, 1973). In this accident, the cockpit crew became distracted from primary flying and monitoring duties while investigating a landing gear warning light. During this period, the autopilot was inadvertently disengaged and the aircraft began a gradual descent from its intended altitude. The controller on duty did not warn the crew in any way of its impending flight into terrain—but merely asked "how are things going out there?" In the cockpit there was a failure to maintain active monitoring of flight controls and in air traffic control there was a failure to share situation awareness that could have prevented the accident.

United Airlines Flight 232

In contrast with the preceding accident, at Sioux City in 1989, the handling of a DC-10 that lost all hydraulic systems and flight controls due to the catastrophic failure of an engine was exemplary (National Transportation Safety Board, 1990). During the in-flight emergency, the flight crew worked effectively with controllers to select an alternate airport and to mobilize emergency units prior to the attempted landing. In this emergency, there was appropriate exchange of information both within the flight deck and between the pilots and the air traffic controllers, and this was combined with sensitivity to workload issues and emotional support needs (Predmore, 1991). The flight crew attributed their successful management of the emergency to formal training in interpersonal human factors known as crew resource management.

Avianca Flight 052

In 1990 a B-707, en route from Medellin, Colombia, crashed near New

York's John F. Kennedy Airport after total fuel exhaustion. The accident followed repeated holds during the flight from South America (National Transportation Safety Board, 1991). Although flight crew coordination and decision making were egregious and were clearly major causal factors, the interactions with air traffic control suggest that cultural patterns in communication may have influenced the crew's behavior with disastrous consequences (Helmreich, 1994). Individuals from highly hierarchical cultures that exhibit high power distance (i.e., high relative difference in power at successive ranks) normatively avoid questioning the actions of superiors (Hofstede, 1980). Demonstrating this style of interaction, the Colombian flight crew maintained a subordinate-to-superior relationship with the controllers and unnecessarily accepted multiple holding patterns and, despite being in an extreme emergency just prior to the crash, accepted instructions that delayed their return to the airport after a missed approach. In addition, the crew was indirect in communicating the urgency of its fuel state to the controllers. This behavior is consistent with Colombian society's high power distance (Hofstede, 1980) and was mirrored within the aircraft in the interactions between subordinate crew members and the captain (Helmreich, 1994). Had the crew declared an emergency or refused a clearance, it could have received immediate assistance with either landing at JFK or diverting to an alternate field.

Controllers have no regulatory requirement to recognize cultural differences in communications, particularly when the crew fails to adhere to standard procedures. However, the accident illustrates the complexity of communications across cultural boundaries and raises concerns about future communications problems in an increasingly global aviation system.

Near-Midair Collision Involving the Traffic Alert Collision Avoidance System

One approach to the reduction of separation errors and midair collisions in terminal areas has been the introduction of an airborne computerized traffic alert and collision avoidance system (TCAS, discussed further in Chapter 12). Despite its important alerting function, TCAS also changes the nature of interactions between controllers and pilots, since the controller is not privy to the cockpit information provided by the system. A particular incident involved a Boeing 737-200 transport plane and a light twin-engine turboprop in a terminal area (S.G. Jones, University of Texas at Austin, personal communication, 1995). The heavily loaded 737 took off on a standard instrument departure (SID) and was cleared by air traffic control to climb to 15,000 feet. At an altitude of 5,500 feet, TCAS provided an aural alert and the 737 crew saw the smaller aircraft ahead at the same altitude and a range of about two miles. Just after the TCAS alert, the controller reported the traffic and instructed the 737 to "maintain visual separation." Immediately thereafter, TCAS issued the commands "Descend, descend now." In accordance with company policy, the 737 commenced a descent and the small aircraft passed approximately 200 feet above the transport. The incident was initiated by an error by the controller, who failed to note the potential conflict between the transport and the light aircraft. However, the critical team issue rests in the fact that the controller did not have access to the TCAS actions and alerts and hence did not share the mental model of the pilots in the aircraft. In this case, the crew was obligated by policy to obey the command that led to a sudden departure from the assigned flight path—a maneuver that was unexpected by the controller. In terms of automation, the TCAS example illustrates how the actions of an automated system can constitute critical items of information required by all team members who must share situation awareness, suggesting that increased verbal interaction may be needed in some cases to share this information.

Team Composition and Operational Errors

Many air traffic control positions are staffed by two controllers who work together, with one handling radar monitoring and communications tasks (R-side position) and the other dealing with flight data (D-side position). Thus a ground-based team manages the aircraft under its control, but a single individual usually communicates with the team's air traffic. This work design not only divides the task but also provides redundancy in the form of additional eyes and ears to maintain situation awareness. However, under low traffic conditions, supervisors frequently elect to increase the efficiency of resource utilization and to combat boredom by combining these duties, thus turning a team activity into an individual one. Although this practice does reduce staffing requirements, it also deemphasizes the team function and can lead to inconsistency in defining duties at a particular position. It is noteworthy that a substantial proportion of operational errors occurs during periods of low workload, when such combined activities are in effect. Investigating changes in task and team resource allocation in a sample of 142 Canadian operational irregularities, Stager and Hameluck (1990) found that more than one-fourth occurred at combined R-side and D-side positions. The authors suggested that changing a team task into an individual one may alter the perception and organization of the task. One of the consequences of this kind of change can be a reduction in situation awareness. There is a need for study of the relationships among workload, teamwork, situational awareness, and operational errors using data obtained at American air traffic control facilities.

TEAM-RELATED RESEARCH IN AIR TRAFFIC CONTROL

Most research on air traffic control communications has been conducted with an individual focus, centered less on the interpersonal component of communication than on its content and form (e.g., Cardosi, 1993; Kinney et al., 1977; Morrow et al., 1993; Nadler et al., 1993). Other research has focused on individual demographic factors such as experience, position, and rated effectiveness (Human Technology, 1991). These lines of research, although valuable, do not deal directly with the team issues associated with controller performance.

Research Specific to Air Traffic Control

Seamster and colleagues (1993) conducted an experiment involving the simulation of operational problems that allowed analysis of performance effectiveness in a controlled context. These investigators grouped subjects into pairs to encourage collaboration on problem solution sets. In some scenarios, the sheer volume of aircraft transmissions prevented controllers from verbalizing their strategies. Experienced controllers simplified these situations and reduced monitoring loads by managing their workload early. Teams that most efficiently handled high workloads did so through the use of situational inquiries, frequent observations, and statements of intent, along with direct responses to queries. Task prioritization, workload management, and contingency planning were most effective if conducted during low workload periods, before periods of high traffic began. Seamster et al.'s (1993) research is important in its demonstration that safe and efficient traffic management requires the practice of effective teamwork skills as well as cognitive ones.

As noted, despite the complex team nature of air traffic control, its organizational culture and procedures have not historically stressed the team aspects of the controller job. Accordingly, team issues are seldom addressed in training or evaluation. The fact that on-the-job training constitutes the major means of socializing and qualifying new controllers as well as maintaining job competence exacerbates this problem. On-the-job instructors are fully qualified in the technical aspects of the controller function, but they do not receive systematic instruction in how to evaluate, instruct in, or reinforce communications and team skills.

Research on the determinants of operational errors in the Southwest Region of the FAA is being conducted by Jones (1993, 1995), as part of an initiative to reduce and contain human errors known as ASSET (Air Safety System Enhancement Team). One of the primary elements of the research is a survey completed without jeopardy by individuals involved in incidents. The survey elicits information on factors surrounding the event.

Three team-related scales were derived from behavioral questions on the survey. These include elements of task management (i.e., planning and workload distribution), information exchange (i.e., description of situation factors and intentions), and interpersonal relations (i.e., interpersonal sensitivity and receptivity). These team scales reflect the concepts included in formal training programs known as crew resource management.

Preliminary analyses contrasted scale scores for operational mishaps with normative and exemplary data from days without incidents. The data suggest that team issues are critical factors in operational errors. There were significant differences between incidents and normative and exemplary conditions on all three team behavior scales, with more positive scores associated with the absence of mishaps.

Data addressing these issues collected across a broad array of air traffic control facilities should prove useful for the determination of critical issues in team training. The data also suggest the potential value of formal programs to train controllers and their supervisors in effective teamwork and communications strategies. It also emphasizes the extent to which effective workload management is as much a team function (knowing when and how to shift task responsibility from one member to another) as it is an individual function, as described in Chapter 6.

Parallel Research in Air Traffic Control and Cockpit Domains

Attitudinal data regarding flight deck management and crew coordination have been collected using a survey that measures the level of endorsement of CRM concepts, the cockpit management attitudes questionnaire (Helmreich, 1984; Gregorich et al., 1990). Data from this questionnaire have been used to isolate training needs and to measure the impact of CRM training (e.g., Helmreich and Foushee, 1993; Helmreich and Wilhelm, 1991). Preliminary data were collected in several air traffic control facilities using an adaptation of the questionnaire and its extension, the flight management attitudes questionnaire (Helmreich and Foushee, 1993). More than 500 controllers from 3 en route facilities and 1 TRACON completed the survey as modified for each environment. Although it would be inappropriate to generalize to the system from this limited sample, the data do demonstrate that the concepts captured in aviation can be reliably measured in the air traffic control environment (Sherman, 1992). Scales that parallel those isolated among flight crews were identified in factor analyses.

The data suggest that, among those queried, there is general acceptance of the importance of team coordination and open communications. However, attitudes regarding leadership responsibilities and the need for interactive leadership differed from those of U.S. pilots. Two items on the revised survey follow a description of four leadership styles: democratic, consultative, directive, and autocratic. These questions ask respondents to identify preferred and experienced leadership style. The majority of controllers would prefer a consultative leadership style, in which leaders seek the opinions of subordinates. However, nearly 60 percent in one facility at which these questions were asked reported experiencing autocratic leadership offering little consultation and explanation for actions. Only 11 percent reported working under consultative leadership.

Scores were also quite low on a scale formed of items showing recognition of the negative effects of stressors on performance, such as fatigue, personal problems, and inexperienced coworkers. Many controllers feel that their decision-making ability is as effective in emergencies as under normal conditions.

Similar denial of vulnerability to stress effects has been found in pilots and medical personnel (Helmreich and Foushee, 1993; Helmreich and Schaefer, 1994). The perception of being "bulletproof" when faced with stressors can result in a failure to use teamwork as a countermeasure to stress.

Another scale, derived in the controller sample, contains items reflecting a willingness to question and disagree with the actions and decisions of others. Not surprisingly, junior (developmental) controllers reported more reluctance to speak up. The results suggest that integrated human factors training with a special focus on leadership and stress issues could have the same beneficial effects measured among flight crews, at least for the groups surveyed.

Because of the paucity of research and empirical data regarding team performance and team training specific to air traffic control, experiences in a related domain, commercial aviation, are discussed in the next section before reviewing the steps that have been taken to introduce team training in air traffic control.

TEAM TRAINING FOR THE FLIGHT DECK

By the late 1970s, research had demonstrated that human error was associated with the majority of accidents and incidents in commercial aviation (Cooper et al., 1980; Murphy, 1980). The same data also indicated that these human failures tended to be in leadership and team communication and coordination rather than technical aspects of flight control. This has implications when deciding where new systems should focus their human support. These findings have remained robust over time in their implication of interpersonal team factors as critical determinants of aviation safety (Helmreich and Foushee, 1993; Helmreich et al., 1993), and hence their implications for the necessity of preserving or enhancing team communication functions in system design changes.

Commercial and military aviation responded to these data, in cooperation with the National Aeronautics and Space Administration, by undertaking the development and evaluation of training programs in interpersonal human factors that were known initially as *crew* resource management (CRM) training. The meaning of the CRM acronym has subsequently changed to the broader designation of cockpit resource management to reflect the fact that team issues extend beyond the cockpit to include interfaces with air traffic control, cabin, dispatch, and ground operations (Helmreich and Foushee, 1993). Considerable empirical data have accumulated over the last decade indicating that CRM training can and does change attitudes and behavior among flight crews and that these changes increase the margin of safety in flight operations (Diehl, 1993; Helmreich and Foushee, 1993; Helmreich and Wilhelm, 1991). However, the data also indicate that some programs have greater impact than others and that a variety of causal factors determine behavioral outcomes and overall crew effectiveness (Helmreich and Foushee, 1993; Taggart, 1993, 1994). A partial listing of factors that influence the acceptance and practice of the concepts provided in this training is relevant to consideration of team issues in air traffic control. In programs with high positive impact:

1. The organizational culture is supportive of human factors concepts and training.
2. Senior management demonstrates its strong support for the program.
3. The program is supported by unions as well as management.
4. Critical role models (instructors and evaluators) and managers practice and reinforce effective team communication and coordination. It is especially important that the concepts taught are evaluated and encouraged under operational conditions and are not expressed as abstract concepts. The failure of many traditional management development training programs (including total quality management efforts) to influence day-to-day behavior comes, at least in part, from a lack of connection with the mundane realities of individuals' jobs. Human factors training is operationally based in the domain, rather than imported and conceptual. Training needs to reflect the organizational culture and to be rooted in operational behavior and situated learning rather than addressed as vague psychological constructs, such as leadership and open communication. Many organizations survey flight crews before developing training to determine critical issues for training, and again following training to measure impact using standard measures of attitudes relating to flight deck management (e.g., Helmreich, 1984; Helmreich and Foushee, 1993; Helmreich and Wilhelm, 1991).
5. Instructors and evaluators receive special training in the evaluation and reinforcement of team concepts. Clearly, if the critical role models in an organization cannot evaluate and reinforce the concepts, the probability is low that they will become embedded in the organizational culture.
6. Human factors training is experiential rather than didactic. Trainees need to practice and experience the concepts being communicated rather than to receive lectures regarding effective behavioral practices.
7. Nonjeopardy simulation is provided to allow team members to practice concepts and receive feedback. Military and commercial aviation has embraced the concept of line oriented flight training (LOFT), in which full mission simulations are conducted under highly realistic conditions to allow crew members to practice concepts without threat to their licenses (Butler, 1993; Federal Aviation Administration, 1978).
8. Human factors data are collected in incidents and operational errors to provide empirical data for training development and evaluation. Instances of both deficient and highly effective team performance provide the most relevant material for both initial and recurrent training. By utilizing reality-based incidents with which participants can identify, the probability of acceptance is greatly enhanced.

The FAA has strongly endorsed CRM training for pilots. A total of airlines, including most of the major carriers, will soon be regulated by a new special federal aviation regulation that allows training innovations and requires both recurring CRM training and LOFT (Federal Aviation Administration, 1990). The FAA has also issued a revised advisory circular that emphasizes many of the points listed above (Federal Aviation Administration, 1992). More recently, the FAA has issued a notice of proposed rule making that will require CRM training for all pilots covered by the Code of Federal Regulations, Volume 14, Parts 121 and 135 operations (applicable to most scheduled air carriers) and will extend the training to flight attendants and dispatchers. The National Transportation Safety Board (1994), in its investigation of the crash of an FAA aircraft, noted that the FAA does not provide CRM training for its own pilots and recommended its implementation.

TEAM TRAINING IN AIR TRAFFIC CONTROL

As CRM training in aviation became widely implemented and enthusiastically accepted, working controllers at several facilities concluded that the issues involved were highly relevant to their duties and operational problems. With the cooperation of airlines having CRM programs in effect, grass-roots training programs were implemented beginning in 1988 at several facilities, including Seattle and Chicago. The locally developed programs were known as controller awareness and resource training (CART). These initiatives were supported by the controllers' union, the National Air Traffic Controllers Association, were well received by participants, and received some support from FAA management. However, these programs were clearly derivative adaptations of airline training, with most examples and exercises focused on cockpit rather than air traffic control issues. The program continued informally until 1993. During its existence, its implementation at a facility was entirely at the discretion of facility management and the controllers' union. Informally, it was widely recognized that many of the facilities with serious human factors problems were most resistant to this type of initiative.

Senior FAA officials in air traffic also became aware of the increasing growth and impact of airline CRM as well as the development of local programs for air traffic controllers. To signal organizational commitment to a system-wide human factors training program, a conference was held in October 1991 in Austin, Texas, to foster the exchange of information among research and operational personnel involved with major airline CRM programs, senior air traffic control management, members of the controllers' union, and facility managers. Following this meeting, in 1992, a steering committee was formed to develop a training program more focused on air traffic control, and a contractor was engaged to develop a national program for controllers in all facilities. This program is known as air traffic teamwork enhancement (ATTE). During 1992 and 1993, 150 workshop leaders (facilitators) were trained.

Following deliberations of a cross-sectional committee composed of labor and management, a revised curriculum for the program was approved in late 1994. Additional facilitators, who are working controllers, were given ATTE facilitator training during the first half of 1995 and nationwide implementation is under way. It is important to note, however, that the program is not mandatory, and the training costs are not a budget item. Funding for ATTE training must be provided by individual facilities. At present there are no data on the percentage of facilities that have initiated ATTE or the number of controllers who have completed the workshop.

ATTE Curriculum and Delivery

The design of the curriculum for ATTE makes it a conventional basic awareness CRM program as defined in an advisory circular (Federal Aviation Administration, 1992). It is implemented as a three-day workshop to be attended by 6-20 controller participants. The curriculum is designed to be presented by two facilitators. The training also uses the same approach adopted by many airlines in having the material presented by facilitators who are working controllers rather than members of management or training professionals. The training strategy is participative rather than didactic. Its goal is to use exercises and experiences to demonstrate the importance of the concepts being presented. In these respects, ATTE complies with the FAA's recommendations for initial CRM training.

The manual for facilitators introduces the concept by pointing out that research into air traffic control incidents has concluded that approximately 70 percent involve human error and that more errors occur when sectors are staffed individually rather than by teams. Thus the framework is set for emphasizing the importance of teams and teamwork. Six major topics are included in the curriculum:

1. *Understanding air traffic teamwork.* This module has the expressed goals of identifying resources available to controllers, demonstrating how they can make better use of resources, and discussing the characteristics of effective teams.

2. *Communicating with others.* This module focuses on teaching skills for communicating effectively and providing feedback. It also discusses barriers to communication and provides practice in applying communications skills.

3. *Being a resource.* In this module, characteristics of controllers who are valuable resources to their teams are introduced and the importance of speaking up (assertiveness) is discussed.

4. *Managing stress.* This segment discusses current stress levels among controllers and the relationships among stress, health, and performance. Methods of reducing stress are discussed.

5. *Managing conflict.* The conflict module discusses how attitudes and values influence ways of dealing with conflict and the outcomes of destructive

and constructive conflict. Styles of conflict management and techniques for effective resolution of interpersonal disputes are also included. In the future, if aircraft become more autonomous in the development and execution of flight plans (for example, free flight navigation using global positioning satellites), the controller will need skills in negotiation, as will pilots, in order to establish optimal, safe routing. Negotiating skills are clearly related to conflict management, and training in this area can be made an extension of the present.

6. *Summary.* Insights and learning from the workshop are summarized and the session is evaluated.

One of the strengths of the program is its attempt to make the experience relevant to the domains of air traffic control. However, the syllabus does not include material on systems issues that may impede team coordination and performance, issues surrounding the interface between controllers and supervisors, or performance evaluation techniques. Although admirable in intent, the ATTE program is lacking in several of the factors discussed earlier that have been shown to influence the success of CRM programs. Specifically, the following concerns can be raised about ATTE:

1. Program development took place in the absence of empirical data regarding controller attitudes and incidents occurring in air traffic control facilities. Facilitators are charged with determining critical issues in their facilities and adapting the curriculum to reflect them. This is an extreme demand to place on individuals who are neither professional researchers nor professional educators.

2. The program does not demonstrate organizational commitment to the concepts by being budgeted and mandated at the national level and integrated into ongoing training and evaluation activities. Indeed, part of the job of the facilitator is to sell the program within his or her facility.

3. The use of peer facilitators has long been practiced in aviation under the working assumption that peers will be the most credible communicators of the nontechnical concepts associated with CRM and will be less threatening to participants. These ideas were certainly relevant to the climate of suspicion regarding psychological training that surrounded the introduction of CRM in the early 1980s. However, the situation has changed dramatically, and it is not uncommon to have the provision of CRM training made part of union contract demands. The unintended negative consequence of using peer facilitators is to dissociate CRM from both formal training and evaluation. In other words, it becomes seen as a training event but not as part of the culture, the "way we do business day to day."

4. Formal evaluation of the impact of ATTE has not been initiated using both behavioral and attitudinal outcome measures, nor is such validation planned at this time.

5. No additional training has been developed for managers, on-the-job training instructors, or evaluators to enable them to provide effective debriefing and

reinforcement of human factors behaviors. With the use of peer facilitators, the formal training and evaluation structure is bypassed and leadership may relate negatively to the program as a result of being excluded.

6. The training is designed as a single-event program without provision for annual recurrent training. Effective programs provide updated annual training that reflects areas of concern isolated from incidents or research.

7. The program does not include nonjeopardy simulation to allow realistic practice of behaviors or to receive feedback on performance. Also, it does not include simulations that reproduce the team environment of facilities.

The strategies that have made CRM programs effective in air carrier operations would seem to apply directly to air traffic control programs. If this assumption is correct, it is unlikely that these programs will achieve their potential unless they are carefully tailored to air traffic control, and unless improved air traffic control programs reflect the concerns mentioned above.

Air Traffic Control Simulation

Because team-centered simulation is at the heart of air carrier CRM programs, it is important to discuss the use of simulators in air traffic control training. Simulation, at least at radar positions, tends to concentrate on the ability of controllers to manage extremely high-density traffic as individuals. An important adjunct to basic human factors training would be simulation that includes coordination among positions and also includes supervisory personnel as active participants. Effective full mission simulation (LOFT), as practiced by air carriers, also involves structured human factors briefings and debriefings to reinforce the concepts practiced.

Extrapolating from air carrier experience, one of the most significant enhancements the FAA could make in its use of simulation would be to initiate team-oriented training utilizing scenarios that involve interactions with supervisors, other sectors, and on-the-job training. However, the agency is faced with a dilemma in implementing standardized simulation training on a system-wide basis because of the idiosyncratic nature of operations at the various facilities. Although the concepts involved are general, the specific cultures at different facilities may dictate differing emphases in training.

IMPLICATIONS OF AUTOMATION FOR TEAMWORK AND COMMUNICATIONS

As the FAA introduces automated systems into the air traffic system, it is essential that the effects of such innovation on teamwork and interpersonal communications be addressed during the design phase. In this section, some of the consequences of an already introduced system are discussed, along with the possible behavioral effects of two other systems.

Historically, it appears that the earliest forms of air traffic control automation, the ARTS and HOST computer systems (discussed further in Chapter 12) that provide more detailed information on target identification and position may have had the unintended consequence of reducing teamwork and acceptance of the importance of team coordination. Before the displays were automated, members of the team had to rely on verbal information transfer to ensure that both controllers and pilots maintained situation awareness. With the additional information provided by automation, understanding of the benefits of sharing mental models may have become lost, fostering the more individualistic role definition found today.

The near-midair collision discussed earlier illustrates unintended consequences of automation. TCAS provides flight crews with a visual display of traffic as well as aural warnings and commands. The warnings and commands it gives are not available to the controller handling the flight. Because of this discrepancy in information, controllers and pilots may not have the same mental model of the situation and the controller may not know in advance that an aircraft will deviate from an assigned altitude or heading. Thus it is possible, as in the example, for conflicting instructions to be issued. Although TCAS is proving to be a valuable tool for collision avoidance, it adds uncertainty to the controller role and may reduce the level of teamwork achieved.

Datalink is designed to provide electronic exchange of information between aircraft and controllers. Providing visual rather than aural information has the potential for reducing misunderstanding of clearances. However, it also may reduce the amount of information available to flight crews, often relayed via nonverbal cues in voice communications, and may have a deleterious effect on the development of teamwork between controllers and flight crews. Furthermore, one of the ancillary benefits of verbal communication between controllers and multiple aircraft on the same frequency is a great deal of information, albeit sometimes ambiguous, about conditions and traffic (Pritchett and Hansman, 1993). The party-line aspect of air traffic control communications provides information on traffic flow, weather, etc. For example, in the Avianca accident described earlier, there was a great deal of information regarding diversions and holding that, had it been processed, could have helped the Colombian crew decide to divert or declare an emergency prior to running out of fuel. The absence of verbal interaction between controller and aircraft may also make it harder to establish an effective team relationship under conditions in which datalink is not working or during emergency conditions when joint decision making is required.

Attitude surveys of flight crew members' reactions to automation have shown wide variability in their liking for automation, in the recognition that team communications requirements are changed by automation (the presence of an "electronic team member"), in perceptions of freedom to adjust the level of automation employed, and in concerns that the use of automation may degrade operational skills (Sherman and Helmreich, in press). These suggest a need to communicate organizational philosophies of automation to personnel and further the need for formal training in the team as well as technical aspects of automation use. These cockpit issues are likely to be mirrored for controllers as automation is increased in the air traffic control environment.

CONCLUSIONS

Teamwork, reflected in verbal communication among controllers and their supervisors and between controllers and flight crews, is likely to be a critical component of air traffic control for the foreseeable future. As in other technological endeavors, a high percentage of operational errors involves breakdowns in communications, coordination, and group decision making. Crew resource management training has proved to be effective in improving team coordination in flight crews and is being mandated on a worldwide basis. Similar training for air traffic controllers and their supervisors and trainers has the potential to provide similar enhancement of teamwork. This potential will only be realized if the necessary commitment by and support from FAA management becomes evident.

The automation of components of the air traffic system may influence team interactions and can, in some circumstances, have a negative effect on teamwork and the ability of controllers to maintain situation awareness. The panel has identified a number of approaches to improving team coordination and communication in the air traffic control system:

1. Making team issues a part of the organizational culture of the air traffic system by defining the nature of team coordination as part of the organization's task description. It is important to include evaluation of team as well as individual skills as part of performance assessment.

2. Focusing on team as well as individual factors in the investigation of operational errors in the air traffic control system.

3. Make team training a centrally funded program required at all air traffic control facilities.

4. Using empirical data, including analysis of team issues in operational errors, survey data on controller attitudes regarding team issues, behavioral measures of team performance in simulation, and participant evaluation of training programs to refine training programs, to ensure that critical issues are addressed in the curriculum, and to measure training impact.

5. Including interface issues (controller to supervisor) as well as controller to air crew as part of team training.

6. Providing additional training in human factors issues should be provided

for supervisors and on-the-job training instructors to allow them to evaluate team performance and reinforce effective behavior.

7. Providing recurrent training in team human factors and using team-oriented simulation as part of training.

8. Evaluating the impact of automation components on interpersonal communication and team performance before adopting systems.

9. Including team-related automation issues in team training.

8

Systems Management

Over the next several years, the Federal Aviation Administration is planning major changes in the air traffic control system through the introduction of more highly automated equipment. In preparation, the agency will want to assess its existing management system in terms of its capacity to effectively support the change process. Our purpose in this chapter is to begin that assessment as a basis for providing managers with a framework for developing strategies, procedures, and organizational structures to help manage the anticipated changes. The discussion focuses on the internal structure and culture factors that form the context within which the air traffic control system operates and on the external factors that influence both the mission and the context. Of particular importance is the interaction between the structure and the culture as it relates to the acceptance of change.

The mission of the Federal Aviation Administration is to promote safety of flight and to foster the development of air commerce (Public Law 85-726). The manner and extent to which the FAA fulfills this dual mandate is scrutinized by constituent users, as well as by a variety of external groups. These groups exert pressures on FAA management to adjust its policies, priorities, procedures, and resources to maximize the efficiency of air travel while maintaining public confidence with respect to the safety of the air traffic control system (Broderick, 1995; Daschle, 1995; Hinson, 1995a). Congress establishes statutes (e.g., amendments to the Federal Aviation Act of 1958) and funding appropriations that constitute limiting conditions within which the FAA management must operate. Local governments also establish regulations—governing, for example, noise abatement and physical construction—that constrain FAA development and operations in their areas. The representatives of air carriers, general aviation, and the military compete for limited resources. The National Transportation Safety Board, other agencies of the executive branch of government, aviation labor unions, representatives of litigants in aviation accidents, environmental groups, and the press also investigate FAA management and operations (Office of Technology Assessment, 1988; Daschle, 1995; Hinson, 1995a).

Pressures are also created by other environmental factors. The projected growth in air traffic yields the prediction that both technological and procedural changes in air traffic control must be made to avert future overload of the system and its controllers—especially because there are practical limits to what can be achieved by adding more controllers. The development of new technologies also pressures FAA management, offering potential solutions to both safety and efficiency concerns while posing challenges regarding the optimal role of controllers and automation and the transition to new technology. The challenges posed by the predicted growth in air traffic, the aging of existing equipment, and the implementation of new technology are compounded by increasingly severe economic constraints, affecting both government budgets and the aviation industry (Office of Technology Assessment, 1988; Broderick, 1995; Daschle, 1995; Hinson, 1995a).

In response to these pressures, the FAA has established formal statements of its vision, mission, and values (Federal Aviation Administration, 1995a, 1995f). These statements (paraphrased here) identify the following goals: make responsibility commensurate with authority; deliver, individually and institutionally, the highest-quality service (including maintenance of safety) on time at the lowest cost; adopt teamwork as the way of doing business and work collaboratively across organizations; empower employees to do their jobs, by providing the resources they require, involving them in decisions that affect their work, and allowing timely decisions to be made at the lowest organizational levels; foster trust, openness, dignity, integrity, and respect for the knowledge and expertise of the workforce; encourage employees to speak out, even when what they say is not popular; and encourage openness to new ideas and new ways of doing business. It is significant that these vision and values statements do not address the question of how trade-offs between conflicting goals are to be made or resolved. Such trade-offs bring management and organizational options into the purview of human factors, which can help resolve them.

APPROACHES TO DESCRIBING THE CONTEXT OF AIR TRAFFIC CONTROL TASKS

Several approaches have been taken to characterizing the organizational structure and culture in organizations concerned with maximizing safety. These approaches are presented in the work of Harss et al. (1990), International Civil Aviation Organization (1993), Moray and Huey (1988) and Reason (1987a,

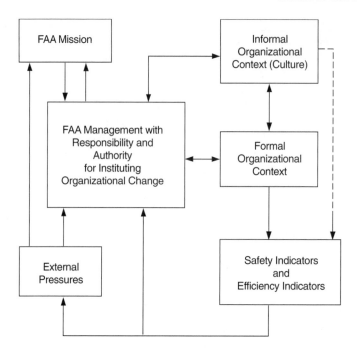

FIGURE 8.1   An air traffic control system management model.

1987b). One compelling approach suggests that air traffic controllers perform their tasks in the context of a "high-reliability organization" (HRO); this type of organization manages an extremely complex set of interacting technologies and is required to perform—and does perform—at an extraordinary level of safety and productive capacity in the face of catastrophic consequences of failure. Examples of other high-reliability organizations are nuclear power plants and aircraft carriers. The performance of operators and technicians in such organizations is affected by organizational context factors such as: operating rules, patterns of interdependency, decision-making and communication dynamics, and norms of behavior. Each of these factors can be defined both formally, for example, by written procedures, and informally, as part of the organizational culture (LaPorte, 1987, 1988, 1996a, 1996b; LaPorte and Consolini, 1991).

The panel's approach is illustrated by Figure 8.1, which depicts the relationship between the various factors contributing to or influencing FAA management. As the figure illustrates, FAA management is influenced by its own stated goals and by the external pressures exerted by interest groups outside the agency. Taking into account these influences, as well as assessments of safety and efficiency, FAA management establishes and adjusts the formal organizational context and also influences the informal organizational context.

LaPorte (1988) and LaPorte and Consolini (1991) reviewed large-scale, high-reliability organizations, including the FAA's air traffic control organization. They asked "How can a large, complex organization like the FAA manage to maintain virtually failure-free (high-reliability) performance despite equipment unreliability and the trial-and-error process characteristic of large organizations?" They found that such organizations, including the FAA, typically establish complex, detailed, hierarchical organizational structures bolstered by equally complex and standardized procedures that usually facilitate operations. These structures and procedures are often effectively bypassed at the field level, however, when planned or unplanned situations arise that require operations in response to a crisis or peak system loading. In these high-tempo situations, authority patterns shift to a basis of functional skill that is situation-dependent. In these modes, task-oriented leaders are spontaneously recognized by their coworkers—the individuals who possess the situational skill requirements become the de facto task leaders. LaPorte and Consolini suggest that this informal authority structure achieves the flexibility, teamwork, communication, and interdependent coordination that is required to maintain the high reliability of the system.

These ad hoc adaptations allow what is normally a formal bureaucracy to function temporally in a highly flexible and inventive manner. The question is how to ensure that such adaptive flexibility is retained in an environment that is increasingly stressed and is increasingly dependent on very advanced technology. Before undertaking any major organizational restructuring, it may be worthwhile to identify the organizational features that tend to ensure the capability to shift into a dynamic problem-solving state when it is needed. Such a study would benefit from a "bottom-up" observation of the informal leadership patterns of the organization (what works in practice), rather than from implementing only top-down reorganizations based on what may work in theory. In any case, both the formal and informal contexts influence the controller's performance. This performance yields safety and efficiency indicators that are monitored by both the FAA management and external constituents. In the sections that follow, we examine each of these interacting factors.

ASSESSING SAFETY AND EFFICIENCY

Safety and efficiency indicators are critical links between organizational context factors and FAA management and between the agency and external groups. Safety and efficiency indicators can both spur changes to the formal organizational context and help to validate its effects. Policies and procedures to manage the simultaneous achievement of these two organizational goals constitute a key feature of the air traffic organizational context.

Accident rates, incident rates (aircraft incidents, near-midair collisions, deviations, runway incursions, operational errors), and hazards are major indicators of the level of safety. The relationship between the three may be described by a "tip of the iceberg" metaphor. Relatively few accidents, visible above the water line, suggest more numerous incidents, which are hidden and often unreported (submerged). However, such incidents may be symptomatic of latent situations that can lead to organizational accidents.

Inadequacies at the organizational level can contribute directly to hazards, incidents, and accidents, for example, by failing to provide adequate resources to controllers, establishing inappropriate or conflicting rules for operations, incorrectly predicting the consequences of using the resources according to the rules, and neglecting to provide processes for reporting and analyzing safety issues and for correcting them. These direct links between antecedent formal organizational context variables and safety outcomes are critically important. The FAA has developed extensive experience with these links; in fact, the air traffic control system has been characterized as a network of resources (infrastructures and operational support equipment) and rules (procedures) that have evolved largely in response to practical lessons learned through accidents and incidents (Planzer, 1995).

The FAA regularly collects safety-related performance measures that include (1) for air traffic control: operational errors and operational deviations, runway incursions, air traffic delays, results of periodic on-site evaluations, and in-flight evaluation reviews and (2) for airway facilities: facility and service interruptions and outages, facility flight checks, and results of on-site inspections and evaluations (Federal Aviation Administration, 1989, 1994a). The FAA also maintains a number of databases in which safety-related data are recorded for analysis:

- The *Air Traffic Activity Data System* (ATADS) is a national system that records facility-specific monthly activities.
- The *Air Traffic Operations Management System* (ATOMS) is a personal computer network linking FAA headquarters, regions, and facilities; ATOMS contains data on operation levels and delays.
- The *Operations Network* (Ops Net) system records monthly data on outages, interruptions, and reduced services.
- The *National Airspace Information Monitoring System* (NAIMS) is an automated data system that tracks reported safety-related incidents such as operational errors, operational deviations, and near-midair collisions.
- The *Aviation Safety Reporting System* (ASRS) includes an FAA/NASA database that stores voluntary reports by pilots, controllers, passengers, and mechanics on safety-related events or conditions. ASRS reports preserve the confidentiality of the reporter.
- The *Facility Flight Check* database stores data from airborne computers and is used to periodically check the performance of NAVAID facilities. The evaluations of field operations performed periodically by the air traffic and airway facilities organizations are reported and stored for later review.

In addition to these data collection efforts, each air traffic control facility records data sufficient to recreate both the state of the system and the operations (and voice communications) of controllers at the time of incidents that require investigation.

The Office of Technology Assessment (1988) points out that the safety data collection system must produce data that are meaningful and can be manageably reviewed and analyzed. The FAA is in the process of developing an automated safety database that unifies the existing independent (and in some cases redundant) databases and provides a more user-friendly interface that supports efficient analysis.

Beyond that, however, individuals must be motivated to report relevant incidents. The ASRS, for example, provides an opportunity for controllers, pilots, mechanics, and passengers to report observations and concerns pertinent to safety; its features of preserving confidentiality and permitting anonymity support the motivation to report information. In addition to espousing its stated goals of open communication, honesty, and trust, the FAA's safety office would benefit from advocating the perception that reporting safety (e.g., error) data represents an achievement—not a failure—and that such achievements are rewarded.

The unique organizational position of the System Safety Office within the FAA provides it with the opportunity to consider the following recommendation, made by both the International Civil Aviation Organization Circular (1993) and Wood (1991): safety analysis should not be limited to consideration of the factors related to the situation, equipment, and personnel that are often emphasized subsequent to incidents and accidents. Rather, safety analysis should include careful attention to organizational context factors and extraorganizational factors that may contribute to safety issues—including factors contributing to "accident waiting to happen" situations. On this account, consideration should be given to enhancing such reporting mechanisms as the ASRS by including prompts that assist reporters in (1) considering concerns related to the formal organizational context and to organizational culture and (2) considering factors that predict future safety concerns.

Indicators of organizational efficiency that are frequently addressed include capacity/demand ratio, delays, flexibility, predictability, access, and monetary cost. Although there is no universally accepted definition of efficiency for air traffic control, the primary definition applied by the FAA is the extent to which the capacity of system resources keeps up with demand, and the traditional measure of efficiency has been traffic delay (Federal Aviation Adminstration, 1996a). The FAA, however, is working with the aviation industry to develop and validate three additional measures of efficiency: (1) flexibility, the ability of the system to meet users' changing needs, (2) predictability, the variance in the system experienced by the user, and (3) access, the ability of users to enter the system and obtain service on demand (Federal Aviation Administration, 1996a) .

The measurement of the efficiency of the air traffic control system is complicated by the need to take into account the dynamic nature of both capacity and demand and the different contributions of various system resources (Federal Aviation Administration, 1993a). Demand, as indicated by the number of aircraft that wish to land at and take off from a given airport at a given time, can change as a function of many things, for example, aircraft fleet conditions and special events. Although ground space and the number of runways at a given airport are often taken as capacity limiters, this capacity can be dynamically altered by such factors as weather conditions, winds, and the mix of aircraft types. Measures of system capacity are insufficient to determine efficiency; the capacity of individual resources must also be considered. These resources, whose capacity can also change dynamically, include airspace subsystems (e.g., surveillance, navigation, communications, flight data processing) and air traffic controllers, whose capacity is often discussed in terms of workload.

The FAA's consideration of efficiency increasingly attempts to take into account the monetary costs to the airline industry associated with the efficiency definitions mentioned above. For example, departure and landing delays translate to sizable monetary costs for the airlines (Planzer and Jenny, 1995). The FAA must also consider its own funding limitations; definitions of efficiency that stress maximizing capacity (Federal Aviation Administration, 1996a) may be inappropriate, because expending resources to achieve capacity that goes unused is not cost-effective. On that account, definitions of efficiency that emphasize balancing capacity and demand—enough but not too much—represent greater efficiency with respect to monetary costs.

FORMAL ORGANIZATIONAL CONTEXT VARIABLES

It is the responsibility of management to develop a formal organizational context for providing adequate resources (including reliable equipment and competent staff), appropriate rules for employing the resources to perform jobs effectively, and mechanisms for instituting changes as needed. The rules established by management constrain both who utilizes which resources and how the resources are to be utilized. Formal structural lines of authority, responsibility, and communication identify individuals and groups authorized to perform specified tasks and to make specified decisions. Policies and procedures (e.g., those governing safety and efficiency), sometimes constrained by legal liability considerations, establish the framework for decision making, priorities for the allocation and scheduling of resources, and approaches to the performance of tasks. The formal organizational context also includes procedures for determining the effectiveness of rules and the adequacy of resources. Processes of change include procedures for assessing what changes to personnel, technology, and formal organizational context are necessary and feasible; for implementing the changes; and for evaluating the effects.

The following discussion focuses on selected formal organizational context factors that may impact controller performance:

1. Policies governing safety and efficiency,
2. Authority and responsibility,
3. Personnel policies and procedures including communication and teamwork, workload, work schedules, performance assessment, personnel development, and legal liability, and
4. Labor-management relations.

Policies and Procedures Governing Safety and Efficiency

The FAA has clearly stated that, when the goals of safety and efficiency conflict, maintaining the safety of air travel is the agency's first priority (Broderick, 1995; Hinson, 1995a; Federal Aviation Administration, 1996a). In an effort to enhance human reliability and contain errors, the FAA prescribes detailed procedures that govern the operational activities of controllers. For example, FAA Order 7110.65J (1995b) prescribes air traffic control procedures, including the instruction to controllers to: "Give first priority to separating aircraft and issuing safety alerts as required in this order. Good judgment shall be used in prioritizing all other provisions of this order based on the requirements of the situation at hand."

The FAA also faces the challenge of improving safety and efficiency while finding new ways to cut costs (Daschle, 1995; Hinson, 1995a). This challenge is made more difficult by the fact that "aviation is so safe today that it takes major efforts to make even minor gains" (Hinson, 1995b). The FAA's strategic plan for 1995 emphasizes that, although resources are scarce (Volume 2, Section 1, "System Safety"):

> FAA will not sacrifice safety because of resources. Instead, FAA must find ways to maintain and even improve worldwide safety without requiring more resources. That means targeting FAA safety efforts where they will be most effective. It means risk assessment that compares the costs of FAA actions to both itself and the aviation community against the reduced risks of people being hurt and killed in aviation accidents and incidents. It means careful assessment of the most cost effective way to achieve an objective.

The achievement of both safety and efficiency goals is complicated by (1) changing criteria of acceptability for both safety and efficiency, (2) possibly complex interactions between safety and efficiency, and (3) possible variance in controllers' interpretations of and responses to these objectives. To allocate resources effectively, to apply appropriate technologies and strategies for improvement, and to validate improvements, the establishment of clear indicators and definitions of acceptability for safety and for efficiency must be coupled with an understanding of the relationship between safety and efficiency and between individual indicators.

Achieving multiple objectives (maintain and improve safety, maximize efficiency, and reduce costs) concurrently is likely to be complicated. A strategy to improve a given indicator may affect a different indicator, and an understanding of these interactions is also necessary if simultaneous objectives are to be achieved. For example, on one hand, introduction of an improved navigation system may improve both safety and efficiency indicators. On the other hand, the introduction of a procedure may improve safety but reduce efficiency by causing flight delays. These relationships also may change at different threshold points. For example, efficiency may be improved without affecting safety by reducing separation cushions to reduce delay (independent relationship) up to some threshold, at which further reduction in separation detracts from safety (negative correlation), particularly if unexpected crises develop in the airspace. Given the possibilities for complicated relationships, the unknowns associated with emerging indicators, and the importance of effective allocation of resources to improvement strategies, the FAA should strive to develop predictive models that establish the indicators, the levels of acceptability, and the relationships between indicators.

Human reliability and good judgment are required to meet safety and efficiency goals. The strategies whereby controllers develop and apply judgments, especially in translating formal procedures to nonstandard situations, and when attempting to maximize efficiency while maintaining safety, are important areas for study. According to LaPorte and Consolini (1991), in high-tempo modes, what works in practice may differ from what works in theory. This supports the suggestion that the development of formal organizational structures and procedures should take into account the successful operations of informal structures and procedures. When formal structures and procedures are defined and promulgated "from the top" without due reference to the interpretations, perceptions, and practical experience of those required to implement them, the following potential problems can occur and can be difficult to resolve: (1) required actions may be inappropriate to specific circumstances; (2) they may be ambiguously specified; (3) their purpose may be unclear; (4) they may be unnecessarily laborious; and (5) they may conflict with other policies or procedures. This is not to say that FAA management does not take into account the experience and knowledge base of controllers when formulating organizational structures and procedures; it does. However, what we are suggesting is that additional study of controllers' informal organizations and procedures is needed to further elicit information from that experience and knowledge base. Moreover, decisions about automation need to take into account the critical importance of designing to enhance human reliability and good judgment.

Other problems may also arise. Procedural failures diminish management's credibility in the eyes of the workers. Bad rules encourage more general contempt for rules. Research has shown that rules and procedures can have an adverse effect when applied to tasks that people are intrinsically motivated to perform (Deci et al., 1989). The more people are managed through external controls, the less likely they are to take personal responsibility for their tasks. This suggests that, to encourage controllers to develop and apply effective strategies to achieve both safety and efficiency objectives, rules and procedures should be kept to the minimum necessary to provide an adequate formal organizational context for safe and efficient working.

Authority and Responsibility

The key features of the FAA's major lines of authority and responsibility include: (1) a hierarchical chain of authority and responsibility organized along functional lines of business, (2) extension of direct chains of command from headquarters through regional and facility levels, and (3) a System Safety Office reporting directly to the FAA administrator. That authority structure reflects a recent reorganization spurred, in part, by external pressures. A 1988 report by the Office of Technology Assessment (OTA), *Safe Skies for Tomorrow*, summarized the major external pressures that suggested the need for change (p. 45):

> Although commercial aviation maintains an enviable safety record, dramatic growth in air travel, major changes in technology and industry operations and structure, the firing of air traffic controllers, and Federal budget constraints have left the FAA scrambling to catch up. Consequently, public attention has again focused sharply on whether the FAA has the institutional capability and resources to carry out its operation, standard setting, rulemaking, and technology development functions effectively and to guarantee compliance through its inspection programs.

The OTA report concluded that a fundamental organizational problem within the FAA's air traffic control organization was the "splintering of authority" between headquarters management and the nine regional offices, which have broad and separate authority (p. 52). Such splintering of authority can result in confusion about responsibilities and in lack of accountability, which becomes especially salient with respect to safety responsibilities.

In response to congressional concerns reflected in the OTA report, increasing industry demands that the efficiency mandate be more forcefully addressed, general federal budget constraints (including the requirement to downsize staff), and increasing public demands for higher standards of safety (fueled by reports of safety incidents and accidents and of the decreasing reliability of air traffic control equipment), the FAA announced in 1994 the restructuring of its organization (Federal Aviation Administration, 1994b, 1994c).

In addition to the overall goal of making accountability commensurate with authority, the reorganization specifically addresses the OTA report's concern about the splintering of authority between FAA headquarters and regional management. Under the reorganization, the nine regional administrators remain responsible for carrying out the policies and regulations set by headquarters management. Each regional administrator reports to the associate administrator for administration, serving as the "eyes and ears" of the FAA administrator. However, regional administrators have no line authority over the operating services located within their regions. The regional operating services (including division managers for air traffic control, airports, flight standards regulation and certification, security, and airway facilities) have line authority over the field personnel and facilities within their respective regions, and they maintain the lines of accountability by reporting directly to their respective associate administrators at FAA headquarters. Another rationale for the reorganization is to achieve a combination of simplicity and flexibility that will allow the FAA to adapt quickly to the constantly changing demands of the aviation industry (Daschle, 1995).

The reorganization undertaken in 1994 to address this problem can be characterized as centralized and hierarchical, a feature of high-reliability organizations. One advantage of this centralization may be the clarification of formal accountability. Another advantage may be the ability to standardize policies and policy interpretation; this would reduce confusion in the field and enhance employee morale. There remain, however, three essential questions concerning the practical utility of the reorganization with respect to meeting those goals:

1. Are the leaders individually appropriate to their roles?
2. Does the reorganization extend appropriately beyond headquarters to the field level?
3. Are the formal structures compatible with the informal structures?

The 1994 reorganization identifies stable offices and clear lines of authority, another feature of high reliability organizations. However, the FAA's frequent reorganizations have combined with a practice of moving managers through a variety of positions as a means of career progression, producing a continuing "musical chairs" among the leadership. Accountability holds more force when the accountable leader expects to remain accountable in the given office. Particularly when career progression involves competition, as it does at FAA headquarters, it is doubtful that either simple or complex reorganization of the formal authority and responsibility chart can produce the hoped-for teamwork, flexibility, and dynamic responsiveness.

The 1994 reorganization primarily addresses headquarters functions. By implication, regional and facility offices participate in the reorganization, but they have not undergone fundamental internal reorganization. The question of possible gaps or discrepancies between the philosophies of headquarters and field management remains. The headquarters organization exhibits a hierarchical structure while espousing a "flat" teamwork philosophy. Which approach is actually implemented and whether the same approach is applied in the regions are questions of practical import that may have safety and efficiency implications.

The OTA's *Safe Skies for Tomorrow* report specifically recommended the establishment of a comprehensive system safety management program. It recommended that the safety program should apply across all phases of planning, data collection, engineering, and operations and that the program should include coordinated data collection, analysis, and reporting of safety issues. In 1994, the FAA created a special System Safety Office reporting directly to the FAA administrator. The System Safety Office serves as a single focal point for safety issues and concerns. The office is responsible for analyzing safety databases, assessing current and predicted safety issues, and providing to the FAA administrator independent advice on safety issues and trends (Federal Aviation Administration, 1994b, 1994c).

The creation of an independent safety office reporting directly to the FAA administrator and assuming authority for safety data collection, analysis, and change recommendations mirrors the standard approach practiced within the aviation industry (Wood, 1991). Our discussion of safety indicators above points out that the FAA is addressing recommendations to improve the integration and usability of the safety databases, and that additional improvements should include prompting of the reporting of data on both formal and informal factors relevant to safety.

Personnel Policies

General Considerations

Formal organizational reporting linkages are necessary but not sufficient elements of organizational structure. Equally important are the channels along which communications—between controllers, between managers, and between controllers and managers—may pass. As mentioned earlier, the FAA has included among its stated goals fostering openness, encouraging communication and feedback generally, and specifically encouraging employees to speak out even if what they have to say is unpopular. Many communication issues reflect the interaction between informal organizational culture and formal organizational context; for example, formal communication channels may be available, but their use may not be informally encouraged.

A key characteristic of air traffic control performance is the sharing of responsibilities among team members, who handle multiple tasks under time pressures as a group as well as individuals. Currently, controller teams are neither assigned legal culpability as team units during accident or incident investigation nor rewarded or penalized as team units under the performance assessment process. It is important to assess the performance of the team unit; it is also important to distinguish and to assess the contributions of individual controllers to both effective team performance and error. The ability to conduct such assessment underlies not only the effective direction of individual and team activities, but also the appropriate administration of reward and discipline.

For achievement of safety and efficiency goals to be validated, it is critically important that performance assessment be conducted thoroughly and carefully, that tasks be assessed with respect to both safety and efficiency, and that controller contributions to system performance be assessed. It is also desirable that the effectiveness of organizational context factors be validated by investigating the effects of these factors on controller performance.

Finally in the assignment of tasks and work schedules, the FAA should be guided by the capabilities and limitations of controllers and the potential negative effects of such factors as sleep loss, overload, and underload.

Legal Liability

Although controllers are indemnified by the FAA,[1] legal liability can indirectly affect them in two ways: (1) operating procedures (e.g., FAA Order 7110, which directs air traffic control procedures) have grown substantially in number and complexity, partly in response to litigation following accidents and incidents, with a concomitant increase in controller workload and training time and (2) internal disciplinary actions, as well as promotional and other reward actions, are largely tied to observance of the operating procedures. Article 62 of the collective bargaining agreement between the FAA and the National Air Traffic Controllers Association (NATCA) union, however, provides for an immunity program that limits the circumstances under which discipline is imposed: disciplinary action is not imposed when the employee's action is inadvertent and does not involve gross negligence or a criminal offense, provided the employee files a report to NASA on the error within the time limits prescribed by applicable regulations and does not otherwise cover up the error.

Labor-Management Relations

Labor-management relations are governed by a contractual agreement reached in 1993 between the FAA and the National Air Traffic Controllers Association (NATCA). Article 7 of this agreement requires the FAA to notify and, upon request of the union, to conduct negotiation with NATCA any time it

---

[1] The Airport and Airway Safety and Capacity Expansion Act of 1987 provides that: "The [FAA] Administrator is empowered to indemnify any officer or employee of the Federal Aviation Administration against any claim or judgment against such person if such claim or judgment arises out of an act or acts committed, as determined by the Administrator, within the scope of such person's official duties" (FAA Order 2300.2A)."

proposes to change personnel policies, practices, working conditions, operational procedures, or procedures related to technological changes. If negotiation does not produce resolution, the agreement provides for escalation of negotiations from facility through regional and national levels, as appropriate. If these negotiations do not produce resolution, the agreement permits the FAA and NATCA to pursue whatever course of action is available to them through the Federal Service Labor-Management Relations Statute.[2]

In addition, both the FAA and NATCA recognized a need for a process of change that highlights the value of trust, communication, and partnership between employees and management, working together to solve problems pertaining to enhancing the quality of service provided by the air traffic control system to its customers (Department of Transportation newsletter, 1993). Toward that objective, in 1991 union and management jointly developed a quality through partnership (QTP) process applicable to air traffic control terminals and centers. The QTP process, defined in FAA Order 3710.16, is structured in tiers, with FAA management and NATCA representatives serving on facility, regional, and national QTP teams. Additional teams may be established to address problems within specific facilities. The QTP process is completely separate from the collective bargaining relationship and is not intended to address matters that are covered under the collective bargaining agreement (e.g., individual grievances). The subjects that may be considered under the QTP process are not otherwise limited. At each level, decisions are based on the consensus of team members. Decisions supported by all team members are binding on all within the jurisdiction of the team making the decision.

The QTP process will take time to institutionalize both formally and informally. At the national level, both FAA management and the NATCA representatives espouse a policy of cooperative teamwork—although they have in the recent past aired differences of opinion regarding the sufficiency of air traffic control staffing and debated the proper role of controllers in the development of new (often automated) equipment. Currently, a full-time NATCA representative works alongside the director of the FAA's Air Traffic Requirements Office, helping to formulate requirements for new air traffic control systems and to monitor systems under development. In practice, at regional and facility levels, there are differences in the style (cooperative versus confrontational) among both FAA managers and union representatives. Such differences reflect the personalities of both individuals and facilities. However, the problem-solving approach

---

[2]Article 8 of the agreement formally recognizes that traditional methods of dispute resolution (e.g., grievance/arbitration and unfair labor practice charges) are not always the most efficient means of problem resolution. It therefore establishes a problem-solving process whereby complaints are submitted for consideration and possible resolution at a meeting attended by representatives of the union, FAA management, and the employee(s), in a good-faith attempt to avoid the grievance procedure. Provision is also included to support training programs in problem-solving techniques.

defined in the collective bargaining agreement and the QTP process represent positive steps to ensure that controllers have a voice in how they work and what tools they use.

INFORMAL ORGANIZATIONAL CONTEXT VARIABLES

The culture of an organization characterizes both its informal workings and its climate and morale. It therefore characterizes workers' perceptions, attitudes, and behavioral norms, which may or may not agree with the formal organizational context. As Figure 8.1 illustrates, informal and formal contexts interact, both are affected by and influence management strategies, and both impact safety and efficiency.

The FAA's strategic plan and vision statements acknowledge the importance of cultural factors and their likely interconnection with formal organizational context factors, as well as with safety and efficiency. Experience and research in the field of industrial and organizational psychology generally suggest that inadequate, inappropriate, or conflicting formal organizational rules, consequences, change processes, and resources contribute to negative organizational cultural climate, and, conversely, can result from them. It is also true that, when adequate and appropriate, formal organizational context factors can help to produce a positive culture, and vice versa. However, a robust set of quantitative and qualitative data is needed to define the extent to which and ways in which organizational culture factors can act as mediating variables, enhancing or detracting from both the formal features of organizational context and the performance of air traffic controllers.

Reason (1987a, 1987b) and Stager (1990) provide detailed discussion of cognitive functions pertinent to such safety- and efficiency-related tasks as air traffic control; their unified conclusion suggests that the playing field on which mediating organizational culture variables exert their influence is the cognitive "environment" (Stager) of the controller. Four key factors that influence controllers' cognitive performance are: (1) the amount of correspondence and conflict between formal and informal norms and rules, (2) subculture differences, (3) job satisfaction, and (4) attitudes toward change.

Informal Norms and Rules

The cultural framework within which the formal organizational context is enacted is comprised of attitudes, perceptions, and behavioral norms that have evolved over a long period of time. A key element of this interaction between culture and formal context is the underlying dynamic of the communication among operators, technical professionals, and managers. Managers and technical professionals often enact formal structures, policies, and procedures to reinforce management control. If these procedures complement cultural norms, the resulting enhancement of job satisfaction, teamwork, and cooperative implementation of changes helps to explain how high-reliability organizations such as air traffic control manage to perform at an extraordinary level of safety and productive capacity in the face of very demanding circumstances. If management enacts formal context features that abrogate the cultural norms, then dissatisfaction and poor performance may result (LaPorte, 1996a).

Westrum (1990) identifies as a link between informal and formal organizational contexts the receptivity of management to the communication of perceived problems. Westrum distinguishes three kinds of organizational culture:

1. In pathological cultures, managers do not wish to know of safety problems, treat punitively those who report problems, avoid responsibility for problems, punish those associated with errors, and actively suppress new ideas.

2. In bureaucratic cultures, managers may not be informed of problems (if the communication process does not involve them), passively receive information, accept responsibility for problems only if it is assigned to them, prefer to localize response strategies, and discourage new ideas.

3. In generative cultures, managers actively seek out information (including reports of problems), train subordinates to seek out and to report information, share responsibility, respond to problems with far-reaching inquiries, and welcome new ideas—including those pertaining to organizational change.

The FAA is attempting to promote a generative culture by endorsing goals of encouraging openness, communication, and uncensored reporting. LaPorte and Consolini point out that high-reliability, high-hazard organizations are especially driven to use a preventive decision-making strategy that encourages the discovery of potential problems and rewards reporting of them.

Subcultures

Although supporting data are anecdotal, subculture differences are known to exist between headquarters and the field, among geographic regions, and among facilities (Breenlove, 1993; panel visits). Mundra (1996) provides an example—the introduction of the ghosting display aid to the radar display—illustrating that subcultural differences in the willingness to adopt a new technology can reflect, in part, real differences between facilities with respect to operating procedures, constraints, and requirements. In addition, the very mission of the FAA may be interpreted differently at different levels of the organization, and these interpretations may receive different emphasis at different geographic regions or facilities. Controllers and local supervisors may focus on the practical aspects of moving aircraft safely and expeditiously in response to tactical conditions. Regional managers may focus on the installation and implementation of equipment and procedures. The focus of headquarters' managers may reflect a broad stance that permits acceptable responses to political questions. The following section describes the results of the FAA's employee attitude survey. In order to verify these plausible hypotheses, data need to be collected and analyzed regarding controller attitudes. If data could be sorted by facility, they could be examined to study the cultural differences between facilities. Such a study could contribute to the development of causal models of facility characteristics that may lead to better or worse cultural climates.

Job Satisfaction

Organizational culture includes the perceptions of the controllers and their managers about the adequacy of resources, rules, consequences, and change mechanisms for performing their jobs—as well as the extent to which job characteristics contribute to the satisfaction of individual needs—and their expectations as to whether these factors will maintain adequacy, improve, or deteriorate. Controllers compare their perceptions of things-as-they-are with their expectations of things-as-they-will-be, producing an overall sense of job satisfaction and morale.

Since 1984, the FAA has conducted biennial surveys of employee job satisfaction. Initially, the survey took the form of the survey feedback action program (SFAP), focusing on employees' confidential reporting of their satisfaction with the styles and behaviors of their immediate supervisors, who discussed the results of survey analysis with their subordinates. The survey included items pertaining to job satisfaction.

In 1991 the survey program was replaced by the currently administered employee attitude survey (EAS), which focuses on employees' perceptions of organizational culture factors. Administered biennially, the EAS addresses a core set of attitudinal variables, including: overall job satisfaction, employee empowerment, employee involvement, communication, recognition and rewards, teamwork, personnel development, and performance appraisal. Each administration of the EAS also addresses attitudes toward current issues of special interest; recent examples are telecommuting and alternative work schedules. EAS results are reported to upper management for each line of business within the FAA. Of course, many of the survey variables relate to the formal organizational context factors discussed in this chapter.

The following discussion summarizes the results of the 1995 EAS for air traffic employees (including but not limited to controllers) and links these perceptions to the formal organizational context factors discussed above. The results for the 1995 EAS were presented in an FAA briefing, *Annotated Summary of the 1995 EAS* (Federal Aviation Administration, 1995c), in terms of the 5-point rating scale used in the survey's administration: 1 = very dissatisfied, 2 = somewhat dissatisfied, 3 = moderately satisfied, 4 = highly satisfied, and 5 = very highly satisfied.

Considering overall job satisfaction, 71 percent of employees reported that they were either highly or very highly satisfied with their jobs. This category includes such factors as satisfaction with pay, benefits, working conditions, the nature of the job tasks, immediate work group and supervisor, and opportunities for development. The factors with which the highest percentage was satisfied are pay, benefits, and the nature of the work itself. In contrast, the factors with which the lowest percentage was satisfied included factors contributing to the environment in which the job is performed: working conditions, the supervisor, the organization, and opportunities to develop potential.

The 1995 reported results were not significantly different from the previous (1992) survey. However, compared with the 1984 survey, the 1995 survey showed a 16 percent increase in the percentage of respondents reporting that they were either highly or very highly satisfied overall. Before concluding that these results demonstrate an improvement in the components that contribute to overall job satisfaction, this comparison should be qualified by three considerations: (1) although both surveys included items on overall job satisfaction, which is still considered a benchmark for agency morale, the surveys were different. The 1984 survey was an SFAP, the 1995 an EAS. (2) Overall job satisfaction is typically higher than facet-specific job satisfaction. (3) Job satisfaction measures are known to be positively skewed because those who are very dissatisfied leave the job, and those who experience dissatisfaction but stay on the job may convince themselves that they are more satisfied, in order to reduce cognitive dissonance.

Any increase in reported job satisfaction since 1984 is tempered by a steady decline, since 1988, in the percentage of survey respondents agency-wide who expressed confidence that FAA management would use the EAS results to improve working conditions and morale. That percentage has declined from 36 percent in 1988 to 19 percent in 1995. Both the results for key overall job satisfaction factors and the response to the question of trust in management suggest that air traffic employees perceive a need for changes in management style and/or structure, particularly relating to specific formal organizational context variables.

With respect to safety and efficiency goals, controllers are governed by formal procedures. They may also be influenced by informal procedures and pressures. In either case, controllers must understand the procedures and, ideally, should influence their development. EAS results indicate that air traffic employees are highly satisfied with their understanding of how their jobs "contribute to the FAA mission." However, they are only moderately satisfied with the adequacy of management's communication of policies and with their opportunities to express concerns openly and with impunity.

We therefore suggest that management should recognize, during the development of new policies or priorities with respect to safety and efficiency, that there may be a discrepancy between the intent of their policy communications and the interpretations drawn by controllers. We question whether the goals of openness and generativity are being met. One method of encouraging both understanding of and compliance with policy is involving the employees in its development; on this account, management should consider in particular employee perceptions pertaining to the open expression of concerns about policy changes.

As noted above, the FAA has both established values statements and instituted reorganizations aimed at aligning responsibility with authority and at empowering employees to make decisions. The agency-wide results of the EAS, however, indicate that employees are only moderately satisfied with the extent to which decisions are made at appropriate levels, employees have authority to make day-to-day decisions, and employees are given the opportunity to contribute to decision making that affects their jobs. Data specific to air traffic functions were not available for these items. Although FAA management has formally expressed these cultural goals, the goals have apparently not been internalized in the culture. The results of the EAS also call into question the effectiveness of the quality through partnership (QTP) process, which was instituted by the union and FAA management to foster employee involvement in decision making.

Communication issues also interact with other considerations. With respect to job satisfaction, we cannot overstress the importance of the extent to which employees feel free and encouraged to communicate their recommendations for change, as well as the extent to which they perceive communications from management to be clear and meaningful. The EAS results show that over 40 percent of surveyed air traffic employees believe that it is safer to agree with management than to disagree and that employees are not encouraged to speak openly. Only 60 percent reported that they feel free to discuss problems with their supervisor. Addressing this perception that communication is inhibited would be a useful means to develop improvements in other reported areas of dissatisfaction.

With regard to the System Safety Office and associated reporting mechanisms, the panel's judgment is that employees should be encouraged to report not only observed incidents and deviations but also their perceptions of latent hazards that might contribute to future accidents or incidents. The substantial percentage of employees who do not feel free to report problems suggests that management should investigate the extent to which these inhibitions apply to the reporting of safety concerns.

The FAA has established a sequenced program of training for controllers, with adjunct retraining when skill decrements are determined and when new technology is introduced. EAS results, however, indicate that employees are only moderately satisfied that they receive the training needed to perform effectively and that the training they receive is applied to their job. With respect to more general opportunities for growth, only about 30 percent of air traffic employees reported that they are satisfied with opportunities to develop their potential.

With regard to the process of performance assessment for air traffic controllers, the EAS results indicate that air traffic employees experience generally low satisfaction with the extent to which recognition and rewards are given for exceptional performance, management responds positively to a job well done, recognition and rewards are administered promptly, and promotions are given on the basis of job qualifications. Assessing proposals for changes to the performance appraisal system, employees expressed high satisfaction with proposals aimed at increasing the timeliness of recognition and rewards, moderate satisfaction with proposals that separate pay from performance and include ratings by coworkers in the appraisal process, and low satisfaction with proposals that rely exclusively on ratings by coworkers. These results suggest that the ongoing process of revising the appraisal process for controllers should include careful consideration, with continued feedback from controllers, of both how performance should be appraised and how rewards and penalties should be tied to the results of appraisals.

The EAS survey results also indicate that air traffic employees report low to moderate satisfaction with the impact of new technologies on their jobs. This topic included such issues as the extent to which new technology is appropriate, sufficient and timely information on the new technology is provided by management, and the organization is generally quick to adopt new work methods. Such results form part of the backdrop of organizational culture against which user involvement (or noninvolvement) occurs during the acquisition of new systems.

There is a lack of research evidence establishing clear causal relationships between formal and informal organizational context factors, and between these factors and the performance of controllers. The EAS, however, does constitute a vehicle for the collection of data on culture, and these data could be applied to the study of these relationships.

Attitudes Toward Change

The general construct of trust is a key variable in the use of automated equipment. Controllers' actions are based not only on formal procedures, but also on the shared subcultural assessment of whether the equipment and the procedures merit trust. The declining reliability of air traffic control equipment, the projected increase in air traffic, pressures to contain staffing levels, and the long-term nature of acquisition processes for modernizing equipment combine to produce a tendency to rely on human controllers to compensate for the deficiencies of other resources and heighten the controllers' concerns about trust.

Controllers' trust (or mistrust) of new equipment and procedures is a function of both the reliability or effectiveness of the changes and the controllers' trust in themselves (Lee and Moray, 1992). Recent press reports of controller reactions to increasing unreliability of air traffic control equipment suggest that, although controllers are becoming increasingly concerned that future equipment failures may exceed their abilities to compensate successfully, they publicly express confidence in their skills and abilities to maintain air traffic safety. However, trust in new equipment and procedures implies a willingness to take reasonable risks associated with adopting the changes; controllers' trust in their equipment, procedures, and one another is also affected by both formal and informal organizational rules pertaining to the taking of risks associated with introducing change.

COORDINATING HUMAN FACTORS RESEARCH ACTIVITIES

It is a central theme of this report that human factors considerations with respect to the development and implementation of automation for air traffic control are critical and broad. Human factors research activities should be applied across all phases of acquisition, from the definition of requirements through test and evaluation, and information about human performance, derived from human factors research, must be available to support all phases of acquisition and implementation. To do this, it is necessary that human factors research activities and the resources that support them be coordinated both within and across research, acquisition, and implementation activities.

Such clear lines of responsibility and authority are not currently evident in the FAA. Human factors research activities are conducted at and/or managed through separate organizational entities. The Civil Aeromedical Institute (CAMI) conducts research internally and manages contracted research pertaining to air traffic control and to airway facilities in such areas as communication, selection, training, performance assessment, information display, workstation configuration, teamwork, fatigue and shift work, and organizational context factors (Federal Aviation Administration, 1996b; Collins and Wayda, 1994; Schroeder, 1996; briefings of panel by CAMI Human Resources Research Division, 1995). CAMI also conducts in-house and contracted research on general aviation, the flight deck, and bioastronautics that support such activities as regulation and certification. The FAA Technical Center (FAATC) applies its Research Development and Human Factors Laboratory, with extensive simulation capability, to the evaluation of advanced concepts and technologies for air traffic control and airway facilities (Stein and Buckley, 1994; Federal Aviation Administration, 1995g).

Within the Research and Acquisition organization at FAA headquarters, the Human Factors Division, the Security Human Factors Branch, and separate integrated product teams sponsor human factors research in response to needs as they emerge. This research is conducted by CAMI; the FAATC; cooperating government facilities at NASA Ames, at the Volpe National Transportation Safety Center, and within the Department of Defense; and university and other contractors. The research entities within the FAA reside within different organizational entities. They do not report through a single chain of responsibility and authority, their resources (e.g., staff and budgets) are separately managed, and their activities are separately evaluated.

Human factors application activities are similarly conducted by disparate organizational entities. The FAA's Human Factors Division occasionally provides support to acquisition programs. However, human factors activities in support of acquisition are performed largely by contract personnel, including human factors personnel working for the design contractor as well as human factors monitors working for FAA program management. Such efforts are currently managed by integrated product team leaders, who determine their own needs for human factors support. Separate human factors support activities, provided largely through contract personnel, are also managed on an as-needed basis within the Air Traffic Services organization (for Airway Facilities and advanced system planning areas) and within the Regulation and Certification organization (for standards and certification areas).

In order to reinforce agency-wide appreciation for the importance of human factors, and in recognition of the need for coordination of human factors activities both across research and acquisition and across disparate organizational lines, in 1993 the FAA promulgated a human factors policy statement (Federal Aviation Administration, 1993b). This order prescribes that "Human factors shall be systematically integrated into the planning and execution of the functions of all FAA elements and activities associated with system acquisitions and system operations." The order also prescribes the composition and function of a Human Factors Coordinating Committee (HFCC).

The HFCC is chaired by the FAA chief scientific and technical advisor for human factors, who is currently located within the Research and Acquisition organization. The HFCC is composed of representatives of several executive directors, associate administrators, assistant administrators, and center directors, all of whom retain authority to represent their organizations in human factors matters. The functions of the HFCC are to: (1) identify research requirements and coordinate research results; (2) foster the dissemination of human factors information across organizations; (3) facilitate the integration of human factors into rulemaking, systems acquisitions, and other activities within the agency; (4) identify the need for changes to existing policies, processes, research programs, regulations, or other human factors activities and programs; and (5) monitor the efficacy of human factors efforts and programs within the FAA. The HFCC, which meets infrequently, communicates largely through a newsletter.

The HFCC represents, in effect, more of an information exchange vehicle than a management vehicle: its members manage (plan, direct, control, allocate resources for, and evaluate) separate human factors activities rather than a unified agency program. In addition, the *National Plan For Civil Aviation Human Factors* (Federal Aviation Administration, 1995d), recently approved by the FAA, outlines a general plan that provides conceptual direction for human factors research and applications, but the plan does not define how those activities will be managed.

In sum, research activities are not adequately coordinated across research centers, research is not systematically performed to support practice needs, and research findings are not systematically applied. Application activities are not adequately coordinated and rely heavily on subcontractor efforts that are not managed by FAA human factors professionals. Given the importance of applying in a timely manner the appropriate skills and knowledge of the multidisciplinary staff that must comprise a human factors program, the importance of maintaining a synergy between research and acquisition activities across all phases of the development of new systems, and the practical constraints of limited budgets that demand effective use of resources, the fragmentation of human factors activities across the agency suggests the need for integrated management. To this end, overall management of human factors research and development activities for the FAA that relate to air traffic control and to airway facilities should be concentrated. Concentrated human factors management should be given authority over staff and budget commensurate with its responsibility and should be assured of the resources required to perform its functions. Furthermore, we urge that the relationship between the System Safety Office and the concentrated human factors management should be a strong one.

It is beyond the scope of this report to consider the location, within the FAA's management structure (e.g., within which line of business or at what level of management), of this proposed concentration of human factors management, how human factors management itself should be organized, and the extent to which this management entity should assume responsibility and authority for human factors activities other than for air traffic control and airway facilities. A subcommittee of the FAA's Research, Engineering, and Development Advisory Council has recommended a plan for human factors within the FAA (Federal Aviation Administration, 1996c:iv):

- Establish an FAA human factors single point manager for all human factors research and application efforts within agency functions for acquisition, regulation and certification, security, and NAS operations, and across agency organizational elements (including the FAA William J. Hughes Technical Center and the Civil Aeromedical Institute):

- Assign authority and resources (people, dollars, and facilities) concomitant with the responsibility and accountability for an effective FAA Human Factors Program for research and applications.

- Designate the FAA Human Factors Division (AAR-100) as the agency's human factors single point manager, and hold that office accountable for the quality of the agency's human factors products and services.

CONCLUSIONS

New technologies are introduced into an existing organizational formal context and informal culture, whose characteristics interact with those of the new technologies to influence subsequent performance outcome indicators, like safety and efficiency. It is therefore critical in planning the introduction of new technologies in the air traffic control system to identify organizational features that tend to ensure the capability for shifting to a dynamic problem-solving state when such is needed. Because it is responsible for instituting organizational change, air traffic control system management at all levels must regard itself as having a potentially strong influence on safety and efficiency, especially by virtue of how it reinforces the processes and values that underlie and support productive change.

It is generally accepted that organizational culture exerts a pervasive influence on performance that is difficult to specify for two reasons: (a) there are distinct cultures and subcultures in different regions, options, and facilities and (b) the variables that comprise culture cannot be exhaustively enumerated. Cultural factors that deserve special attention because of their potential to influence the success of efforts to introduce new technologies into the air traffic control system include a positive change orientation, the perception that change can and will be accomplished effectively, and the belief that the new technologies will serve important performance needs or goals. Managing the technological change process well—from planning and procurement through implementation and full incorporation into the air traffic control organization and culture—is a significant way to promote positive performance effects.

More research is needed on the multiplicity of conditions and variables that mediate the effects of formal organizational context and informal organizational context (culture) on technological change and performance (safety and efficiency) in the air traffic control system.

When indicators and criteria of acceptability for safety and for efficiency are unclear or not accepted, proposals for technological or procedural improvements, including automation, may constitute solutions in search of a problem. To allocate resources effectively, to apply appropriate technologies and strategies for improvement, and to validate improvements, the establishment of clear indicators and definitions of acceptability for safety and for efficiency is necessary but not sufficient. An understanding of the relationship between safety and efficiency and between individual indicators is also required. To be effective, standard post facto analyses of the factors contributing to safety (accidents, incidents, and hazards) and to efficiency (e.g., delays) should be complemented by predictive risk assessments. Comprehensive predictive assessments should include the development and application of models that: (a) identify indicators, measures, and levels of acceptability for safety and for efficiency; (b) assess the interaction of safety and efficiency factors; and (c) assess the contributions of controller cognitive tasks, including decision making, to outcomes. Predictive assessments should precede the acquisition of technologies proposed as solutions to safety or efficiency concerns and should include assessment of the proposed technologies.

The restructuring of the air traffic control organization, the development of policies and procedures governing operations that address safety and efficiency, and the development or acquisition of related technologies need to take into account both the formal and informal processes judged by controllers to work in practice. When necessary, it is important to study systematically the informal processes, especially those by which controllers respond to high-tempo contingencies.

Personnel are critical resources that must be properly developed and maintained. The FAA is currently relying on the skills of its controllers to compensate for the deterioration of equipment resources. Examining the impact of staffing on opportunities for training would be useful. New or altered policies governing controllers' actions to maximize efficiency while maintaining safety need to be accompanied by requisite training. Existing safety databases and their proposed integration should be complemented by efforts to encourage the perception by controllers that reporting safety data represents an achievement, not a failure, that such reports will be rewarded, and that reports should include concerns relating to organizational factors.

The FAA's biennial Employee Attitude Survey generally indicates that air traffic employees are dissatisfied or only moderately satisfied with management practices and the organizational context within which they perform their jobs, although the jobs themselves are reported as satisfying. Addressing their perception that communication is inhibited holds great promise as a means to develop improvements in other reported areas of dissatisfaction. If the existing EAS data can be sorted by facility, the data could be examined to study the cultural differences between facilities, which could contribute to the development of causal models of facility characteristics that may lead to better or worse cultural climates.

Human factors activities within the FAA, including both research and practice activities, are fragmented. Research activities are not adequately coordinated across research centers, research is not systematically performed to support practice needs, and research findings are not systematically applied. Application activities are not adequately coordinated and rely heavily on subcontractor efforts that are not managed by FAA human factors professionals. Overall management of human factors research and development activities for the FAA that relate to air traffic control and to airway facilities should be concentrated. Concentrated management should be given authority over staff and budget commensurate with its responsibility, and it should be assured of the resources required to perform its functions. The relationship between the System Safety Office and the concentrated human factors management should be a strong one.

9

Human Factors in Airway Facilities

The human factors aspects of Airway Facilities operations, equipment, job classifications, selection, training, and performance appraisal are critical to the impacts of current automation. In this chapter we discuss the effects of increased automation on Airway Facilities operations and staffing and the state of research in the human factors of Airway Facilities. This chapter builds on the overview of Airway Facilities presented in Chapter 4.

EFFECTS OF INCREASED AUTOMATION

It is noteworthy and perhaps paradoxical that, when new components or systems are introduced, the impact of automation is often experienced more by Airway Facilities than by Air Traffic Services. The new components or systems occasionally include increased automation of air traffic control functions; often they represent modernization of aging equipment without significant change to the human-machine interface for air traffic controllers. In either case, the new systems increasingly include automation of such functions as diagnostics, fault localization, status and performance monitoring, and maintenance logging. These automation enhancements are usually transparent to the air traffic controllers, but they can impose on Airway Facilties specialists the requirement to learn new and often complex functional and human-machine characteristics of the modernized equipment.

It is difficult to make generalizations about current components and systems; about the procedures and activities associated with their operation, monitoring, and control; and about the associated personnel selection, staffing, training, and performance appraisal procedures because: (1) there is variation across FAA regions, sectors, and facilities with respect to equipment, systems, and personnel considerations and (2) the current process of modernization involves the piecemeal introduction of new technologies, with associated changes to operations and personnel activities, in a manner that places the Airway Facilities domain in a state of flux.

As new technology tends toward more software-intensive and automated functions, network linkages, space-based systems, and highly reliable distributed architectures—introduced at different times in different places—several changes are occurring within Airway Facilities (Schroeder and Deloney, 1983; Reynolds and Prabhu, 1993; Federal Aviation Administration, 1991c, 1993a,.1995b, 1995c). There is a continuing trend toward increased automation of such functions as data acquisition and storage, diagnostics and fault localization for modularized equipment, reconfiguration through the use of redundant software as well as hardware elements, and maintenance logging. At the same time, automation support for such higher-level cognitive functions as system-level diagnostics, trend analysis, decision making, and problem solving is reserved for longer-term development. Maintenance philosophy is turning from an emphasis on corrective and regularly scheduled preventive maintenance to an emphasis on performance-based maintenance that takes advantage of automated trend analyses to identify the most efficient scheduling for maintenance to prevent failures. Maintenance philosophy is also turning away from concentration on on-site diagnosis and repair of elements of equipment toward more centralized and consolidated operational control centers (OCCs) that monitor and control equipment and systems at unmanned facilities, accompanied by automated localization of problems to line replaceable units that are replaced and sent to contractors for repair. The focus on "systems within one's jurisdiction" is being replaced by a focus on sharing of information, resources, and responsibilities across jurisdictions.

Airway Facilities job classifications have traditionally stressed specialized knowledge of hardware for specific equipment or systems—knowledge that is still required to keep the current system operational. However, new job classifications are placing much more emphasis on knowledge of and responsibility for monitoring and controlling interacting systems, on management of software-intensive, distributed networked resources, and on application of systems engineering methods to provide system services to users. Selection procedures, which previously encouraged the hiring of military personnel with electronics backgrounds, are placing more emphasis on the hiring of personnel with skills and abilities related to systems engineering, computer science, and automation. Training programs have traditionally involved strings of courses that develop expertise with single items of equipment or single, independent subsystems; there is increasing demand for programs that develop expertise in diagnosing and responding to system-wide difficulties, including understanding of the interactions between systems that cooperate to provide national airspace system services. There is a corresponding change in the way the performance of Airway Facilities specialists is appraised, away from "fix the box" tests and observations toward evaluation of specialists' abilities to diagnose and respond to system problems.

In this chapter the panel makes some generalizations with respect to such moving targets. In discussing components of the national airspace system, equipment supporting Airway Facilities activities, and its operations, our approach is to suggest sets of items that may have a bearing on automation issues, rather than attempt to distinguish the many site-specific variations.

One generalization that has received recent attention in the national media is that the equipment and systems that support air traffic control represent obsolete technology. Triggered by the occurrence of major equipment and system outages at busy air traffic control facilities, media reports have reflected concern that the current components collectively represent a museum of electronic (including computer) equipment whose increasing unreliability may be approaching the point of overwhelming the efforts of Airway Facilities specialists to maintain its operation and of air traffic controllers to cope with its degradations.

A highly significant but less broadcasted fact is that the Airway Facilities technical workforce has aged along with its equipment. The FAA estimates that, within 10 years, up to 35 percent of this technical workforce will be eligible for voluntary retirement (Federal Aviation Administration, 1993b). This has raised two questions: (1) If the experienced technicians, who are now relied on to maintain the aging equipment, retire before the equipment is modernized, who will keep the system operational? and (2) If efforts to replace the aged equipment with modernized equipment succeed before the experienced technicians retire, who will be trained to ensure the proper operation of the new equipment?

Given the many changes taking place, the near future is a critical period for Airway Facilities, during which the decisions made with respect to automation will be extremely significant. A significant contributor to the success of the FAA's modernization efforts will be the extent to which the application of automation reflects appreciation for the human factors issues associated with each of the following questions, as well as for the fact that answers to each question must properly *interact*:

1. To what extent do the systems for which Airway Facilities is responsible perform automated functions to maintain their operation and to recover from degradation or failure of their components?
2. What are the automation and human-computer interface characteristics of the monitoring and control tools provided to Airway Facilities?
3. What are the job classifications, descriptions, and qualifications for Airway Facilities technicians?
4. What critical operations do technicians perform?

5. How do Airway Facilities and Air Traffic cooperate as a team to respond to problems?

6. What are the selection procedures for and the characteristics of technicians?

7. How are technicians trained?

8. How is the job performance of technicians appraised?

9. What roles do Airway Facilities representatives play during the equipment acquisition process? What roles do human factors representatives play with respect to the design, development, and test of equipment and systems used by Airway Facilities personnel?

10. What are the effects of the agency's organizational structure and culture on the job satisfaction of Airway Facilities personnel?

Automation at the Maintenance Control Centers

As explained in Chapter 4, each Airway Facilities sector is supported by a centralized monitoring and control workstation suite, which functions as the command center for the technical support of en route centers and terminal facilities. Automation and computer assistance (with respect to both information display and control) are applied to different levels in the different systems and equipment monitored and controlled through the maintenance control center (MCC). In general, automation and computer assistance are applied more often to support such sensing, calculation, data searching, and control actuation functions as information retrieval, alarm reporting, remote control, and data recording. Automation is rarely applied to perform or to support such higher-level cognitive functions as trend analysis, failure anticipation, system-level diagnostics and problem determination, or certification.

In fact, FAA Order 6000.30B, which establishes a long-term policy for national airspace system maintenance, recommends that automation be applied to repetitive maintenance tasks and that the Airway Facilities specialist be left "free to accomplish higher level, decision-oriented work" (Federal Aviation Administration, 1991c:5). FAA Order 6000.39, which defines the MCC operations concept, summarizes the current philosophy of automation (Federal Aviation Administration, 1991a:5):

> The MCC will implement automation to the degree that tasks can and should be automated. Advances in expert systems and artificial intelligence will be applied where possible to automate tasks requiring a small degree of human intervention. . . . The following functions shall be automated: . . . (1) Routine functions requiring little or no human intervention, such as diagnostic report generation . . . (2) Data gathering not requiring narrative or human interpretation. (3) Administrative paper documentation.

Given the fact that modernization is accomplished through many different programs within the FAA and involves many different vendors of equipment and systems, and given the fact that the national airspace system is the focus of rapidly advancing technologies, Airway Facilities specialists can find themselves faced with a variety of new technologies, provided by different vendors, with varying levels of automation and different human-machine interface design strategies, at the same time that the procedures and the human-machine interface for air traffic controllers experience more controlled growth and change. As a rule, air traffic controllers are provided with integrated workstations whose display/control logic and formats are carefully guarded and monitored during the development of new supporting systems. In contrast, the technicians who monitor and control the supporting equipment are typically provided with new monitoring and control devices that are tacked onto the array of such devices for other equipment in a loosely arranged combination that lacks integration (Theisen et al., 1987).

The Airway Facilities community has specified standardized protocols and data acquisition and processing requirements to ensure that new components and systems will provide data to and accept control commands from the centralized monitoring and control workstations (Federal Aviation Administration, 1994c). However, these and other (Federal Aviation Administration, 1991a) recommendations address only the lower-level automation tasks mentioned above. They do not address the proper allocation of higher-level tasks between human and machine, the integration of automation functions across disparate systems, or the integration of the associated human-machine interfaces at the MCC. Without an adequate attempt to understand, clarify, and standardize the automation requirements and associated human-computer interface for its workstations, Airway Facilities cannot play an effective role in the systems acquisition process that determines to a large extent whether automation of various systems by various vendors under various program management personnel will contribute to or detract from its ability to fulfill its obligations.

Supervisory Control and Automation

Investigation of supervisory control and automation for air traffic control should not be limited to examination of the air traffic controllers' workstations. In the process of monitoring and controlling air traffic patterns and activities, air traffic controllers do monitor the apparent quality of the data appearing on their workstations and the performance of their display and control devices. Air traffic controllers will question, for example, the quality of radar-provided data, and they do have limited control over the selection of radar parameters for display. However, it is the responsibility of Airway Facilities to monitor and control all equipment that ultimately supports the controllers, to inform the controllers of the status and performance of equipment and systems on which their tasks depend (including the controllers' workstations), to reconfigure and maintain degraded or failed equipment in a manner that minimizes interference with air traffic control tasks, and to respond to requests for service from controllers. Therefore, the air traffic controllers are not the only supervisory controllers of their equipment; the air traffic supervisory control tasks must be viewed as cooperative efforts of both controllers and technicians. On that account, many issues pertaining to the automation of air traffic control functions, discussed in Chapter 12 of this report, require examination of the activities of both Air Traffic Services and Airway Facilities that contribute to the performance, monitoring, and control of those functions. As equipment and systems evolve, these roles and responsibilities may change with respect to responding to degradation or failure of the software and hardware that support the automated functions, and with respect to performing supervisory control tasks for air traffic control when monitoring and control of automation equipment is included.

A word of caution is offered concerning the general use of the term *automation*. Airway Facilities has been increasingly confronted with new equipment and systems labeled "automated"; this trend helped prompt the creation of the job classification "automation specialist" electronics technician; a similar job classification once existed within the Air Traffic staffing organization (this job classification recently became the responsibility of Airway Facilities). The activities and expertise of automation specialists have largely addressed computer hardware diagnosis and maintenance, as well as computer program analysis and development tasks. Automation was generally understood to be associated with computers, and particularly with computer software, as distinguished from the activities associated with maintaining the other electronic hardware components of radar, communications, and navigation systems. Although this distinction did recognize the technological trend toward software-intensive and software-modifiable computer-based systems, it masked two aspects of the difference between "automation" and "modernization." First, many modernization efforts within the national airspace system involve application of computers without significant changes to the allocation of functions between humans and machines; automation is not a necessary corollary of computerization. Second, automation is not a discrete attribute of a new system; automation may be applied in degrees or in levels. New systems often do introduce changes in the level of automation for some tasks. However, these changes generally apply to such tasks as data gathering, calculation, rule-based determination of the status of components, comparison of performance against thresholds, and automated switching to redundant backup components when primary components fail. Computerization has not generally introduced to Airway Facilities tasks the automation of such problem-solving functions as diagnosing faults from patterns of failures, predicting faults from trends in data, reconfiguring systems (as opposed to single components) in response to system failures, or certifying systems and services.

Masking of the distinctions between automation and modernization and between automation and computerization has significant impact on the tasks expected of automation specialists, and on the selection, training, and performance appraisal procedures associated with this job classification. The FAA has implicitly begun to recognize these distinctions with the introduction of the GS-2101 job classification, the airway transportation specialist. The position description of this job classification helps to distinguish among automation, computerization, and modernization. The GS-2101 incumbent must understand and will rely on increasingly automated computer functions and must also possess knowledge and skills representative of systems engineering: how components and systems interact to produce services.

Almost all former automation specialists have been reclassified as GS-2101s, along with most other electronics specialists who have mastered interacting systems. On this account, it should be made explicit that, within Airway Facilities, "working with automation" is no longer recognized as a special job classification; it is rather a set of tasks that virtually *all* Airway Facilities specialists perform while working with systems. In this connection, it would be useful if new systems introduced to the national airspace system included precise descriptions of their automated functions rather than the ubiquitous "automation" label. The distinction required now is not whether systems are or are not automated, but which *functions* of systems are or are not automated.

OPERATIONS

General Responsibilities

The duties most germane to the current concern with the implications of automation for air traffic control are monitoring and control, diagnosis of systems (as opposed to equipment components), certification, and restoration of services. The general applications of automation to monitoring and control have been mentioned in Chapter 4 in connection with maintenance control centers. Automation has been applied to system diagnostics in connection with both the MCC capabilities, discussed above, and the processes of certification and restoration, discussed below.

Certification

The increased reliability of computer-based systems and the automation support for diagnostics that are often embedded in such systems currently suggest the following trends in certification: extension of the acceptable certification intervals; increasing reliance on the results of built-in diagnostics that can support certification while the equipment remains in operation; more performance of remote certification, replacing the need to examine the equipment directly; and more automated maintenance logging and equipment performance recording.

However, these trends and the application of automation to the certification process must be considered in the light of the following current formal procedures for performing certification, defined in FAA Order 6000.15B (Federal Aviation Administration, 1991b): in addition to the procedures recommended in technical handbooks, instructional books, and other technical documentation that accompanies the delivery of equipment and systems, the certifier should use methods that include: direct measurement of certification parameters; monitoring status indicators; analyzing technical performance; performing a comparative analysis of flight inspection data with previous results; visual and aural observations of data, extraneous noises, excessive heat, and questionable odors; reports by pilots and controllers; and diagnostic testing. It is important to note that the certification process is not governed by standard algorithmic procedures. The FAA emphasizes that the choice of methods used for certification determination is left to the professional judgment of the certifying technician.

The automation applied to the certification process should support the judgment strategy of the certifier. Since the FAA orders governing certification suggest that certifiers are free to—and must—develop their independent judgments, the question of automating certification should be associated with attempts to analyze the entire set of certification tasks, including the judgment process. As with the general tasks of monitoring and control, automation is currently applied to lower-level certification tasks of data acquisition, data calculation, and simple rule-based diagnostics, but not to higher-level judgment tasks.

A significant practical consideration with respect to the automation of certification functions is: How does the automation of certification affect legal liability? If automation is relied on for certification and it errs, is it appropriate (legally) to blame the machine or to blame the certifier whose judgment accepted the machine's error?

With respect to the issue of user trust in automated systems, it is important to emphasize that, within the FAA, the certification process represents the formalization and operationalization of trust. When an Airway Facilties specialist certifies a system, that specialist formally and legally expresses the FAA's conclusion that the system is trustworthy. When the specialist ceases to trust a system, he or she formally decertifies the system. Therefore, when a certified system fails, the issue of trust extends through multiple orders: the controllers may question not only their trust in the system and its equipment, but also their trust in the individual who certified the system. This introduces the possibility of mistrust in the qualifications of the certifier (and therefore in the process by which the certifier was "certified to certify") and in the process of equipment/system certification, which ultimately and formally relies on the professional judgment of the certifier. One response to these concerns has been the suggestion that the certification process should be as automated as possible—in which case the question arises: Who will certify *that* certifier?

Restoration to Service: An Example of Teamwork

The contributions of both automated equipment and its human users to the cause, complication, or resolution of major outages should be considered. Human error, particularly within the staff who control the automation equipment, can cause or contribute to outages. One option frequently considered by Airway Facilities specialists when complex systems demonstrate performance decrements is to do nothing, since experience has shown that frequently performance decrements are transient, and complex systems sometimes salvage themselves. One general rule followed by experienced specialists is: analyze before you act. This suggests that important features of automation are the extent to which it contributes to system self-stabilization, the extent to which it supports system analysis, and the extent to which it discourages human error.

Troublesome system problems are not restricted to the catastrophic. Small, infrequent problems can sum to produce an unstable system. These problems can be produced by software bugs, errors in data transmission or storage, timing errors, and subtle design deficiencies not detected in formal acquisition tests. Johannssen (1992) estimated that between 1987 and 1992 there were approximately 4,000 reported software problems in the national airspace system, of which about 1,600 were not resolved by 1992. The FAA specifies a procedure for filing, maintaining, and resolving program technical reports in response to such problems.

A significant question regarding the application of automation is: Will Airway Facilities specialists be able to effectively restore equipment and systems to service when (1) the equipment or systems that have failed contain automation on which air traffic controllers rely heavily to perform their duties and when (2) Airway Facilities itself relies on automation to perform the restoration, but the automation has failed or is difficult to work with? Improper design or application of automation to both Air Traffic Services and Airway Facilities can produce a double indemnity situation that complicates extremely any problems relating to failure of the automation supporting air traffic tasks that are the focus of this report.

Airway Facilities has always shared with controllers the responsibility for and the philosophy of maintaining the safe and efficient flow of air traffic; it is open to question whether the Airway Facilities roles within the team will actually increase with increased automation, or whether increased automation will require that the controller roles expand into supervisory control functions currently performed by Airway Facilities. The function of maintaining automation software, which traditionally resided within the Air Traffic Services organization, has recently been transferred to the Air Facilities organization. This has helped to eliminate the traditional difficulties associated with assignment of responsibility for hardware and for software to different organizations, especially when diagnosis of problems is at issue. The new Airway Facilities job classification emphasizes that the GS-2101 specialist must be able to understand and work with new automated technologies from a systems engineering perspective.

The responsibility for restoration also highlights the need for teamwork within the Airway Facilities sector itself. Despite the recent reclassification of most electronics specialists to the GS-2101 classification, Airway Facilities remains staffed with technicians whose knowledge base represents depth in specific systems and their associated items of equipment (e.g., radars, communications, navigational aids, computers). The workforce does include operations managers whose understanding spans the systems, but they typically rely on the in-depth knowledge of system specialists. The FAA recognizes that, since facilities may not be staffed with representatives of all disciplines on a 24-hour basis, a frequent requirement during restoration will be to call back needed off-duty specialists. The GS-2101 classification seems to rely heavily on the assumption that the anticipated systems will contain embedded automation that will relieve these system-level specialists of the requirement to understand their in-depth functioning. If that assumption is incorrect, the utility of the GS-2101 classification with respect to the callback problem must be questioned. In addition, since Airway Facilities personnel currently tend to become individual specialists in particular areas (e.g., radar, communications, navigation, or equipment or subsystems within these areas), they informally rely on one another's expertise to solve problems. A move toward more breadth of responsibility may affect the dynamics of Airway Facilities teamwork.

Teamwork is the focus of recent study by the FAA. The FAA Civil Aeromedical Institute is conducting research in the following areas: knowledge and skills that predict successful membership in and leadership of self-managed teams; tools to assess the progress of work teams; organizational culture factors that inhibit or facilitate acceptance of new technology by the Airway Facilities workforce; and methods for introducing new technology (e.g., quality circles, town hall meetings, goal setting, teaming).

Workload

In Chapter 6 we discussed issues relating to workload in general and to air traffic control in particular. Airway Facilities is also subject to the problem of sudden workload transition from low troughs to high peaks. Scheduling of preventive maintenance and certification tasks is currently a commonly applied method to average workload. Airway Facilities also schedules tasks that affect controller operations in coordination with Air Traffic Services, taking into consideration controllers' workload. The most significant workload challenge for Airway Facilities personnel occurs when multiple critical elements fail, creating or threatening service outage. Under these situations they are faced with the complex task of rapidly diagnosing the cause from the pattern of failures, simultaneously assessing the progress of the diagnosis, logistics support factors, and the utility of applying alternative solutions to maintain or restore service.

Acquisition of Airway Facilities Automated Systems

Chapter 11 of this report describes the FAA's procedures for development and acquisition of new systems and makes recommendations pertaining to the consideration of human factors issues and the appropriate involvement of human factors professionals during all phases of system development, acquisition, implementation, and test and evaluation. *Those recommendations apply with equal force to Airway Facilities equipment.*

The application of human factors to the process of designing and developing the human-computer interaction for the air traffic controllers' workstation during acquisition of automated systems benefits from two facts: these workstations are highly visible concerns during the acquisition process, and it is widely appreciated that the introduction of new systems must fundamentally preserve the current integration of these workstations.

The situation with respect to Airway Facilities has been quite different. As mentioned previously, the maintenance control center (MCC) is a concatenation of disparate workstations without an integrated human-computer interface. Simply expanding the MCC to assimilate additional workstations for new systems will foster idiosyncratic human-computer interface designs that may exhibit internally consistent application of human factors principles but fail to integrate with other MCC designs. The FAA has recently required that all new systems provide data in standard formats to the remote monitoring system that feeds the MCC, but this still allows for idiosyncratic design of the MCC human-computer interface and of automation for each new system. There is a significant need for the specification of an MCC human-computer interface into which all new designs must fit well, and a corresponding need for an overall MCC automation strategy against which proposed automation designs can be evaluated.

The FAA has recently produced the *Human Factors Design Guide (HFDG) for Acquisition of Commercial-off-the-Shelf (COTS) Subsystems, Nondevelopmental Items (NDI), and Developmental Systems* (Wagner et al., 1996). This design guide is intended to overcome the limitations associated with using commercial and military human factors design standards within the FAA environment. The current version of the design guide focuses on ground systems and equipment managed and maintained by Airway Facilities. The FAA plans to expand the design guide to address air traffic control operations, aircraft maintenance, aircraft and airborne equipment certification, and regulatory certification for aviation personnel. The design guide is not intended to be a substitute for indepth professional practice.

The design guide should be maintained and updated in accordance with results from a systematic, continuing program of human factors research. Efforts, including the use of such guidelines, to standardize the application of new technologies should include attention to issues pertaining to the application of automation to support Airway Facilities tasks—especially monitoring, certification, and restoration tasks—as well as issues pertaining to the human-computer interface characteristics of associated equipment, including maintenance control centers, off-line diagnostic tools, maintenance logging tools, and software development tools.

STAFFING

Selection and Demographics

The GS-2101 job classification, which now covers the majority of Airway Facilities electronics technicians, is likely to require change in the population from which hirees are selected. There is no known FAA documentation of the strategy for identifying this population or for determining the precise relationship between selection criteria, performance during training, and on-the-job performance for the GS-2101 specialists or for any other electronics specialists. The same concern for identifying relationships between automation factors, selection criteria, and the GS-2101 knowledge and skill requirements will apply if the FAA decides to pursue the notion of developing a GS-2101 selection test.

As discussed in Chapter 4, demographic data suggest that the Airway Facilities workforce will see, within 10 years, a simultaneous retirement of significant percentages of its experienced technicians and the equipment on which they have developed their experience. This suggests that the introduction of the GS-2101 job classification is quite timely—fostering the hiring and training of new types of people for new types of equipment—but it also adds to the urgency of validating the GS-2101 hiring and training devices and procedures.

Training

It is noteworthy that the training track specified for automation specialists not only includes instruction in general hardware and software aspects of computers and in computer programming but also emphasizes knowledge of specific computer-based systems such as the HOST computer, display channel processors, and MCC operations. There is currently no training track that specifically addresses the position descriptions of the GS-2101; therefore, GS-2101 trainees currently receive tailored instruction selected from among the pool of instructional sources that were developed to train the specialists in radar, navigation, communications, and automation (computer systems). The GS-2101 position descriptions and qualification standards emphasize knowledge and skills characteristic of systems engineers, with a focus on how systems interact to produce services. In contrast, existing training materials have been developed to effectively train technicians in the operation and maintenance of specific systems and equipment. There is a clear need for the development of a training program and associated course content for the GS-2101 job classification; the current courses of instruction do not address the systems engineering requirements of the GS-2101 qualification standard.

An associated concern looks in the opposite direction. The rationale for the GS-2101 job classification relies partly on the expectations that new systems are likely to automate current equipment- and subsystem-level monitoring, diagnostic, and reconfiguration functions. It also relies on the expectation that these systems will be modularized to permit automatically failed components to be removed, replaced with equivalent modules (pull-and-replace maintenance), and returned to the manufacturer for repair. The concern is that these assumptions may lead to the conclusion that the GS-2101 can focus on system- and service-level activities, relying on automation to monitor and control lower-level functions—and that, consequently, training for these lower-level functions can be eliminated. Such "dumbing down" of training would be suspect in the light of two questions: What will the GS-2101 do when the automation fails, and how will the GS-2101 maintain proficiency in these automated tasks?

There is currently no discernible consistent philosophy within Airway Facilities that governs the maintenance of skills and proficiency during the use of automated systems. It is also noteworthy that despite the heavy reliance on teamwork during its operations (i.e., reliance within Airway Facilities on the cooperation of domain experts and general reliance on cooperative work between Airway Facilities and Air Traffic Services), and despite the emphasis in performance appraisal on performance within the team context, training for Airway Facilities team operations is not in evidence in the FAA *Catalogue of Training Courses*.

Performance Appraisal

The development of new GS-2101 course work will require associated development of new examinations. The GS-2101 position descriptions will have to form the basis for the development of any associated personnel certification examinations and performance ratings. In each of these enterprises, there will be the challenge of addressing the question: How can the performance of the technician be distinguished from the performance of the machine when tasks are automated? This question currently applies especially to such tasks as certification, monitoring, and control. Detailed analyses must be performed for each system to determine how to characterize the GS-2101 supervisory control tasks to permit effective performance appraisal.

Job Satisfaction

In Chapter 8 we discuss the interaction between organizational culture (which includes the perceptions and attitudes of employees) and formal organizational context factors (formal structure, policies, and procedures), as well as their combined influence on job performance, noting the importance of job satisfaction as a key element of organizational culture. Chapter 8 also describes the FAA's employee attitude survey (EAS), which includes indicators of job satisfaction, and reviews their results for air traffic control employees.

Key results of the 1995 EAS for Airway Facilities employees are summarized here; these results, and related conclusions, are very similar to those reported for air traffic control employees. The reader is encouraged to refer to the discussion of the EAS in Chapter 8 for more detailed considerations relating to interpretation of the EAS results and related conclusions.

Sixty-eight percent of Airway Facilities employees reported that they were either highly or very highly satisfied with their jobs. The 1995 survey showed a 22 percent increase, since 1988, in the percentage of Airway Facilities respondents reporting that they were either highly or very highly satisfied overall; however, reported overall job satisfaction dropped 6 percent from the previous (1992) survey. For reasons detailed in Chapter 8, such findings do not indicate conclusively a net increase in satisfaction with all of the key facets that contribute to overall job satisfaction. In 1995 Airway Facilities employees were especially satisfied with pay, benefits, and the nature of the work itself, but less satisfied with factors contributing to the job environment, including working conditions, the supervisor, the organization, and opportunities to develop potential. These results, combined with a steady decline in reported confidence that FAA management would utilize the EAS results to improve working conditions and morale, suggest that, like air traffic controllers, Airway Facilities employees perceive a need for changes in management style and/or structure.

Some suggestions for those changes emerge from examination of the results relating to specific organizational context variables. EAS results indicate that Airway Facilities employees are highly satisfied with their understanding of how their jobs contribute to the FAA mission. However, they are only moderately satisfied with the adequacy of management's communication of policies and with their opportunities to express concerns openly and with impunity. These results occur against a backdrop of agency-wide EAS results indicating that employees are only moderately satisfied with their involvement in decision making, and that many employees do not feel free to discuss problems with their supervisors.

Chapter 8 details conclusions related to similar EAS results for air traffic controllers. These conclusions, which apply as well to the results for Airway Facilities personnel, include, in brief: (1) Management should recognize possible discrepancies between the intent of their policy communications and the interpretations drawn by employees, especially with respect to policies that address safety and efficiency. (2) Management should involve employees in policy development and encourage open expression of concerns. (3) Employee empowerment is a means of aligning responsibility with authority, and empowerment must be internalized within the culture. (4) Employees' perception that communication is inhibited should be addressed as a means to develop improvements in other reported areas of dissatisfaction. In the discussion of training and personnel development above, it was reported that the FAA has established a tailored program of training and retraining, as required, for Airway Facilities specialists. EAS results, however, indicate that Airway Facilities employees are only moderately satisfied that they receive the training needed to perform effectively, and only about 40 percent of Airway Facilities employees reported that they are satisfied with opportunities to develop their potential. The process of performance assessment for Airway Facilities specialists is also discussed above. The EAS results indicate that Airway Facilities employees experience generally low to moderate satisfaction with the extent, timing, and appropriateness of recognition and rewards for exceptional performance. The ongoing process of revising the appraisal and training processes for Airway Facilities specialists should include careful consideration, with continual feedback from Airway Facilities specialists, of how performance and potential should be trained and appraised and how rewards and penalties should be tied to the results of appraisals.

Airway Facilities employees reported low to moderate satisfaction with the impact of new technologies on their jobs—including both appropriateness and timeliness of the introduction of new technology. This lack of satisfaction forms part of the cultural backdrop against which user involvement (or noninvolvement) transpires during the acquisition of new systems, as discussed in Chapter 11.

HUMAN FACTORS RESEARCH

Until very recently there has been an extreme paucity of human factors research pertaining to Airway Facilities at either of the FAA's major human factors research organizations, the Human Factors Research Laboratory at the Civil Aeromedical Institute (CAMI) and the Research and Development Human Factors Laboratory at the FAA Technical Center in New Jersey (Collins and Wayda, 1994; Stein and Buckley, 1994; Human Factors at the FAA Technical Center: Bibliography 1958-1994).

A notable exception is Blanchard and Vardaman's (1994) development of an outage assessment inventory to study a broad range of factors relating to equipment and system outages. Blanchard and Vardaman's study concluded that adequate understanding of the factors contributing to outages must include attention to the following variables: system and equipment design factors (reliability, accessibility, level of automation and built-in testing, and degree of automated switching to redundant components); human behavioral processes (information gathering and interpretation, knowledge base, problem-solving and decision-making strategies, and planning); personnel factors (training and experience level, skills, availability and assignment of personnel, staffing levels, shift scheduling, and management); logistics factors (maintenance philosophy, parts availability, job training aids, and availability and operability of test support equipment); and physical environment factors (travel time to equipment and physical characteristics of the maintenance environment).

Blanchard and Vardaman suggested that these factors should be studied in relation to each of the following tasks required for restoring failed equipment: detection of the outage, scheduling of the maintenance, assignment of the maintenance technician, traveling to the equipment, preparing maintenance tools, diagnosing the faults, repairing or replacing components, verifying and certifying the equipment or system, and logging the maintenance report. Blanchard and Vardaman's conclusions represent one promising framework for investigating, within the context of a standard task sequence, variables that may interact with automation to mediate the effectiveness of automation applied to Airway Facilities.

Very recently the FAA has developed plans for, and initiated on some fronts, human factors research pertaining to anticipated changes in Airway Facilities job tasks, workstations, skill requirements, demographics, selection procedures, training needs, and organizational structure and culture. The FAA is developing plans for human factors research on automation in Airway Facilities. The FAA's *Plan for Research, Engineering, and Development* (1995b) proposes that a plan be developed to conduct and apply research in the following areas: task analyses to provide the necessary data for developing knowledge, skills, abilities, position descriptions, and training criteria for current and future positions that work with automated systems; guidelines for the development of human-computer interfaces for Airway Facilities workstations; assessment of factors that affect human performance of Airway Facilities activities; development and validation of selection criteria for Airway Facilities; criteria for effectively using intelligent systems in Airway Facilities maintenance; analyses of organizational effectiveness for work with current and future systems; and analysis of the workload associated with Airway Facilities tasks. The plan proposes extensive use of rapid prototyping and simulation. It suggests a consistent long-term evaluation of proposed new technologies that Airway Facilities must monitor and control and that can be applied to assist its operations.

The FAA's *National Plan for Civil Aviation Human Factors* (1995e) proposes the following avenues of research related to both Air Traffic Services and Airway Facilities: develop concepts and guidelines for applying human factors to the design of human-machine interfaces for automated systems; identify the workload and performance implications of applying automation; investigate transitions between low and high workload; analyze new classes of error that result from new technology and procedures; examine methods to articulate and coordinate a human-centered automation philosophy; investigate how overautomation and lack of appropriate feedback to the operator can create performance problems; identify the conditions that may lead to overreliance or underreliance on automation; resolve issues related to the degradation of basic skills with associated performance implications should the automation fail; study situation awareness during the use of automation; and identify selection, training, and performance requirements associated with new systems.

Recently initiated research at the FAA supports, in part, the research agenda outlined in the *National Plan for Civil Aviation Human Factors*. It focuses on the impacts of anticipated technology, including automation, and includes: human-centered automation studies of human factors considerations in advanced operations control centers (OCCs); development and validation of selection, training, and assessment methods for Airway Facilities; and study of the organizational impact of new technologies on Airway Facilities performance (Federal Aviation Administration, 1996).

The FAA Technical Center is applying the resources of its Research and Development Human Factors Laboratory to study human factors considerations in advanced OCCs. Using its OCC test bed, rapid prototyping, and operational scenarios the FAA Technical Center is studying the operational suitability of the human-computer interface characteristics of proposed OCC concepts and designs, including those that involve the introduction of intelligent systems. The laboratory is also applying virtual reality technology to the visualization and analysis of candidate layouts for advanced OCCs. As the OCC concepts develop, the FAA's plan is to apply the resources of the laboratory to human factors evaluations of conceptual designs and prototype systems, translate research findings into human-computer interface requirements and specifications for modifications to existing systems and for future systems, and perform human factors acceptance testing of OCCs.

Related efforts include the development of job task analyses, the identification of knowledge, skills, and abilities, the construction of human performance and workload models, and the definition of team structures—all of which characterize anticipated changes to Airway Facilities personnel and activities. The FAA's plan is to maintain feedback between these efforts and the Technical Center's efforts to develop and evaluate OCC workstations, so that design requirements will reflect consideration of personnel, team, and procedural factors.

Toward the development and validation of selection, training, and assessment methods for Airway Facilities, the CAMI human factors researchers are collecting data on job tasks, biodemographics, personality characteristics, and results of post-hire assessments, Academy training examinations, field equipment certification tests, and on-the-job performance evaluations. The immediate goal of this research is to validate methodologies for the post-hire assessment of Airway Facilities specialist knowledge, skills, and abilities (used to decide whether specialists may bypass training courses), as well as other characteristics relevant to training placement and job performance. The longer-term goal of this research is to develop new or additional tests incorporating innovative assessment methodologies, such as computer-adaptive testing. This research also contributes to the wider, long-term refinement of the Airway Facilities systems model for assessment, recruitment, and training (Airway Facilities SMART) program, whose goal is the development of improved assessment and placement tests, instructional technology, and recruitment strategies appropriate to the anticipated OCC environment. CAMI researchers are performing studies to identify knowledge and skills that predict successful membership in and leadership of self-managed teams and are developing tools to assess the progress of Airway Facilities work teams.

CAMI human factors researchers are also developing and validating an organizational culture survey that will be used to study the organizational impact of new technologies on Airway Facilities performance. The FAA plans to include in this study examination of various methods for introducing new technology, including quality circles, advanced training, management town hall meetings, goal setting, and teaming arrangements. Dependent variables for the study include direct measures of performance (e.g., time to identify defective equipment, time to begin repairs and to restore equipment, and number of equipment failures per period) as well as culture variables such as morale and attitude. The goals of the study are to assess attitudes toward new technology; identify relationships between culture factors, organizational structure, and performance; and evaluate methods for the effective introduction of new technology into the Airway Facilities work environment.

The general plans for human factors research pertaining to Airway Facilities, described above, suggest the need to develop knowledge based on empirical findings, and those plans recommend the application of that knowledge to the design, development, and evaluation of new systems. The high level of detail at which the plans are discussed, the general issues planned for consideration, and the instances of ongoing research confirm that: (1) attention to the human factors of Airway Facilities is at its inception, and (2) the FAA has recognized the value of investigating several of the Airway Facilities human factors issues identified in this report. Although ongoing human factors research is consistent with several areas of concern identified in this report, the scope of Airway Facilities human factors research remains very small by comparison with the FAA's air traffic control research efforts. It is a central theme of this chapter that ongoing and impending changes to Airway Facilities technology and personnel are dramatic and require significant human factors attention. Ongoing research should continue and expand according to the plans stated in the *Human Factors Research Project Initiatives* (Federal Aviation Administration, 1996). Additional research should be conducted across the spectrum of research areas defined in this report and in the *National Plan for Civil Aviation Human Factors* (Federal Aviation Administration, 1995e).

CONCLUSIONS

The impact of automation when new equipment or systems are introduced is often experienced more by Airway Facilities specialists than by air traffic controllers. Even at centralized maintenance control centers, Airway Facilities specialists can find themselves faced with a variety of new technologies, provided by different vendors, with varying levels of automation and inconsistent human-machine interface design strategies. The current process of modernization involves the piecemeal introduction of new technologies into the national airspace system, with associated requirements for changes to operations and personnel activities, in a manner that has placed Airway Facilities in a state of flux. The workforce has been aging with the equipment it maintains, and the FAA is faced with the prospect of hiring "modernized" Airway Facilities specialists in conjunction with modernized equipment and systems. The near-term challenge will be to maintain operation of existing systems while phasing in new ones whose complexity and increased automation are likely to demand new job skills.

Automation has not been applied on a large scale to support decision-making and problem-solving functions such as system-level diagnosis of faults from patterns of failures; trend analysis of system performance; prediction of failures; certification of equipment, systems, and services; and the development of restoration strategies—which are critical functions performed by Airway Facilities in support of air traffic control.

Changes in the definition of Airway Facilities job responsibilities and associated qualification standards, combined with a tendency to label new computer-based systems as "automated," suggest that the FAA has not established clear and distinctive definitions for the terms *automation*, *modernization*, and *computerization*. Masking of the distinctions between these terms has been one factor detracting from detailed analysis of new systems with respect to: the precise allocation of functions between human and machine that each new system introduces; the resulting impact on the performance of job tasks; associated requirements with respect to selection, assignment, training, maintenance of proficiency, and performance appraisal of Airway Facilities specialists; appropriate criteria and methods for the evaluation and testing of the systems; human-computer interface requirements; and avenues of research required to support effective system design. The term *automated* should not be used as a label for systems that are not fully automated. Definition of new systems should include clear identification of which tasks are automated, which tasks are performed by the human, and which human-performed tasks involve the use of automated support. The description of new systems should not focus on whether they contain automation, but, rather, on which specific functions they automate.

Detailed human factors guidelines (such as the *Human Factors Design Guide for Acquisition of Commercial-off-the-Shelf Subsystems, Nondevelopmental Items, and Developmental Systems*) should be maintained and updated in accordance with results from a systematic, continuing program of human factors research. Efforts, including the use of such guidelines, to standardize the application of new technologies should include attention to issues pertaining to the application of automation to support Airway Facilities tasks—especially monitoring, certification, and restoration tasks—as well as issues pertaining to the human-computer interface characteristics of associated equipment, including maintenance control centers, off-line diagnostic tools, maintenance logging tools, and software development tools.

There is no clear strategy evident for the application of automation to the systems management tasks of GS-2101 specialists. The GS-2101 job classification will require revision to the procedures for selection, assignment, training, and performance appraisal of Airway Facilities specialists. Specific requirements for work with automation should be defined for the GS-2101 job classification. Team training should be developed to support both Airway Facilities team activities and cooperative Airway Facilities and Air Traffic teamwork.

The Employee Attitude Survey (EAS) results indicate that, whereas Airway Facilities personnel are highly satisfied with the nature of their work, they are only moderately satisfied with their working conditions, opportunities for training and for developing potential, processes for appraising and rewarding performance, opportunities to communicate concerns with impunity, and the ways in which new technologies have been selected and applied. Management should address these perceptions by facilitating communication and by effectively involving Airway Facilities specialists in the development of new equipment and procedures.

10

Strategies for Research

The fundamental methodological challenge for human factors research in air traffic control involves the central requirement to find a cost-effective means of generating valid human factors information and design recommendations. Resources for human factors research programs are usually limited, and as a result early planning decisions must be made to determine which research topics can be included in a program and which ones cannot. An awareness of alternative research methods, including their strengths and weaknesses, can assist in determining which research topics are the most likely to produce valid conclusions with the resources available.

Two salient questions in the design and development of complex human-machine systems, including the modernization of the contemporary air traffic control system, are: "What methodologies will identify the variables that influence operator performance and thus yield the information required for system design?" and "What are the appropriate methodologies for evaluating existing or new systems?" Both of these questions revolve around two requirements: (1) the specification of behavioral or performance criteria and (2) the determination of valid measurement procedures. This chapter focuses primarily on the different means by which the relevant human factors variables are identified (Dennison and Gawron, 1995; Wickens, 1995a). Although cautions for evaluation methodology have been provided by several authors in a recent volume on verification and validation of complex systems (David, 1993; Hancock, 1993; Harwood, 1993; Hollnagel, 1993a; Jorna, 1993; Woods and Sarter, 1993), we also provide an overview of measurement issues in complex systems.

Throughout this chapter, references are made to the concepts of validity and validation. Both are integral to discussions of research and system development. Campbell and Stanley (Campbell, 1957, 1969; Campbell and Stanley, 1966) have distinguished between the concepts of internal validity and external validity.

Internal validity means that the findings (e.g., the observed cause-and-effect relationships) of a particular investigation follow logically and unequivocally from the way the investigation was designed and conducted. An investigation would be internally valid (and would constitute a controlled study) if no contaminating factors (i.e., confounding of variables) undermined the conclusions.

External validity refers to the generalizability of the findings of an internally valid study to other situations. Generalizability refers to the assumption that a finding will hold or apply in situations other than the one in which it was observed (Chapanis, 1988; Locke, 1986; Sherwood-Jones cited in Taylor and MacLeod, 1994). Internal validity is a prerequisite for external validity or generalizability.

Validation refers to the determination that a system design is appropriate for the intended purpose (i.e., that the system, when implemented, will provide the necessary functionality and that it will allow the articulated operational goals to be achieved, presumably in a safe and reliable manner).

This chapter describes the research methodologies available for collecting human factors data, as well as the relative strengths and weaknesses associated with each approach. We consider:

- human engineering databases and literature,
- analysis of controller responses,
- computer simulation and modeling,
- design prototyping,
- real-time simulation, and
- field studies.

Each of these methodologies is reviewed; we then discuss how the different methodologies can be combined in the investigation of a particular topic (e.g., operational errors in air traffic control). The chapter ends by summarizing the human factors measurement issues associated with the design and evaluation of complex systems.

What we are describing is a series of strategies for research on air traffic control. Many of these strategies involve collection of data from past (through accident and incident analysis), present, and projected users of the system. However, other strategies, particularly those involved with modeling and computer simulation, examine issues in the absence of new data, for example, when a model is run to predict trade-offs between the safety and efficiency of a new system innovation, such as free flight. Appropriate marshaling of the full arsenal of research strategies requires understanding both the strengths and weaknesses of each, as detailed in the sections that follow.

HUMAN ENGINEERING DATABASES AND LITERATURE

The application of human factors research to potential problems in air traffic control is sometimes prompted by the identification of generalizable human performance issues in other complex systems and at other times by direct analysis of the operational environment of air traffic control, including accident and incident analysis. Similarly, the relevant human factors literature and engineering databases may generalize across particular domains of human behavior but may also be the subject of tailoring for application in specific contexts. The breadth of engineering databases and the human engineering literature reflects the diversity of the human factors needs, not only in aerospace but also in surface transportation, weapons, medical, and other types of systems.

Literature Sources

Researchers in aviation systems are likely to be aware of specialized periodicals such as the *International Journal of Aviation Psychology* and the proceedings for the biannual meetings of the International Symposium on Aviation Psychology and the International Conference on Experimental Analysis and Measurement of Situation Awareness (Garland and Endsley, 1996). The *Proceedings of the Human Factors and Ergonomics Society* also publishes several papers on aerospace systems, including air traffic control. The same *Proceedings*, together with the society's journal, *Human Factors*, publishes many papers on human performance issues (e.g., workload, models of human error, perceptual processes, decision making, shift work, workspace design) that are directly relevant to problem areas in air traffic control as well as other human-machine systems.

One of the most comprehensive references for the human factors database is Boff and Lincoln's (1988) *Engineering Data Compendium: Human Perception and Performance*. Although it has the appearance of a design handbook, in fact it can be valuable as a resource for air traffic control research questions as well. Handbooks such as Boff et al. (1986), *Handbook of Perception and Human Performance*, and Salvendy (1987), *Handbook of Human Factors* (with Volume 2 in press), provide research reviews (and design implications) of the major human factors issues.

Several texts on aviation human factors are available, including Cardosi and Huntley (1993), Fuller et al. (1995), Hawkins (1993), Hopkin (1982a, 1995), Jensen (1989), Maurino et al. (1995), McDonald et al. (1994), O'Hare and Roscoe (1990), and Wiener and Nagel (1988). Texts by Cardosi and Murphy (1995) and by Hopkin are specifically addressed to air traffic control. Others, like the Jensen and the Wiener and Nagel texts, include chapters on air traffic control, and the research on many human performance issues in the cockpit are applicable to the operational controller as well. For example, the human factors issues in cockpit resource management described in Wiener et al. (1993) and their applicability to air traffic control environments are relevant to our discussion of teamwork and communication. A series of documents published by the International Civil Aviation Organization focuses on human factors issues in air traffic control (e.g., International Civil Aviation Organization, 1993, 1994), including workspace, automation, communication, navigation and surveillance, and management systems.

Another collection of texts by Wise and his colleagues (Wise and Debons, 1987; Wise et al., 1991, 1993, 1994a, 1994b) provides a wide-ranging survey of human factors problems and needs in aviation, with a particular emphasis on air traffic control. These texts report on papers presented and discussed at working conferences on contemporary challenges to the development and certification of complex aviation systems.

Through the Civil Aeromedical Institute (CAMI) in Oklahoma City, the FAA Office of Aviation Medicine has compiled a listing of aviation medicine reports (Collins and Wayda, 1994). This list and the FAA Technical Center bibliography of human factors studies (Stein and Buckley, 1994), completed at the center over the last 35 years, together provide another significant component of the aviation human factors database. More recently, the Armstrong Laboratory at Wright-Patterson Air Force Base has published a bibliography of 50 years of human engineering research in the Fitts Human Engineering Division (Green et al., 1995).

Transferability of Human Factors Data

Caution must be exercised before lessons and recommendations obtained from databases and literature are transferred to air traffic control system design. As an example, Sperandio's (1971) early descriptions of how controllers use adaptive strategies to maintain performance in the face of increases in task load was one of the first indications that existing models of workload and performance could not be generalized for the direct prediction of controller behavior. Another indication is provided by the findings that operational errors tend to be associated with moderate or low workload (Stager and Hameluck, 1989, 1990; Rodgers, 1993). Some research has been done in developing cognitive models of controller processes (discussed below), but there are not yet normative models of controller behavior or controller performance by which to ascertain the validity of transferring findings from another task environment (within the broader aviation context) to air traffic control. Substantial work in completing cognitive task analyses for different types of air traffic control sector operations is required to help adapt existing human factors literature to air traffic control needs.

At the same time, some literature would appear to be more or less directly applicable. For example, descriptive models of human error (discussed below in the section on combining sources of human factors data) may be directly applicable in the post hoc analyses of operational errors and in the design of more effective controller-system interfaces.

Limitations

The data and recommendations contained in the human engineering literature frequently have not been tailored for specific applications. Expert interpretation is often required to determine the applicability (particularly without further validation) of data to a specific research question. Although it is often possible for human factors specialists to extrapolate from the literature to a design application, whenever possible, usability testing (i.e., for user acceptability) should be conducted in a rapid-prototyping or other simulation environment (see below).

ANALYSIS OF CONTROLLER RESPONSES

In this section, we describe three sources of human factors information: (1) incident analysis, (2) reporting systems, and (3) subjective assessments and verbal reports—a collective expression for a number of independent methodologies that depend on the subjective responses or comments of controllers rather than on their performance within the system. Our emphasis in this section is on the working system and the user experience (and performance) within the air traffic control system. Subjective assessments, however, are also an integral part of the rapid-prototyping process.

Incident Analysis

Because of the multiple causes of most accidents in highly redundant systems, such as those involved in aviation, accident analysis is often ambiguous in revealing human factors causes (Diehl, 1991). The occurrence of aviation accidents that are directly attributable to an air traffic control system error, such as the runway collision at Los Angeles International Airport (National Transportation Safety Board, 1991), is extremely rare. Incidents (which are referred to as operational errors), such as a loss of the required separation between aircraft, are more common but still relatively infrequent (Rodgers, 1993). The low frequency of incidents (versus the occurrence of errors that do not result in incidents) imposes particular constraints on the observation of precipitating conditions and statistical inference. By definition, incidents are concerned with either system or operator error or with a procedural deficiency.

McCoy and Funk (1991) recently attempted to develop a taxonomy of operator errors based on a model of human information processing using NTSB aircraft accident reports. They found that the air traffic control system was a contributing or probable cause in 6 of the 38 accidents they reviewed for the 1985-1989 period. When the search was extended back to 1973, they found a total of 29 examples of air traffic control involvement. The errors were related to attention, memory, perception (i.e., the validity of the controller's world model), and response selection (including the issuing of a clearance, coordination, and a variety of other procedures). From an analysis of operational errors, Redding (1992) reported that failure to maintain adequate situation awareness was the likely cause of most errors. As a result of their own review, McCoy and Funk argued for the design of error-tolerant systems (see Wiener, 1987, 1989) while still trying to prevent errors.

Stager and Hameluck (1989, 1990) reported that, in their analysis of 301 Fact Finding Board reports, the occurrence of incidents was not related directly to rated workload. Operating irregularities were associated with conditions of moderate or low workload, normal complexity, and intermediate traffic volume and complexity. Allowing for the fact that more than one cause could be assigned by a review board to the same incident, both attention and judgment errors were cited as the cause in more than 60 percent of the cases examined. In a related study, at least half of all system errors were found to have "causal or contributory factors which are directly attributable to breakdowns in the information transfer process—usually in oral communications" (Canadian Aviation Safety Board, 1990:6). Similar findings were reported by Rodgers (1993) for an analysis of the operational error database for air route traffic control centers. In a second analysis, neither controller workload (number of aircraft being worked) nor air traffic complexity was found to be related to the severity of the operational errors.

Incident analysis is a post hoc process, and the data that are available for analysis have frequently been filtered through a conceptual system that is reflected in the classification structure of the database itself. What data are collected at the time of a given incident are determined largely by the questions posed during the gathering of evidence. Rodgers (1993) has indicated that it is necessary to be able to review the dynamics associated with the air traffic situation (and not just the error-related event itself) when examining operational errors (Rodgers and Duke, 1994). Consequently, an analysis of operating irregularities can sometimes provide insight into the patterning of the occurrence of incidents without clearly identifying the underlying causal factors involved (Stager, 1991b; Stager and Hameluck, 1989, 1990; Stager et al., 1989). Still, from a procedural perspective, it is important to identify controller or system errors that can impact system safety (Rodgers, 1993; Durso et al., 1995).

By focusing on the purely operational factors that are associated with an accident, it may be that the higher-level management and organizational factors are overlooked. Maurino et al. (1995) have recently tried to extend the scope of analysis beyond the individual to the system as a whole (see also Reason, 1990).

Reporting Systems

The aviation safety reporting system (ASRS), coordinated by NASA in the United States (Nagel, 1988; Reynard et al., 1986; Cushing, 1994; Rosenthal and Reynard, 1991; Wickens and McCloy, 1993), the confidential human incident reporting programme (CHIRP), under the auspices of the Civil Aviation Authority in the United Kingdom, and comparable focal points in a few other countries provide confidential and anonymous channels for reporting actual occurrences or potential sources of error in the interests of aviation safety. Both pilots and controllers are able to report situations in which there has been a breakdown in standard procedures or errors in behavior have been observed. They are guaranteed anonymity, provided that the error has not been previously reported by others and a formal loss of separation did not result. The confidential reporting procedure allows appropriate follow-up action, including interviews, to be taken. This kind of reporting facility benefits safety by tapping evidence not otherwise available, complementing rather than replacing other more traditional means to improve aviation safety. The ASRS can sometimes provide a means of documenting not only a particular problem area in air traffic control operations (Monan, 1983) but also the impact of system changes, such as the introduction of the collision avoidance system, TCAS II (Mellone and Frank, 1993).

The investigation of controller errors has often relied on ASRS data. For example, Morrison and Wright (1989) grouped controller errors within two broad concepts: control (monitoring, coordination) and communications (clearance composition, read/hearback errors). Rosenthal and Mellone (1989) investigated anticipatory clearances (e.g., fast sequence clearances to expedite traffic flows in high-volume situations).

Although the value of the ASRS for aviation safety has long been acknowledged, there are inherent limitations with the reporting system as a research methodology (Prinzo and Britton, 1993; Wickens and McCloy, 1993). For example, the language in which events are described by the participants does not necessarily reflect the same concepts used by human factors personnel to define causal relationships (e.g., mental workload, perceptual failure, inappropriate mental model). The controller or pilot may not appreciate the need for a description of (or may not be able to articulate precisely) the antecedent conditions. Some clarification can be achieved through the follow-up interviews by ASRS personnel and the use of keywords for incidents.

There is always the concern in reporting systems (and in descriptions of operating irregularities prepared for boards of inquiry) that data can be constrained if not specifically determined by a predetermined conceptual structure. Questionnaires or lists of keywords prepared by persons with an operational background may not capture those aspects of an event that the psychologist or human factors specialist needs to interpret an incident within a valid framework.

Harwood et al. (1991) have suggested that a relational schema (based on the controllers' conceptualization of the air traffic control domain knowledge) overlaid on the ASRS data would be a helpful organizing tool. Conceptual structures found in their analysis of controllers' representations of relationships between concepts could provide a means of drawing together seemingly disparate incidents.

Finally, the very significant volume of incident data that is collected each year itself imposes a constraint on use of the reporting system as an effective research methodology. The ASRS staff are able to follow up only a fraction of the reported incidents and to encode them in the appropriate psychological language. Moreover, a fully user-friendly means of exploring the ASRS database in order to generate hypothetical causal relationships is not yet available. Consequently, this information resource has been underutilized as a methodology.

Subjective Assessments and Verbal Reports

Subjective assessments are frequently used in the air traffic control environment (Hopkin, 1982b). They are convenient, inexpensive, and always available as an option; in some circumstances, they may yield data that can be obtained in no other way (Manning and Broach, 1992). For example, Harwood (1993) has described the role of subjective assessment methods in identifying, first, the human-centered issues associated with air traffic control system upgrades and, second, the required criteria and measures that are applicable during system transition.

Subjective assessments are integral to the rapid-prototyping process (see below) and are commonly obtained in the laboratory investigations, real-time simulations, and field studies that provide a context for measures of controller performance as well as physiological and biochemical indices. The verbal reports can be very helpful in supplementing and explaining other measures, although they are not adequate as a substitute for other measures. For example, subjective comments and ratings are often collected to supplement behavioral, physiological, and biochemical measures of workload, effort, fatigue, and stress in air traffic controllers (Costa, 1991; Melton, 1982; Moroney et al., 1995; Smith, 1980; Stein, 1988; Tattersall et al., 1991). One of the most common applications of subjective assessments and verbal reports is in subjective workload assessments, which we discuss in a later section.

In the discussion of subjective assessments in this section, however, the focus is on the use of controllers' own responses as a means of making inferences about their cognitive structure and information processing.

Modeling Controller Processes Through Verbal Reports

From a general systems design perspective, it is understood that the displays and inherent functionality of an operator's workstation (e.g., the nature of computer-human interaction) must be compatible with the operator's mental model of the system characteristics (Edwards, 1991; Hollnagel, 1988; Lind, 1988; Van der Veer, 1987; Waern, 1989) and that the nature of the displays must match the nature and level of information processing at which the operator is working (Moray, 1988; Rasmussen, 1985; Rasmussen and Vicente, 1989).

In air traffic control research, the subjective assessments of controllers coupled with both structured and unstructured measures of their information processes provide a means of gaining insight into their cognitive models (Leroux, 1993b, 1995; Mogford, 1991, 1994; Murphy et al., 1989). In some instances, the subjective assessments may depend simply on verbal reports (Whitfield, 1979; Whitfield and Jackson, 1983); in others, on the use of psychometric analyses, such as multidimensional scaling and cluster analysis, of subjective assessments (Kellogg and Breen, 1987; Stager and Hameluck, 1986).

For example, verbal reports can be used to reveal the cognitive processes underlying a controller's performance during the management of traffic scenarios (Amaldi, 1994; Endsley and Rodgers, 1994) and the parameters that are considered by controllers in their decision making and that contribute to perceived airspace complexity (Mogford, 1994; Mogford et al., 1994a, 1994b).

Limitations

Subjective assessments can be useful, yet verbal comments reflect what individuals think they do or what they are supposed to do, not always what they actually do. It is often advocated that videotaped records be made in order to ensure completeness of the information obtained through direct subjective assessments and verbal reports. Assessments and verbal reports are subject to error through distortions of individual emphasis and the fallibility of human memory. On some occasions, users may voice subjective preference for systems that do not support the best performance (Andre and Wickens, 1995; Yeh and Wickens, 1988; Druckman and Bjork, 1994). If subjective measures are in disagreement with other measures, this does not justify discarding either type of measure, and the disagreement need not imply that one or the other measure is wrong (Muckler and Seven, 1992). Agreement with other measures may support and help to validate both the subjective and the objective measures. Finally, it is important to emphasize that with any use of subjective data, whether ratings collected in the laboratory or opinions collected in surveys, a good deal of expertise is necessary in order to design the instrument in such a way that the data obtained will be unbiased.

Workload Assessment

One of the most crucial functions of subjective reports in human factors has been to provide estimates of mental workload. However, this is an area that must be considered very much in concert with other workload assessment methodologies. Four main classes of workload measurement procedures have been proposed and used: primary task, secondary task, physiological, and subjective measures. In addition, modeling, or predictive workload assessment, has also been proposed. Only the major aspects of each method and its associated strengths and limitations are considered here. For additional details on each of these classes of workload methods, consult O'Donnell and Eggemeier (1986), Lysaght et al. (1989), and Wickens (1992).

Primary-Task Measures

This method involves measurement of performance on the primary tasks of interest. The techniques available for assessment of controller performance, particularly as they pertain to selection and training, were discussed in Chapter 3. Given that reliable performance assessment methods are used, the performance of the controller on a particular task is assumed to reflect directly the mental workload associated with achieving that level of performance.

Primary-task measures such as the number of aircraft per unit of time, number of control actions, and mean aircraft proximity (Rodgers et al., 1994) have the merit that, if shown to be valid, they can be directly related to operational performance, which can be an advantage in relating workload to system performance, thereby aiding in system evaluation. However, there are at least two nullifying disadvantages: (1) Primary-task performance can be dissociated from mental workload; that is, the same output level of performance may be associated with different degrees of controller workload. Sperandio (1971) showed that controllers often respond to an increase in imposed task load (e.g., increased traffic density) by subtle variations in operating procedures (e.g., shortening the length of verbal messages to pilots) in order to regulate their performance. More generally, controllers may use a variety of strategies to maintain a certain criterion level of performance in response to an increase in task load. However, this could come at the cost of higher mental workload, leaving potentially little margin for dealing with emergencies or additional tasks. The primary-task workload index is insensitive to this potential problem. (2) Primary-task performance measures may be difficult to obtain in practice, particularly for cognitive activities such as planning and decision making, during which the controller may make very few if any overt responses that can be measured. Furthermore, the overt response represents the end product of a number of considerably demanding information-processing activities and as such may provide only an incomplete index of the workload associated with these processes.

Secondary-Task Measures

In the secondary-task procedure (Brown and Poulton, 1961; Garvey and

Taylor, 1959), the operator is asked to concentrate on performing well on the primary task and to allocate any residual attentional resources or capacity to the secondary task. The basic premise is that performance of the secondary task reflects the workload demands of the task to be assessed, given that primary and secondary tasks both make demands on the same information-processing resources.[1] Early studies of controller workload used the oral and written communications of the controller as an embedded secondary task and found that, as task load increased, verbal communications became shorter and more stereotyped (Leplat and Browaeys, 1965) and handwriting deteriorated in form and content (Kalsbeek, 1965). More recent studies have used various secondary tasks drawn from the experimental psychology literature on dual-task performance, for example, probe-reaction time, rhythmic tapping, Sternberg memory search, random number generation, time estimation, and combinations of tasks (see O'Donnell and Eggemeier, 1986, for a review).

One of the advantages of the secondary-task procedure is that the choice of secondary task can be theory-driven (e.g., the multiple resource theory of Wickens, 1984) and therefore potentially diagnostic of the source of workload, rather than simply providing an estimate of overall workload. This method is also one of the few workload techniques that can reveal the upper limit of a controller's capability and hence can be potentially valuable for estimating controller response to emergency events. One of the major disadvantages of the technique is its relative obtrusiveness, particularly when even transient diversion of the controller's resources away from the primary task may compromise safety. The use of secondary tasks that are embedded as a natural but lower-priority element within the main tasks that the controller has to perform (e.g., removal of a flight strip that has been handed off to another controller) may partially overcome this problem.

Physiological Measures

Physiological measures that reflect aspects of mental workload have been a focus of continuing interest for many years, and early applications included assessment of controller workload (e.g., Kalsbeek, 1965). Although different classification schemes can be used to describe the various physiological measures that are available, in general they cluster around two types: (1) background measures that are not specifically linked to ongoing task events or the timing of controller activities or responses. Measures in this category include the sponta-

---

[1]Various other assumptions must also be met for the secondary-task method to yield interpretable results. For example, both tasks should be resource sensitive and not data limited (Norman and Bobrow, 1975); the secondary task should not be capable of being performed purely automatically; and primary-task performance should not vary with the introduction of the secondary task or with different secondary tasks (see Fisk et al., 1986, and Wickens, 1984, for further discussion).

neous electroencephalogram, heart rate and heart rate variability, skin conductance, urinary catecholamine output, etc., and (2) event-related (or task-related) measures that are linked to specific task events, controller responses, or irrelevant (secondary-task) stimuli, such as event-related brain potentials to secondary-task or embedded stimuli, evoked pupillary responses, eye movement scan paths, task-related heart rate changes, etc. Of the background measures, heart rate and heart rate variability have been used in a number of studies of simulated air traffic control and appear to provide a sensitive index of the overall workload level of the controller (Byrne and Porges, 1994; Rose and Fogg, 1993). Of the event-related measures, evaluation of eye movements and of scan paths would seem to be of potential importance in understanding controller workload and vigilance. In one recent study, experienced controllers were found to have longer fixation durations than trainees, and saccade durations decreased as workload increased (as assessed by traffic load and by observer rating) (Stein, 1992). However, little is known about other characteristics of controller scan paths and their relation to workload; as Stein (1992) pointed out, additional studies are needed in this area.

Evaluations based on the two classes of physiological measures, background and event-related, may provide complementary information about controller workload within the same study. For example, Hilburn et al. (1995) carried out a study with experienced en route controllers using a high-fidelity simulation of Dutch airspace. They used heart rate variability measures to show that the use of a new automated decision aid—the descent advisor of the center-TRACON automation system (CTAS), described in Chapter 12—reduced overall mental workload compared with normal operations. Eye movement measures were also used and indicated that the momentary workload associated with handling particular display events (e.g., datalink information) was also reduced.

The primary advantage of physiological measures is their continuous availability, or high bandwidth (in comparison to discrete operator responses, which are relatively infrequent in most tasks except sustained data entry). Physiological measures (particularly background measures) can also be obtained with only minimal disruption of the controller's primary tasks. Another advantage is that physiological measures are potentially sensitive over the entire range of workload, that is, to both underload and overload, whereas other measures may "bottom out" at the low end of workload or provide results that are difficult to interpret (e.g., secondary task performance may be insensitive to the primary task, whose performance paradoxically may improve with the addition of the secondary task; see Poulton, 1973). This feature may be a particularly useful advantage in the assessment of workload in highly automated systems, in which underload can be a problem (Byrne and Parasuraman, 1996). Disadvantages of physiological measures include possible discomfort and lack of acceptance by the controller (although this is becoming less of a problem with the development of lightweight sensors, helmet-based recording, telemetry, etc.) and the lack of known baselines against which increases or decreases in workload can be judged. Also, it is rare that any physiological measure by itself can provide information that can be used to infer workload, without reference to behavior or task events. Physiological measures can sometimes provide unique information about controller state not available from any other source, so that the usefulness of these measures may be best realized when they are combined with other workload metrics. For a review of the range of physiological measures available in workload assessment, see Kramer (1991). For a discussion of how one class of physiological measure, event-related brain potentials, can contribute to mental workload assessment and other ergonomic problems, see Parasuraman (1990) and Wickens (1990).

Subjective Measures

Assessing the subjective perceptions of controllers regarding their workload offers one of the most convenient and cheapest methods for workload assessment. Several subjective measures have been proposed, and they generally differ in whether they assume a unidimensional or multidimensional view of subjective workload. The two most commonly used indices are multidimensional, the NASA-TLX (Hart and Staveland, 1988) and the SWAT (Reid and Nygren, 1988). Each provides for the rating of sources of workload on different scales (temporal demand, effort, etc.) and a weighted overall score. The NASA-TLX has been extensively validated in several studies using laboratory and simulator tasks in which subjects were required to perform tasks singly or in combination (Hart and Staveland, 1988); it has also been found to be an effective workload index in comparisons of the reliability, validity, and other psychometric properties of different subjective scales (Hill et al., 1992). Another way to validate a subjective scale is to examine whether it is sensitive to manipulations of perceptual factors that are known to affect task difficulty (Natsoulas, 1967). This strategy attempts to establish the validity of subjective scales by investigating whether psychophysical factors that degrade performance increase subjective workload, whereas factors that enhance performance diminish perceived workload. Using this basic strategy, Warm and colleagues (1996) have reported a large series of studies in which the TLX was found to be very sensitive to a variety of psychophysical and task manipulations during vigilance performance.

Subjective measures tailored more specifically to controller activities have also been developed. One such method is the air traffic workload input technique or ATWIT (Stein, 1985). In this procedure, the controller is probed for a subjective workload estimate at preset intervals during normal operations. This method produces a workload profile that is thought to reflect accurately variations in task load related to airspace variables and communications.

Subjective scales have some advantages. They are easy to use, cheap, and easily accepted by controllers and other system operators. However, subjective workload measures, like many subjective scales, are potentially sensitive to various sources of bias (the halo effect, under- or overreporting of sources of workload, etc.). Other disadvantages include uncertainty as to whether verbal reports fully reflect all aspects of mental workload, particularly those information-processing activities that are not accessible to working memory, and the fact that subjective measures and primary task performance measures may sometimes disagree (Hart and Wickens, 1990; Yeh and Wickens, 1988; Andre and Wickens, 1995).

MODELING AND COMPUTER SIMULATION

Modeling and computer simulation of the human operator represent important research tools in the study of human factors issues in air traffic control. This importance is manifest in two generic contexts. First, if valid models of either the human or the airspace are developed, then such models can be "run" subject to changing assumptions about the nature of the human-system interface (e.g., the replacement of human functioning by automation for certain tasks, the shift in responsibility of certain monitoring functions from the ground to the air). In this way, predictions of the feasibility of such changes can be made early in the design process (Elkind et al., 1990), thereby potentially saving time and effort if such proposed changes are predicted to be safety-compromising or to lead to capacity losses. Second, such models can often be employed to replace more complex electronic or human components in complex simulation experiments with the human in the loop (see the section below on real-time simulation). For example, an accurate computer model of pilot response to air traffic control commands could greatly simplify the task of running experiments on controller-pilot communications with controller subjects.

Features of Models

Models of all kinds, and particularly those relevant to air traffic control, can be characterized by a wide number of different features. Four of the important features are described in the following paragraphs.

1. *The kind of entity that is modeled.* A model can be developed of either the human operator or of the system. System models in turn can focus (with increasing complexity) on individual elements, such as a single airport or sector, on regional elements, such as an airport and the surrounding airspace, or on networks, such as the network of major airports in the United States (Odoni, 1991). Human operator models can focus on either the individual operator (Elkind et al., 1990) or on groups of operators, such as a pilot-controller link. Generally, system models make either no assumptions or very simplified assumptions about the human element in the system. For example, pilots are assumed to fly at constant speeds, controller delays in managing traffic are assumed to be negligible, and all human elements are assumed to be error free.

2. *Computation of the model.* Typically, a distinction is made between analytic models and simulation models. Analytic models are relatively straightforward equations, in which a predicted output can be rapidly generated from the mathematical analysis of a set of specified inputs. An example of a human analytic model would be one that predicts sector air traffic complexity (and therefore cognitive workload) as a function of a number of parameters of the situation (Rodgers, 1993; Pawlak et al., 1996). An example of a system analytic model would be one that predicts airport delay as a function of a number of parameters such as traffic arrival rate, runway configuration, miles-in-trail, etc. Analytic models are often simple regression equations.

Simulation models are more complex. They are "runnable" in the sense that a number of operating rules are specified in computer language, inputs and changes (events) are delivered to the model over time (e.g., aircraft push back from the gate), and the computer then generates a set of performance outputs, typically probabilistic ones. For example, a recent simulation modeling exercise predicted the implications to traffic flow and density of implementing the free-flight procedure (Planzer and Jenny, 1995), in which pilots are responsible for maintaining their separation in the en route sectors (MITRE, 1995). Simulation models are often necessary in systems like air traffic control, both because the great complexity of the system prevents its behavior from being captured by analytic equations, and because the inherently dynamic behavior of the airspace is well suited for a dynamic simulation (i.e., one that can produce event sequences as outputs). Simulation models also become particularly valuable when the modeled variables have a range of values (i.e., the distribution of air speeds, turn radii, or separation requirements between a fleet of aircraft) and system performance must be evaluated with elements that vary across this range. As a consequence, the simulation may run many times over, and a distribution of output parameters can be provided.

3. *Output of the model.* Most air traffic control system models provide as outputs measures of unit capacity or delay (Odoni, 1991). Other critical parameters of separation loss have not been adequately modeled. Human operator models typically focus on modeling operator processing delays or some aspect of workload (Burbank, 1994; Elkind et al., 1990; Corker and Smith, 1995). Generally, human operator models have not been developed to predict operator error.

4. *Goal of the model.* Odoni (1991), distinguishes among three classes of goals for air traffic control models.

- Policy analysis models are designed to predict the safety or economic impact of major structural changes in the airspace, such as implementing a free-flight regime, building a new airport, or implementing certain new levels of automation that may impact spacing requirements. These models may be analytic, and they are often expressed at a macro level, meaning that they do not need to model processes in fine detail.

- Detailed planning models focus more directly on addressing design and research questions; addressing such issues, for example, as how the interaction between pilots and controllers will be changed with the introduction of the datalink system (Kerns, 1991; Parks and Boucek, 1988). Such models of necessity operate at a more micro level, examining detailed aspects of performance and timing. They are usually simulation models.
- Operational support models (or fast-time simulation models) are intended to provide air traffic controllers with real-time support on their tasks; for example, working out solutions "on line" to help manage traffic and prevent conflict. They are, by nature, models of more microscopic processes.

In fast-time simulation, the normal variables are system ones and not human factors ones. For example, advisory programs like COMPAS (Völckers, 1991) and the final approach spacing tool, FAST (Lee and Davis, 1995) could be tested in fast-time simulation, but the dynamics that would be tested would not bear directly on human factors questions. Some of the difficulties in integrating fast-time simulation and real-time simulation findings occur because the former has excluded human variability and the latter has included it. Sources of human variance and of system variance are not usually expressed in similar terms and therefore resist integration. Practitioners of both real-time and fast-time modeling seem to ignore some variables that to the other are important. This is regrettable, because in much air traffic control research a combination of fast-time simulation and real-time simulation could be much stronger than either is alone.

Because of these differing goals, the three kinds of models have somewhat different requirements. As noted, because it is at a macro level, the policy analysis model can afford to make simplifying assumptions about the fine details of human and system performance. Furthermore, both policy analysis and operational support models must run relatively rapidly; for the former, the need for speed results because vast numbers of runs are typically needed to generate enough data across various possible configurations that reliable solutions can be estimated; the need for speed for the latter results because the controller in the field cannot afford to wait for a long time, while a computer generates a predicted (modeled) output.

All three goal-defined categories of models, whether applied to the airport, the region, the network system, or the human operator have a strong need for validation data. These data should come from databases on delays, operational errors, etc., characterizing the frequency with which such incidents occur and the various conditions in place at the time of their occurrence.

Challenges to Modeling

Air traffic control modeling efforts at all levels are presented with three distinct challenges. The first of these comes from the complexity of the process that is modeled, a complexity resulting from the fact that there are multiple entities and that these entities interact. For example, a policy analysis model of flow within a network of 20 airports, in which each airport can be characterized by three levels of capacity (i.e., associated with varying weather conditions), will have 3^{20}, or 3.5 billion different possible states (Odoni, 1991). Even the model of a single airport, with four different runways, each of which can be configured in one of two directions, can have a phenomenal number of different capacity levels.

The second challenge is to extend models, of both the system and the human, to predict errors and separation violations. This extension becomes considerably more difficult to do for typical capacity and delay systems models, because of the need to incorporate characteristics of the human element and of faulty equipment. It becomes more difficult to do for typical workload models of the human controller, because of the uncertain relation between workload and error.

The third challenge, related to the second, is to develop valid human controller and pilot models that can fit as modules into the more broadly encompassing system models. How do limits on human operator workload influence the capacity of sectors to deliver traffic? How do limits of controller perceptual and predictive resolution limit the allowable spacing? Most critically, what are the sources of human error, which may have catastrophic influences on capacity and flow in worst-case scenarios?

Existing Models

Odoni (1991) has reviewed the state of the art of system-level models, identifying certain versions that appear to be useful and promising. For example SIMMOD (Federal Aviation Administration, 1996) provides a good candidate for modeling flow at a regional level. The NASPAC model (Frolow and Sinnott, 1989) is a network simulation model that includes 58 U.S. airports and is capable of integrating both ground (i.e., airport) operations with airborne operations. TAAM is another model developed in Australia that consists of three modules, representing different operational areas: airport, TMA, and en route. The airport module covers all ground movements at an airport. The TMA module covers runway operations plus the terminal airspace operation at an airport or a cluster of airports. It may be used to assist in the (re)design of standard instrument departures and standard arrival routes and also in solving runway capacity problems. The en route module concentrates on higher-altitude operations worldwide. It may assist in the identification, qualification, and solution of problems with airways congestion. The FAA has developed and is refining a national airspace system simulation model (NASSIM) of ground-based air traffic control equipment and tasks. NASSIM permits identification of bottlenecks (e.g., in controller task loading and communication equipment) that can result from equipment failures (e.g., failure of a HOST computer or radar).

We can distinguish among three classes of models that have been applied, with varying degrees of development, to the air traffic control domain. Most mature perhaps are the analytic sector or traffic complexity models (Rodgers, 1993; Pawlak et al., 1996), which have been designed primarily to predict controller workload. At a different level are operator parameter models, whose intent is to provide somewhat valid estimates of operator performance characteristics, which can then fit into general system level models. Examples include models of typical conflict situations, of the time required for communications interchanges between pilots and controllers, and the response time of pilots to a TCAS alert.

At the most complex level (but least mature in terms of validation) are models of the full array of controller perceptual/cognitive/motor performance (Elkind et al., 1990; Corker and Smith, 1995; Pawlak et al., 1996; see McMillan et al., 1989, for a review). Some of these models have already been applied to pilot performance (e.g., Baron and Corker, 1989; Loughery et al., 1989). To the extent that they can or will be applied to controller performance (and efforts to do so are under way at NASA Ames and the Canada Institute of Environmental Medicine—Burbank, 1994), they should be able to address issues such as the change of controller response time imposed by the datalink system, and the change in controller workload imposed by implementation of a particular automation device. Once validly modeled, these features can then be incorporated, as modules, into the systems-level models.

Limitations

Human simulation research is complex and costly. As an alternative, computer simulation represents an essential research methodology to answer design questions for air traffic control. But models will be only as useful as their ability to accurately capture (and predict) the main sources of variance in system performance (i.e., establishing their external validity is necessary). To achieve this goal, the collection of performance data necessary to evaluate the existing models must be viewed as a number one priority. A second limitation of some models is their complexity and the technical expertise necessary for them to be run.

The air traffic control research community needs to begin efforts to integrate the two levels of models—of the airspace system and of the human component—so that more accurate estimates of controller processing latency and workload can be fed into the simulations of the national airspace, as design changes are made. Although a complete mega model with accurate estimates of all parameters could reach an unwieldy level of complexity, such a research effort could specify the level of modeling precision of each component (human or airspace) that might be appropriate for answering different kinds of questions. For example, in some circumstances, such as those involved in en route traffic management, estimates of controller response times may not need to be very precise. Under others, e.g., response to a sudden conflict avoidance in a terminal area, much greater precision would be needed. This effort would allow the model to be considerably more flexible.

DESIGN PROTOTYPING

A significant methodological concern in human factors research has been how to bridge the gulf between laboratory and field (Chapanis, 1967; Dipboye, 1990). In response to this concern, advances in contemporary computers and software have contributed to the development of rapid prototyping in a laboratory environment.

Description of Prototyping

Rapid prototyping is a technique for gathering the comments and impressions of future users and others regarding the capabilities and limitations of a simulated air traffic control workstation or workspace while it can still be changed. It is not a substitute for testing, and its functions are more analogous to planning, since it encourages the formulation of alternatives but cannot usually yield quantitative evidence, for example, about capacities or characteristic human errors. Prototyping can be a useful technique to discard fundamentally flawed options quickly, to identify crucial combinations of circumstances that require testing, and to discover main topics of agreement and of disagreement among controllers.

The display of dynamic and interactive representations of proposed operational controls and displays (i.e., the computer-human interface) on computer screens enables several alternative design concepts to be quickly evaluated by research, design, and operations personnel (Dennison and Gawron, 1995). Suggested modifications can be made to the prototype displays, often on line, with the observers providing further input to the revised interface and the proposed functionality. Contemporary prototyping provides a means of simulating many of the elements of the displays, including the input mechanisms, the visual display characteristics, and the data processing functionality that will ultimately comprise the nature of the controller's interaction with the fielded system.

Task analyses can provide the basis for a workstation design for an air traffic control tower, for example, in order to meet design requirements (e.g., Miller and Wolfman, 1993), but whether the specifications translate into a usable interface in the operational environment has to be determined through prototyping and evaluation by experienced users (Fassert and Pichancourt, 1994). There is a need, for example, to promote human factors experimentation and user involvement during the design process in order to ensure design validity. One of the objectives for design programs in several air traffic control projects has been to ensure active participation by experienced controllers in the developmental work (Day, 1991; Dujardin, 1993; Leroux, 1993a; Simolunas and Bashinski, 1991; Stager, 1991a). After a review of the procurement process for the advanced automation system (AAS), Small (1994) forcefully recommended more effective use of prototyping and more appropriate use of controller teams in design programs.

Limitations

Much of the prototyping process depends on subjective assessments. In some instances, subjective assessments may be the only type of data collected during rapid prototyping. For that reason, as noted above, the concept of performance-preference dissociations (i.e., subjective preferences that are not supported by measures of performance) are discussed in this context (Bailey, 1993). Wickens and Andre (1994) and Andre and Wickens (1995) have reviewed evidence concerning the dissociation between preference and performance in terms of design factors such as display interfaces and color applications (Hopkin, 1994; Narborough-Hall, 1985). Druckman and Bjork (1994) discuss other aspects of this dissociation. The potential for the dissociation of preference and performance measures argues for the regular use of performance measures to augment preference ratings during prototyping and usability testing (Bailey, 1993).

Practical constraints on training do not enable controllers to become fully proficient with the prototyped interface. The objectives in prototyping are therefore often limited to demonstrating feasibility and viability rather than aspiring to determine ultimate performance capacities and quantification.

As an evaluation methodology, rapid prototyping is constrained in the evidence that it can afford the investigator, but it is capable of demonstrating that a certain design response to a given requirement is feasible and warranting of further development. Given the apparent fidelity of the computer-controller interaction that can be presented and evaluated in the rapid prototyping environment, however, there is a concern that acceptability in the design laboratory can be too quickly equated with task validation for the operational environment. Hopkin (1995a) has observed that it is appropriate to use the prototyping environment to establish the feasibility of design concepts, provided that it is recognized that, in part, rapid prototyping is actually a surrogate for the planning, thinking, and formulation stages of system evolution and cannot function as an alternative form of validation. Similarly, Jorna (1993) has cautioned that subjective opinions (during prototyping) are not necessarily sensitive to all design factors and can actually lead to nonoptimal designs, and that relative ease of prototyping can lead designers to skip or postpone the research phases of more systematic, integrative performance evaluations (i.e., evaluations that carefully consider a full range of operating conditions). Such evaluations are best accomplished through real-time simulation.

REAL-TIME SIMULATION

Contemporary human factors research, whether directed toward problem solving in current air traffic control operations or toward system development, usually involves system simulation and the measurement of operator performance in real time (i.e., with the tempo of actual or live operations). Early human factors research, however, depended either on laboratory experimentation (with the expectation that the observations could be extrapolated to operational settings) or on the use of simulators with controls and displays that could not be readily changed (e.g., Fitts, 1947; Green et al., 1995; Taylor, 1947). Even the early attempts at simulating larger parts of systems (e.g., Fitts, 1951; Parsons, 1972; Porter, 1964) depended on electromechanical and hardwired equipment and were closer to complex laboratory experiments than the advanced high-fidelity simulations familiar today.

Uses

The use of simulation as a research methodology is usually undertaken with the intent to validate human factors design decisions or tentative conceptual models. As a technique for collecting human factors information, real-time simulation has many strengths (Sandiford, 1991). It can also be useful in predicting training requirements, identifying sources of potential error or failure, validating planning decisions (Beevis and St. Denis, 1992), and exploring interactions between human and machine roles (Hollnagel, 1993a).

There are three kinds of approaches to real-time simulation for human factors air traffic control research and development. The first uses a complex air traffic control simulation facility built to replicate the functioning of major air traffic control regions. As many as 20 or 30 staffed control positions may be studied concurrently, together with much of the interaction and communications between them and pilots in the simulation as well as with adjacent sectors and other agencies (Stager et al., 1980a, 1980b; Transport Canada, 1979). The purpose is to test and quantify the viability of forms of air traffic control proposed for the region simulated or for other regions of airspace that it typifies. Many system measures as well as human factors ones are normally taken, because the simulation is not exclusively or even primarily a human factors evaluation (Stein and Buckley, 1994). Sometimes the purpose is exploratory, to establish feasibility and quantify capacities or to compare options by simulating them. Measures include system functioning, communications demands and characteristics, system errors or deficiencies, and controller performance (and its costs in terms of workload, effort, motivation, attitudes, and team roles and relationships). Particular expertise may be required to determine the validity of the behavioral data collected from a given simulation environment (Narborough-Hall and Hopkin, 1988). Such simulations are seldom continuous; each exercise typically lasts for an hour or two, although simulations of longer duration are not uncommon. Questions that would require continuous running that would involve gross fluctuations in traffic demands or shift characteristics, for example, are rarely undertaken.

In the second approach, simulation experiments represent an integral part of the design process in order to provide iterative validation assessments of the proposed operational system (see also the section below on when the measurement process should occur). For advanced complex automated systems, Taylor and MacLeod (1994) advocate that progressive acceptance testing and evaluation, using real-time simulation, should be embodied throughout the different stages of system design and development.

The third kind of approach to real-time simulation is usually simpler than the first two approaches and often deals more exclusively with human factors. Typically it identifies the question first and simulates whatever is thought to be necessary within the resources and time scales available to answer it validly (Albright et al., 1994; Bussolari, 1991; Collins and Wayda, 1994; Mogford, 1994; Parsons, 1972; Stager and Paine, 1980; Stein, 1985, 1988, 1992; Stein and Garland, 1993; Vortac et al., 1994, 1992).

Traditional laboratory experimentation on a wide range of perceptual-motor and cognitive processes relevant to the air traffic control environment is usually required to evaluate and document new technology (e.g., display systems, input devices) (Stager and Paine, 1980; Vortac et al., 1993; Vortac and Manning, 1994). When a new technology is considered for air traffic control (e.g., electronic flight strips), researchers may employ quite a rudimentary air traffic control simulation to discover its potential strengths and weaknesses and to ascertain whether it is a serious candidate for application. The terminal airspace simulation facility (TASF) at the FAA Technical Center provides an example of such a facility to examine controller issues with the introduction of automation in both terminal and en route environments (Benel and Domino, 1993). If the question is simple, the simulation might be as well. Comprehensive simulation of every aspect of air traffic control is not sought; the aim is to impose sufficient control to disentangle the effects of the main variables and to ascertain the most sensitive measures.

Limitations

Simulation Fidelity

There has been a tendency in real-time simulation to equate fidelity with validity. Strenuous efforts may be made to replicate faithfully air traffic control workspaces, tasks, equipment, and communications so that even an informed visitor may not recognize at once that the air traffic control is not real. There is no adequate theoretical or empirical basis, however, for prescribing which aspects of air traffic control must be simulated with what fidelity in order to yield findings with a specified validity. In the absence of such guidance and in order to provide assurance that the desired level of predictive validity will be achieved, the practice has been to strive for similarity between the experimental simulation (i.e., subjects, equipment, tasks, and test environment) and the intended application (Chapanis, 1988). There is no established procedure for measuring similarity, however, or knowing how similar is similar enough (Chapanis, 1988; Druckman and Bjork, 1994; Meister, 1991).

A knowledge of human capabilities and limitations, and especially of principles of learning and the transfer of training (Singley and Anderson, 1989; Holding, 1987), often helps to provide an insight into which aspects of a system are crucial for the human tasks (and must be simulated faithfully) and which are not.

Most simulations impose restrictions on their usage, and the methods and measures employed must take account of this (Buckley et al., 1983). All participants in simulations receive specific instructions. Initiatives, nonstandard practices and short cuts, the development of professional norms and standards, and team and supervisory roles are typically curtailed in simulation. Many organizational, managerial, and scheduling features of air traffic control, its work-rest cycles, its working conditions, and the interactions between work and domestic life are absent altogether from simulated air traffic control.

If simulation exercises run too smoothly, the time absorbed in communications, coordination, liaison, delays, and difficulties in conducting dialogues or in reaching agreement is underrepresented. The roles of teamwork and supervision can be underestimated in evaluations if these elements have not been fully developed. In addition, real-time simulations tend to underplay individual differences between controllers as unwanted sources of variance in relation to the simulation objectives.

Practical Constraints on Data Collection

Experimenters frequently attempt to enhance the face validity of experimental simulation by using highly trained operators as subjects. Baker and Marshall (1988) suggest that experienced operators are probably more highly motivated to do well, but they will also tend to perform well on any reasonable system. As a result, it is difficult to find differences between alternative designs even though, in operational environments, it is the differences experienced by the less well trained (or less skilled) controller that are most likely to compromise safety (Wickens, 1995b).

In many human-machine experiments, the combination of shorter work periods and high motivation can lead to artificially high levels of operator performance or, simply, invalid estimates of human behavior in the planned system (Baker and Marshall, 1988). A restricted test population often requires elaborate repeated-measure designs, with their attendant problems of fatigue, and practice. The restriction to use highly trained operators may also provide an inadequate sample size for stable performance estimates.

One of the primary sources of difficulty for human-machine experiments is the restricted time scale under which most system development projects must operate. For this reason, Baker and Marshall have expressed the concern that experimental factors may be manipulated more from considerations of expediency than from validity and that the experiments provide an overly optimistic expectation concerning the input of factors related to shift work, fatigue, and boredom. There is an obvious need to look at longer experimental sessions, low activity periods, and transitions from inactivity to peak loads (e.g., Hancock, 1987; Huey and Wickens, 1993; Smolensky and Hitchcock, 1993).

In real-time simulations, it is often difficult to establish the validity of data concerning absolute capacities, workloads, strategies, and error rates because of the constraints outlined here. Specialized knowledge is usually required to evaluated their generalizability to actual operations.

FIELD STUDIES

Field investigations enable operational evaluations of elements of a system to be conducted on integral equipment that is being used on-line in the control process (Moody, 1991). In addition, field studies are likely to be undertaken when the interactions of a comprehensive set of variables can be observed dependably only in the actual operational environment (i.e., the limitations of real-time simulation are most evident). These studies represent a cost-effective method of obtaining human factors information provided that the constraints imposed on the data collection process do not invalidate the observations because of sampling restrictions (e.g., limited parameter variation and less experimental control of the relevant variables).

Uses of Field Studies

One of the advantages of collecting human factors information in a field setting rather than in real-time simulation is that the live operational environment provides a means of capturing the subtleties in operational practices and work habits that may not carry over into a simulation environment. Direct access to the operational personnel allows the discovery of unexpected feature use and an assessment of the extent to which a proposed tool or functionality will support the controllers. The ability to capture their experience with new technology, for example, is especially important for complex automation in which the implications of the interactions between system components are largely unknown prior to implementation (Harwood, 1994).

In field studies, the methodology followed for data collection can be quite varied from one study to another, depending on the objectives of each study. In some cases, unobtrusive measures are used to collect data to avoid contamination through the observation process itself, but the validity of the data are highly dependent on the sampling procedures. In their review of the air traffic control communications literature, Prinzo and Britton (1993) point to the advantages of audiotaped databases (e.g., objective, reliable, and verifiable real-time records) versus the ASRS database. The studies of air-ground communications reported by Cardosi (1993) and by Morrow et al. (1993) were based on off-line analyses of audiotaped samples. Similarly, the cross-validation of the type and frequency of communications errors observed in the bilingual simulations (Transport Canada, 1979) described by Stager et al. (1980a) involved the analysis of audiotaped communications from the actual operational sectors that were also being simulated in the laboratory.

V.D. Hopkin (personal communication, 1995b) has described a field study undertaken to determine the effect of lowering the altitude for the base of the holding patterns used at Heathrow airport. In order to evaluate the impact of the change on adjacent en route sectors, concurrent simulation of Heathrow and adjacent sectors would have been required—a requirement beyond the available simulation capability. Because the field trial involved live traffic, it could only be conducted when the controllers and their supervisors agreed to try it. One of the best and most sensitive measures proved to be of circumstances when they were willing to try it and other circumstances when they were not willing to go on trying it any longer. Some of the main findings concerned variables that would have been held constant in a simulation but could not be controlled live and therefore had to be measured as and when they occurred.

Harwood and Sanford (1994) and Scott (1996) describe recent field evaluations of an element of the center-TRACON automation system (CTAS) (Erzberger et al., 1993; Lee and Davis, 1995), undertaken at Denver International Airport. Harwood and Sanford suggest that early field testing during the development cycle can provide both an insight into how the system elements will function in the operational environment and an opportunity to capture and refine meaningful requirements for system certification. The development and evaluation of the CTAS was undertaken at two operational field sites, applying a field development and assessment process to one of the CTAS tools, the traffic management advisor (TMA). Context-sensitive data collection techniques (i.e., techniques based on observation and interpretation in the context of the user's work environment—Whiteside et al., 1988) were used in the evaluation.

Direct behavioral observations in field settings can also provide the data for a research method called the critical incident technique (Flanagan, 1954; Meister, 1985). Researchers have traditionally relied on the critical incident technique to determine measures of proficiency or the attributes relevant to successful performance in complex operations. With the increased emphasis on cognitive activities in operational environments, however, behavioral observation has become more difficult, if not irrelevant for understanding operator performance; Shattuck and Woods (1994) have cautioned that a new set of principles is needed. This is particularly the case in air traffic control, in which observable behaviors do not adequately reflect what the controller is doing at any given moment.

Limitations of Field Studies

In field studies, there are inherent constraints on variable or parameter control (e.g., traffic volume or complexity) that limit the conditions over which the fielded system can be evaluated. There may also be restricted on-line access to controllers. These constraints, however, have to be weighed against the advantages that have already been described and the fact that the limitations cited for real-time simulation are less likely to apply in the field environment.

COMBINING SOURCES OF HUMAN FACTORS DATA

Although each source of human factors data has been described independently of others, it is usually preferable that one source or research setting supplements another. For example, Seamster et al. (1993) have described their cognitive task analysis of expertise in air traffic control using a combination of methods, including simulation exercises, structured and unstructured interviews, critical incident interviews, paired paper problem solving, cognitive style assessment, structured problem solving, and simulated performance modeling. The designs that evolve through rapid prototyping are customarily validated in real-time simulation. The effects of certain constraints in real-time simulation can be evaluated only through systematic observation in supplemental field studies. Sarter and Woods (1995) provide an excellent example of the convergent use of real-time simulation, verbal protocols, and accident and incident analysis for understanding mode errors in flight management systems in transport aircraft.

One of the best examples of using different methodologies to expand the array of evidence on a human factors problem is the investigation of human error in aviation (Nagel, 1988; Wiener, 1980, 1987, 1989), and particularly in air traffic control (Danaher, 1980; Stager, 1991b). The description and analysis of the variables associated with human error tend to lie in the domain of descriptive models of human error and cognitive processes (Hollnagel, 1993a, 1993b; Rasmussen, 1987; Reason, 1990, 1993; Reason and Zapf, 1994; Senders and Moray, 1991; Stager, 1991b; Woods, 1989; Woods et al., 1994).

Nagel (1988) suggests that there are four approaches to gathering evidence about errors (already reviewed in this chapter): direct observation of the operational environment; incident analysis (Baker, 1993; Rodgers, 1993); use of the ASRS system (although the methodology of using self-reports need not be limited to formal aviation reporting systems and can involve diary studies within one or more control centers; Empson, 1991); and the observation of operator behavior in simulations. In spite of the inherent difficulty of acquiring low-frequency error data, real-time simulation probably represents the only potential source for such data and therefore the best means of validating models of human error. Even with the operational fidelity of real-time simulations, however, the validity of any error data can be brought into question by the assumptions that are made concerning how real-world failures occur (Hollnagel, 1993a; Maurino et al., 1995; Reason, 1990, 1993; Rubel, 1976). For this reason, observations in the simulation environment need to be seen as complementary to structured observations in the actual operating environment (Hollnagel, 1993a); in fact, observations in either context are influenced by information drawn from incident analysis and reporting systems.

MEASUREMENT IN COMPLEX SYSTEMS

All of the methodologies that have been described in this chapter raise issues of measurement, including the associated concepts of measurement validity and system validation. This is particularly true for real-time simulations and field studies, for example, when the studies are undertaken to evaluate proposed design changes. The investigation of human factors questions and design validation in air traffic control requires that three questions be addressed:

1. What is to be measured?
2. How should the measurement be done?
3. When should the measurement process occur?

All three aspects can affect the external validity (i.e., the generalizability) of evidence that has been gathered. Contemporary human engineering design (and ultimately system validation) is challenged by the requirement to accommodate and to predict the variance in human behavior in complex human-machine systems, in spite of the practical constraints that can be placed on studies of operator behavior (Stager, 1993).

Validation can easily be seen as a matter of measurement (Kantowitz, 1992) with the concomitant concerns of what one measures as criterion variables (Harwood, 1993) as well as how one measures the behavior of human-machine systems (Hollnagel, 1993b; Reason, 1993; Woods and Sarter, 1993). A distinguishing feature of many performance measures (Hopkin, 1979, 1980, 1982a) is that they are not simply direct measures of controller behavior and that the same measures are often taken, for example, to be indices of system safety.

What Is to Be Measured?

One of the issues in measurement might be called the criterion problem. Current engineering requirements, as outlined in MIL-H-46855B (U.S. Department of Defense, 1979), call for any contractor to establish and conduct a test and evaluation program to ensure fulfillment of the applicable requirements. Section 3.2.3 states that human engineering testing is to include the identification of criteria for acceptable performance of the test.

The criterion measures that are associated with the operational requirements and detailed in system specifications define only a part of the evaluation process that will be required in system development. The challenge for human engineering evaluation is created by the criteria that are left unspecified and/or are to be identified in the human engineering program (Harwood, 1993). Often the unspecified criteria will relate to error rates, workload, maintenance of situation awareness, and the potential for performance decrement associated with prolonged operational stress or fatigue. The issue of defining criterion measures for the human-system component in complex systems therefore impacts system safety and efficiency (Christensen, 1958; Harwood, 1993).

Criterion measures of performance are also required in order to evaluate the relative effectiveness of system design concepts. Only when there is a standard of required performance that is compared with actual performance can one talk about measures of effectiveness (Meister, 1991).

When there are insufficient grounds for extrapolation of performance standards from an existing system, human performance models and models of cognition may provide estimates for the level of performance that can be anticipated (Bainbridge, 1988; Pew and Baron, 1983; and Rouse and Cody, 1989).

Although there are human performance models for basic cognitive processes (Elkind et al., 1990) and for task allocation and workload analysis (McMillan et al., 1989, 1992), there is a fundamental requirement for the development of human performance models that are directly applicable to the air traffic control environment.

How Should the Measurement Be Done?

External validity and thus generalizability can be viewed as having three major components: representativeness of subjects; representativeness of variables; and representativeness of setting (i.e., ecological validity) (Kantowitz, 1992; Westrum, 1994). Test situations have to be representative of those encountered in the operational environment, and the functions provided by the interface have to be sufficiently complete if valid measures are to be obtained (Hollnagel, 1993a). The validation methods chosen have to be sufficiently sensitive to detect design errors (Woods and Sarter, 1993).

When Should the Measurement Process Occur?

An encouraging aspect of the trend toward including cost-effectiveness as a criterion in judging the efficacy of research is that it may force more critical consideration of the reliability and validity of air traffic control research and its outcomes (Westrum, 1993).

Validation (as iterative evaluation) has to be an integral part of system design rather than a "fig leaf" at the end of the process (Woods and Sarter, 1993). The objective of system evaluation should be to help the designer improve the system and not simply justify the resulting design (see Chapter 11). When system designers are properly seen as experimenters, the measurement process becomes a critical element in the design and interpretation of the converging evidence on system performance (Woods and Sarter, 1993).

CONCLUSIONS

A number of research methodologies are available to obtain human factors data, each with relative merits as well as constraints and limitations. Our review of human factors methodology has identified the following critical needs for air traffic control research:

1. Systematic efforts are needed to make access to the aviation safety reporting system database more user-friendly, both to encourage exploratory data analysis and to enable specific questions on human performance to be asked and answered more quickly. The constraints on the usability of the data contained in aviation reporting systems work against an early focus on potential safety areas within the air traffic control system.

2. Systematic work is needed to formalize the role and to enhance the contribution of rapid prototyping to the process of determining the characteristics of computer-human interaction. Careful application of current computer technology to the methods, standards, and objectives of rapid prototyping, particularly for multitask evaluations, could significantly advance the activity as a integral methodology for human factors research.

3. A cost-effective simulation capability is needed, within each system design program, that will support progressive acceptance testing. At present, there is a tendency to equate acceptability in the design-prototyping laboratory prematurely with task validation for the operational environment.

4. There is a need for the development and validation of human performance models applicable to air traffic control research, as well as for approaches to integrate human performance models with system models.

5. There is a need for universally recognized quantifiable dimensions of controller performance. Dependent variables that define controller performance across the spectrum of operational contexts that are sensitive to variation in determinants of performance (including, for example, the cognitive variables of workload and situation awareness) are needed for human factors research. In the absence of a commonly accepted set of measures, the articulation of the critical (but measurable) variables is likely to be undertaken anew in each project.

6. Additional human engineering standards and guidelines are needed for design validation (i.e., beyond the current MIL-H-46885B), that will be applicable to research undertaken to support system development.

11

Human Factors and System Development

There are three modes by which the human factor enters into the system development process. The first mode is based on existing data. As described in Chapter 10, research reports, textbooks and handbooks are the sources of such data. In many cases, the first mode involves the application of standard human factors principles from such sources when design decisions are being made. Since these principles incorporate the physiological and psychological capabilities of humans, system performance is not likely to be impaired by faulty design when this mode is properly carried out.

The second mode is called advanced applications. In this mode, judgments are based on psychophysiological theories, and findings from non-system-specific research are used to inform the design decisions. A specific example is the design of the garments worn by astronauts in extravehicular activities. The designs had to be completed before any actual experience could be accumulated. Consequently, the arrangement of environmental status controls was determined on the basis of experiences with other types of protective garments. The values realized from the employment of this mode can be substantial (Karat, 1992).

The third mode is more speculative. It comes into play when technological advances or changes in the operational situation are so large that conclusions about design options cannot be based on data or experience. The problem for the human factors expert in this mode is often that of a choice between extrapolating from what is already known or calling for focused research to resolve the uncertainty about the decisions to be made. Although the research option generates costs of time and funds, the answers to design questions are likely to be much better than those obtained by extrapolation—particularly if the research involves extensive transactions with users (Bikson, Law et al., 1995).

This chapter examines the possibilities for ensuring the safety and efficiency of the air traffic control system and related systems by the inclusion of human factors—using whatever mode is appropriate—during the system development process. The basic warrant for such an enterprise is that the current national policy stipulates that, no matter how sophisticated the air traffic control system becomes with respect to its inclusion of extensive computer capabilities, there will remain a human presence with operational decision making and management at its core (Federal Aviation Administration, 1995). Even if such a doctrine were not in effect, insofar as the system is operated to achieve human purposes and to serve human clients and customers, information about human capabilities, limitations, and preferences must be included in the decisions about the design and development of the system (*Air Traffic Management*, 1995).

The challenge of incorporating human factors in system development is particularly important because the air traffic control subsystems that will form the national airspace system are likely to incorporate increasingly advanced technology that will have the potential to do much more of the controller's job. In this chapter we highlight the positive advantages of early and sustained inclusion of human factors in the sequence of system development. We also attempt to characterize the barriers to such inclusion—and the means by which such barriers might be overcome. We provide a series of route markers that, if followed, should help ensure that the human factor is considered early in the system development sequence and that this aspect is sustained whenever design decisions are being made.

HISTORY, ORIENTATION, AND RATIONALE

Since the early months of World War II, the inclusion of human factors in system development has been at least partly assigned to people specially trained in psychology or physiology or both (see e.g., Fitts, 1951a, 1951b). However, professional engineers who typically were concerned with other technical issues were most often the people given the primary responsibilities for system development. As larger and larger programs of system development were launched, it became evident that success was more likely if the program participants from various disciplines worked as a team (Chapanis, 1960). This observation, in turn, led program managers to designate specific roles for team members. One consequence was the emergence of systems engineering as a special field that incorporated particular responsibilities for system integration and program management. In short, systems engineers became the formal leaders of the team. Simultaneously, the human factors role also became more clearly defined as a distinct professional specialty.

The teamwork approach has been effective when the system engineer, the human factors specialist, and other members of the team have shared a set of objectives and have sought to produce a system that performs its functions at the highest levels of effectiveness—at the least aggregate cost (Goode and Machol, 1957).

In 1974, Singleton proposed the concept of human-centered design in recognition of the fact that most of the variability in system performance derived from the responses of the human operator (Singleton, 1974). The concept has been elaborated in the intervening years. It is the human who becomes fatigued. It is the human who makes errors if the displayed information is ambiguous or incomplete. Humans can become distracted or frustrated with features of the work setting; machines show no such tendencies. Insofar as the human user is a likely source of significant variance in system performance, it is better to control this source of variance early. Failure to do so will almost always result in a system that is inconsistent in its operation and less than fully dependable.

Human-centered or user-centered design is partly a reaction against conditions in the recent past in which the technology drove the constraints on design options. It is easy to forget that early computer utilities demanded that the user be adept at programming—or that users employ the services of an intermediary. It was only when the technology became more versatile that it was possible to make advanced computer-based systems that readily accommodated human characteristics. The present emphasis on usability is a welcome shift—away from the exclusivity of the computer technicians and in the direction of responsiveness to a broader user population (Lingaard, 1994).

Likewise, it should be noted that, in systems that rely on computers, the software engineers may not be able to anticipate all event contingencies. The human operator is usually expected to compensate for such gaps. Consequently, it is usually a good idea to use human factors concepts to direct the design in ways that will facilitate operator flexibility.

The implementation of the concept of human- or user-centered design—or its friendly competitor, "total systems design"—requires designers to have as much knowledge about human physiology and psychology as possible (Bailey, 1982). It is this knowledge base that the human factors specialist is responsible for bringing to the design team. Ideally, the human factors specialist would not only be a carrier of such knowledge but also would share it—in an educational sense—with the other team members.

In any case, when existing human factors knowledge is applied, more effective designs can result. An example of the successful application of human factors principles outside the field of aviation can be found in the design of agricultural equipment such as tractors. For many years, the designers ignored operator comfort. Then human factors specialists proposed new seating designs and moved the whole industry toward the adoption of enclosed cabs in which noise, vibration effects, dust, and temperature could be controlled for the improvement of total system performance—not to mention user acceptance (e.g., Hornick, 1962; Woodson and Conover, 1964; Sahal, 1975).

In a second example, human factors research has made major contributions to the effectiveness of maintenance work. Specifically, the utility of various formats for maintenance manuals was compared, and the results pointed to fully proceduralized job aids as the better way to support the maintenance worker. The use of this step-by-step approach, combined with clearly intelligible graphics showing such acts as tool positioning, made large differences in performance— even when the workers were given only modest training in working on the test systems (Duffy et al., 1987).

In air traffic control systems, human factors studies done in the 1950s revealed that the use of radar greatly improved the capabilities of the human-machine system. When radar was augmented by target identification, and other information such as altitude was presented adjacent to the radar target, performance improved even more (Schipper et al., 1957). It might be argued that these innovations would have come about without the attention given by human factors researchers. However, the history of developments in air traffic control, particularly the resistance of operators to the first use of radar in civilian air traffic situations, suggests that progress in these ways would have been far slower if the human factors research had not been done and the findings had not been widely distributed throughout the technical communities responsible for air traffic control system development.

FORMAL ARRANGEMENTS FOR INCORPORATING HUMAN FACTORS

The classic method of imposing design standards on the system development and procurement processes is the issuance of formal specifications. Such specifications were put in place by the armed services in the early 1950s and have been updated since. Current human factors standards for design of military equipment are incorporated in U.S. Army TM 21-62, U.S. Navy MIL-H-22174(AER), and U.S. Air Force MIL-STD-1472 and MIL-H-27894. However, these dicta have tended to generate a ritualistic response from systems contractors, who were not interested in spending more than they had to on what some designers still considered to be a frill. To counteract such resistance, the U.S. Air Force attempted to strengthen the inclusion of human factors in the early stages of system design by means of a standard operating procedure labeled "Qualitative and Quantitative Personnel Requirements Information." In general, the idea was to give an institutional role to human factors people on the government's side of the systems acquisition process. If an Air Force human factors specialist was present to impose design requirements on contractors from the outset of the design/development activity, then integration would take place in spite of the contractor's lack of enthusiasm (Demaree and Marks, 1962; Eckstrand et al., 1962). This effort was pursued with some vigor so long as top Air Force officers endorsed it, but it withered when they retired.

In the late 1970s, a joint Army-Navy-Air Force project was initiated at the urging of high civilian officials in the Department of Defense. The lead agency was the U.S. Army Research Institute for the Behavioral and Social Sciences. The conceptual goal was to show key agency people, such as the top technologists at the U.S. Army's Training and Doctrine Command (TRADOC), that their efforts at developing superior weapon systems would be furthered if they embraced human factors and included it at the earliest stages of the development process (i.e., when requirements from the field are interpreted in the form of new mission analyses by TRADOC staffers). The avowed goal was to be achieved by the preparation and publication of a how-to manual for the military systems engineering community (Price et al., 1980). This effort was the precursor of the Office of Management and Budget's Circular No. A-109, which is now the main source document for procurement procedures related to the inclusion of human factors for the Air Force, the Navy, and the Federal Aviation Administration (see Order 1810.1F, *Acquisition Policy*, DOT/FAA, March 19, 1993.).

Meanwhile, in the late 1980s, high-level Army officers again became dissatisfied with the degree to which human factors were included in the development of their systems. New reforms were initiated. The focus this time included the Army Materiel Command as well as the Training and Doctrine Command. The ambitions were loftier as well—including the idea that human factors might become a driving force in technological innovation—not merely a means of guiding a development process driven by advances in technology. The program that emerged was called MANPRINT (Booher, 1990). The concept is still in effect in the U.S. Army procurement activities and has been copied abroad (by the British military) but has not yet been adopted by the Air Force, the Navy, or the FAA (U.S. Department of the Army, 1994).

From the issuance of specifications and monitoring schemes that were to be imposed on systems contractors, the goal of MANPRINT was that someone who was qualified by particular educational accomplishments would examine the proposed configuration of a system under development at specified stages in the development process to make sure that no violations of human factors principles were taking place.

The presence of these specifications also served to ensure that the human factors profession would mature in particular ways. For example, all system contractors of even moderate size were expected to employ a human factors specialist—either as a full-time employee or as a consultant. Academic programs were established to meet the demand for qualified people and, within such academic programs, research projects were promulgated. The results of this research, along with the like products of government laboratories, enlarged the base of established principles that contractors were required to follow. Thus, the field of human factors became a self-amplifying discipline. It also became institutionalized under various names (e.g., ergonomics, human engineering) throughout the industrialized world.

Despite of the increasing stability of the field, however, its impact on system development programs has remained modest. Although the managers of system development programs and human factors specialists share crucial goals, there is also some major divergence of viewpoints. For example, engineers tend to be rewarded for being inventive and often seek to incorporate the most recent technology in their plans, whether it is best for the functions of the system or not (Carrigan and Kaufman, 1966; Kidd, 1990). Human factors specialists can appreciate the potential advantages of advanced technologies but often worry about the degree to which the form of a given technology is compatible with the general capabilities and limitations of prospective human operators (Fanwick, 1967).

Those representing the human factors viewpoint will sometimes assert that a particular design decision should not be made until more information on the question of human compatibility is available. If the information must come from research that takes some time to complete, the engineers tend to see in this scenario the prospect of missed deadlines and cost overruns. All this is not to condemn project managers' attitudes toward human factors. They tend to avoid *all* inputs that might add uncertainty to design decision making or that might add to their investment in the information-gathering phase in such deliberations (Rouse and Cody, 1988). There is a tendency on the part of project managers to minimize human factors in system development projects. Such a tendency is probably aggravated by a lack of appreciation of technical issues on the part of some human factors specialists. Some system design efforts incorporate high levels of human factors participation, and some do not. This variability in the amount and quality of the human factors contribution is a problem for those who have high-level administrative responsibilities for system acquisition activities. Since most senior officials in the agencies that initiate and fund system development efforts are aware that, when human factors are ignored, serious system failures can be the consequence (Busey, 1991; Del Balzo, 1995), administrative rules and standard procedures have been promulgated over the years since World War II as means to ensure that the human factor would indeed come into the deliberations on system development whenever such inputs were needed. The issue that remains is that these arrangements can serve as a means to give the superficial appearance of incorporating human factors when the real impact is minimal. Even when MANPRINT is nominally adhered to by contractors, the process can appear to be more of a ritual than a rigorous procedure (Government and Systems Technology Group, 1995).

UNDERTAKINGS WITH RESPECT TO AIR TRAFFIC CONTROL

As suggested above, the FAA generally follows the Department of Defense protocols on human factors. These practices tend to distribute responsibility between the FAA and its contractors. The situation is somewhat complicated for air traffic control because of the ways in which its operations are managed, the relative heterogeneity of its operational settings, and the distribution of research, development, test, and evaluation activities among internal laboratories (i.e., the Civil Aeromedical Institute, the Atlantic City Technical Center), Transportation Department facilities (i.e., the Volpe Center), other government agencies such as NASA and the Air Force, nonprofit organizations (e.g., MITRE and certain academic centers), and commercial contractors. The need for better coordination among these organizations is discussed in Chapter 8.

Currently, two forces for conceptual integration are provided by the FAA headquarters units called the Directorate of Air Traffic Plans and Requirements and the Research and Acquisitions organization. These offices oversee the full array of FAA system development programs. The FAA is also in the process of reviewing the question of how best to structurally organize the human factors effort within the agency (see Chapter 8). The goal is to achieve a better level of mutual support between in-house research, contracted research and development, and system installation and operations.

The Acquisition of Automated Systems

Further formal procedures for the inclusion of human factors in system development can be found in FAA Order 1810.1F, which describes the FAA's acquisition policy. The policy document stipulates the following as critical steps:

- Performance of the mission need analysis and preparation of the mission need statement, which describes the required operational capability of the new system and explains the deficiencies that the new system will rectify.
- Analysis of the trade-offs among alternative concepts and preparation of an operational requirements document to define the system-level functional and performance objectives of the new system.
- Preparation of a maintenance requirements document (FAA Notice 6000.162).
- Preparation of the system-level specification that defines in detail the system-level functional and performance requirements of the new system.
- Preparation of the request for proposal that defines for potential bidders the work requested by the FAA toward the design, development, fabrication, and delivery of the new system.
- Evaluation of bidders' proposed designs.
- Evaluation of the selected bidders' designs at preliminary design reviews, critical design reviews, and any other program-specific design reviews.
- Testing of the design at various levels (e.g., developmental tests, operational tests, acceptance tests, and field shakedowns).

These steps represent the FAA's interpretation of the standard sequence for systems development projects. They scarcely convey the true difficulties of doing real system design. For example, it is extremely rare that a new system does not have a predecessor. The characteristics of the predecessor constrain what the new system can be. This is clearly exemplified by the fact that the current air routes were laid over the routes developed when pilots flew at night from one searchlight beacon to another. System developers must find ways to fit the most exotic new technologies into frameworks laid down by traditions—the very roots of which may now be entirely forgotten. Furthermore, the designers must cope with the fact that the system objectives can reflect competing values. In air traffic control, this is clearly the case when attempts are made to reconcile the desire for denser traffic flows with the overriding goal of safety.

In addition to these steps, which reflect a systems approach that is fully congruent with human factors participation, there are some even more specific provisos that reflect continuing concerns on the part of top-level administrators. Thus, FAA Order 1810.1F stipulates that:

- Human factors shall be applied to the development and acquisition of national airspace systems software, equipment, and facilities.
- Human factors engineering shall be integrated with the system engineering and development effort throughout the acquisition process, including requirements analysis, system analysis, task analysis, system design, equipment and facilities design, testing, and reporting.
- An initial human factors plan shall be developed prior to the finalization of the system level specification.

These instructions represent a long-standing set of practices and are the relatively routinized part of the human factors contribution. However, even such well-established guidelines have not been adequate to ensure that human factors have been fully incorporated in some design processes. A prime example of a missed opportunity is provided by the actual configuration of equipment in the typical Airway Facilities maintenance control center, discussed in Chapters 4 and 9. Such centers are an assembly of disparate workstations that do not exhibit an integrated human-computer interface. Although the FAA-MD-793A specification (Federal Aviation Administration, 1994) represents an attempt to require that all new systems provide data in standard formats to the remote monitoring system that feeds into a center, idiosyncratic designs are still being generated. There appears to be a need for an overall maintenance center automation strategy as a baseline for evaluating proposed designs. These same needs apply to the design of tools that support other Airway Facilities activities, such as off-line diagnosis of equipment, maintenance logging, and maintenance of software (Simms, 1993).

Limitations in the Formal Order Procedure

Although FAA Order 1810.1F recommends the use of military standard MIL-H-46855 as a guide for the human factors plan, the plan may be tailored by the systems contractor to the scope and level of specificity judged appropriate to the new system's complexity and its major attributes. The acquisition policy also prescribes involvement by both human factors and operational representatives (users). The policy implicitly relies on the author of the human factors plan to identify methods for addressing the issues especially pertinent to automated systems (the policy itself does not identify these issues and does not address special concerns for automated systems). Other issues not included in the statement of acquisitions policy are the distinctions between system acquisition procedures at the national level and the acquisition support activities in the various regions. Each region has its own culture, and each has some unique features, such as the configuration of airways and the locations of terminals within the region's geographic boundaries. New subsystem acquisitions must be designed so that they can be adapted to all these distinctive settings.

Proposed Reforms

Recent additional modifications of the FAA's acquisitions procedures include the establishment of integrated product teams (IPTs) to smooth the discrepancies between the perceptions of FAA officials, suppliers, and users—and to generally expedite the actual fielding of new subsystems. The IPT concept follows the line-of-business approach to acquisitions that is provided to establish clearer assignment of decision-making authority and responsibility.

The variations in culture among the regions have not been clearly designated as one of the problems to be solved by the IPTs, but this is an area in which these new organizational units could make useful contributions to the total modernization process. Such contributions would be facilitated by a provision for the full- or part-time presence of a fully qualified human factors specialist on each IPT. This practice might recapture the advantages that characterized the QQPRI (qualitative and quantitative personnel resources inventory) approach and that are now sought in the MANPRINT procedures. It would give the FAA a means of enforcing the protocols regarding human factors contained in Order 1810.1F with regard to contractors' adherence to human factors principles.

A second reform being sought is an increase in the utilization of equipment for which no special development effort is required—nondevelopmental items and commercial off-the-shelf equipment. The advantages in time and dollar costs are self-evident. However, there can be hidden costs with respect to the human factors question. Simply depending on the manufacturer to have carried out adequate human factors engineering—in the initial interest of furthering commercial market appeal—is not prudent. The point is that each item to be procured by a straight commercial transaction should be subjected to a strict human factors review, in much the same mode as would be used on a system under contractual development. A good analogy is that of consumer protection in the conventional, competitive marketplace.

THE IMPLEMENTATION OF INNOVATIONS

Acceptance by System Operators

The historical pattern of technological development of many advanced systems reveals a number of pitfalls on the road to modernization. One such pitfall is the lack of user acceptance. New systems might represent real technical advances, but they will serve no good purpose if their use is forestalled by their unacceptability to operational personnel. For example, there was strong initial resistance to the use of radar by the FAA—despite the favorable operational examples provided by the Air Force use of this technology in directly parallel operations such as the Berlin airlift (Fitts et al., 1958). Likewise, the airborne automatic conflict warning system (TCAS, discussed Chapter 12) is receiving a negative reaction from some controllers. This recent innovation has great significance because it can directly influence the relationship between the controllers and the pilots. It has been noted that pilots have changed course independently when under positive control. When the automatic warning system was the source of steering instructions, the pilots sometimes have delayed reporting to the controller what is happening (U.S. House of Representatives, 1993).

A positive instance of user acceptance is provided by the success story of one significant subsystem's development for the national airspace system. The example is a computer-based aid for the proper spacing of aircraft on converging approach paths used in landing at a terminal with convergent active runways. Significant improvements in traffic flow rates at night or during adverse weather conditions have been achieved in field tests of this subsystem. Moreover, because it was presented to users in advanced prototype form in a laboratory setting and then field tested at many TRACONs in different regions, it seems likely that the subsystem would generate little resistance if it were to be installed at all the appropriate terminals across the country (for details see—Mundra, 1989; Mundra and Levin, 1990).

There is a substantial body of knowledge on the problems of innovation acceptance and adaptation (see e.g., Rogers, 1983). Acceptance is largely dependent on the user's subjective cost-benefit calculations. However, several other factors are also important. Studies of workers' reactions to changes in work procedures date from the research of Alex Bavelas on the responses of the workers in a factory in Marion, Virginia, to different levels of involvement in the decision to implement new cloth-cutting equipment (Lewin, 1947). This work was followed by specific verifications (Coch and French, 1948) and by major philosophical and practical guidebooks (McGregor, 1960; Bennis et al., 1976). These works and others make it quite clear that adoption of an innovation is strongly influenced by the collective responses of the rank-and-file workers.

When relatively cohesive work groups are brought into the decision-making processes related to distinctive modifications of work procedures—well in advance of the implementation of the change—a positive work climate is maintained and productivity is good. If the workers are excluded from the decision processes and simply told, arbitrarily, that a change is to take place—even when the reasons for the change are explained thoroughly—negative reactions of many kinds, including production suppression, can be expected. This does not mean that managers or systems engineers need to abandon their prerogatives or that workers or groups of workers need to be involved in every decision made by management in an organization. Vroom has shown in repeated studies that workers fixate only on functions that affect their daily tasks and their status in the workplace (Vroom and Jago, 1987). For example, workers are perfectly willing to forego consultation on company investment strategies. They apparently concede that financial experts in the management cadre should make such decisions—not lay people at the worker level.

Acceptance at Different Organizational Levels

Operational personnel are not the only ones who can accept or reject innovations. Resistance to change has been experienced at all levels. For example, despite the presence of a union representative in the office that provides key oversight of all systems development decisions, it is evident that some union leaders still feel themselves to be excluded from the modernization deliberations (Thornton, 1993).

Supervisors, managers, and administrators must also provide support if the adoption of an innovation is to be successful. One approach to dealing with this prospective problem is to try to assess the economic, political, and psychological issues in addition to the technological factors that are driving the change. When all these conditions are identified and understood, managers can assess the specific requirements for organizational adaptations and move to correct any gaps in the support functions.

New Approaches

When the principles of innovation promotion and the integration of human factors in system development are put together, the conclusion that emerges is that extensive user participation is a key to success in both sets of activities. In fact, if there is very extensive user participation throughout the design and development process, there is a strong possibility that the resultant system will yield superior performance and will also be more readily acceptable by its prospective adopters.

The main barriers to the use of user participation are logistical and economic. That is, it takes time and money to provide large numbers of users with exposure to prospective design features, and it can be very expensive to create conditions that lead to incremental changes in design. Known as "requirements creep," such changes are a notorious problem for the managers of large-scale system development efforts. There are also difficulties in the area of logical rigor, in the sense that it is difficult to obtain objective, quantitative conclusions from what can be a series of informal preference statements from prospective users.

Some of these difficulties can be overcome by techniques invented by researchers in the fields of marketing and advertising psychology. Examples include the use of opinion surveys and focus groups for product assessments prior to mass production. During the development of the first versions of the Apple Computer, Steve Wozniak, who was responsible for the technical development of the product, initiated a focus group by recruiting high school students who had formed a "hackers" club in San Jose, California. New system features were reviewed informally by club members as they were instantiated by Wozniak in his garage workshop. Club members vigorously debated the pros and cons of the design and then arrived at a reasonable consensus about the attractiveness of each feature. Apparently, Wozniak never—or very rarely—went against the collective judgments of the club members. The remarkable market penetration of the early versions of the Apple Computer lends credence to the proposition that user participation should guide development and that, when it does, an appealing product is the likely outcome (Byte, 1984, 1985).

Similar successes have been attained by air traffic control subsystem developers who have used various methods of simulation to give users an opportunity to experience the operational features of a particular configuration while major modifications could still be made at low cost (see, for example, Lee, 1994; Erzberger et al., 1993). This procedure can give the designer confidence about the design approach being used, and it also appears to serve as a way of avoiding major design errors by giving the programmers useful clues about what features are attractive (Baldwin and Chung, 1995). For an earlier exposition of similar concepts, see Sinaiko and Belden (1961). They show that important information can be obtained to guide the design engineer without resorting to formal experimentation.

The techniques based on focus groups have also been used to solve some problems related to air traffic control that do not involve advanced technologies to any great degree. Specifically, a team of researchers from two universities and NASA Ames convened a focus group made up of airline dispatchers, the coordinators who negotiate with FAA intermediaries at the Air Traffic Control System Command Center and traffic managers from en route control centers. There are many causes of friction between these entities, and the focus group was empaneled to determine if there were some solid common ground on which improved coordination procedures might be built. The process was successful with the proviso that some continuing follow-through would be needed (McCoy et al., 1995).

Exploitation of the computer as an aid in system design in conjunction with user participation has been tried in Sweden (Akselsson et al., cited in Karwowski and Rahimi, 1990). These investigators used a local area network to link a group of 15 stakeholders in a project to redesign the materials flow and workspace layout in a factory. The network facilitated communication, and problem solving was enhanced. (However, there were some real stresses within the group associated with the lack of uniform levels of skill in the use of the computer as a communication tool.)

On an even more positive note, rapid prototyping, discussed in the previous chapter, has been shown to be successful in certain areas of system development. Rapid prototyping can be regarded as a sophisticated version of basic trial-and-error methods of problem solving (Connell and Shafer, 1995). Another way of describing the process within the system development sequence is the test-and-adjust approach. Human factors specialists were not long in adopting the structured procedures of rapid prototyping for their own purposes. In fact, Gould and Lewis (1983) found themselves admonishing programmers to make sure that they, the programmers, actually follow some form of rapid prototyping in every major software development project.

Variations on the theme of rapid prototyping are exemplified by work on development and installation of a new information system in a manufacturing firm (Mankin et al., 1996). The procedure, called mutual design and implementation, is intended to facilitate user acceptance of computer-based systems; it incorporates the techniques of user participation into a full-scale organizational adaptation program. The basic message is that a technological innovation must be correctly perceived as being congruent with both the overall objectives of the organization and the individual objectives of the users of the technology. The approach moves away from traditional bureaucratic norms but, as such, is not incompatible with the streamlined FAA acquisition process (Donahue, 1996). The approach also recaptures the idea of organizational dynamics found in the early Air Force studies of human-machine systems (Porter, 1964).

If, in the instance of developing a crucial subsystem, costs are seen as a strongly limiting boundary to extensive user participation, compromise protocols do exist. Among the most attractive is the sequential experimentation protocol laid out by Williges and colleagues (1993). Their approach is a variant of the successive approximation strategy in system design. The scheme works in steps toward the initiation of a large-scale, multivariate, factorial experiment—the results of which should resolve the total configuration issue near the end of the design process. Rapid prototyping is restricted to the second step of an 11-step sequence. Its role in this scheme is simply to screen out extremely unacceptable design options after brainstorming and other inclusive techniques have generated a large set of candidates. The next sequence of steps are relatively small-scale experiments that test the surviving design options separately or in limited factorial combinations. At about step 8, the relative strengths of single variables in driving critical performance measures will be known, gaps in the database will be evident, and clues will be present about the presence of significant interactions among the independent variables. The crucial multivariate experiment (or experiments) is to be conducted at step 9. Williges et al. recommend that step 10 be devoted to the construction of an abstract model of the relationship between the tested design options and performance and that step 11 consist of the actual promulgation of the newly determined optimal design configuration. This methodology represents a major commitment to a mode 3 approach. It also provides the means for engaging users early in the design process.

The engagement of a relatively large cross section of operational personnel would yield the added advantage of providing a core of individuals in each FAA region who could become influential in the acceptance of the system upon its installation. The very fact that colleagues had their say in the final configuration of the system would be a positive factor among the rest of the workforce. Moreover, those operators recruited for evaluation episodes would have acquired skill in the use of the system and consequently would be able to help explain the procedures and train their coworkers in the actual employment of the new tool.

The final and perhaps the most important advantage to the gradual approach represented by prototyping as a tool in the development sequence for air traffic control modernization is the prospect that the installed system could vary in some modest ways from region to region and even from site to site. The specifications for such variants would involve the determination of local requirements and the actual inclusion of the variant features in each subsystem when it is installed. Such steps would take some time but are commensurate with a test-and-adjust approach to system development.

CONCLUSIONS

In general, system developers adopt a top-down approach to the design process. The crucial move in this approach is the delineation of functional requirements. The assumption is that, if these functional requirements are fulfilled within the constraints of the technology and time and money expenditure, the system is thereby successful. Traditional human factors contributions have been shaped to fit into this approach. For example, user task analyses are intended to contribute greatly to the specification of functional requirements. Also, since the top-down approach can be neatly divided into discrete steps, it is easy to stipulate at what points design review should take place in order to ensure that no human factors principles have been violated.

More recent experiences with systems designed exclusively by top-down procedures indicate that, in the more complex systems in which cognitive behaviors and strong affective elements come into play through the human user/operator, serious deficiencies can become apparent after the system is delivered and is put to work. One means to avoid such deficiencies involves the early fabrication of prototypes and their evaluation by user groups. Prototypes that could be regarded as virtual representations of an operational system can also serve these purposes. When the end user is so engaged in the design process, it becomes bottom-up (e.g., problem-driven or scenario-based design) rather than top-down. Integration of top-down and bottom-up procedures will probably be needed to achieve optimal cost-effectiveness for the national airspace system of the future.

In addition to bridging the barrier that is likely to exist between ordinary users and design engineers, human factors specialists bring to design deliberations knowledge about human capabilities and limitations that has been acquired by rigorous scientific research. The human factors specialist is also expert in identifying and seeing the implications of subjective user attitudes, opinions, and tastes.

The integrated project team for every major subsystem in the advanced automation system should contain at least one full-time human factors specialist who would have the authority and responsibility to ensure that (a) user participation is timely and extensive, (b) human capabilities, limitations, and values are considered as part of every design decision, and (c) gaps in the knowledge base that could compromise the quality of the resultant system are identified and rectified by appropriately rigorous research.

It is important to consider specific interests of the government as system procurer in the allocation of human factors resources. That is, the final design of a major subsystem should not be the exclusive prerogative of a contractor. A way should be found to ensure that human factors/human engineering is not slighted by contractors as an arbitrary cost-saving ploy. Authoritative oversight is essential in this matter.

Systems that enjoy intensive and extensive user participation in their development are generally more likely to be more usable, effective, and acceptable than systems that are thrust on users after development has been completed.

However, user participation can be expensive and time-consuming—and can lead as easily to ambiguity as to clarity with respect to the choice of design options if good care is not exercised. In particular, users' perceptions can change while the development process is still under way, and user demands can expand over time. The resultant "requirements creep" can seriously disrupt the procurement process. There is a variety of strategies for minimizing the costs, delays, and ambiguities that can come from extensive user participation. Such strategies look at simplification of the procedures of rapid prototyping and the limitation of the use of such procedures to stipulated stages in the system development sequence.

12

Automation

Automation technology for air traffic control has steadily advanced in complexity and sophistication over time. Techniques for measurement and control, failure detection and diagnosis, display technology, weather prediction, data and voice communication, multitrajectory optimization, and expert systems have all steadily improved. These technological advances have made realistic the prospect of revolutionary changes in the quality of data and the aids available to the air traffic controller. The most ambitious of this new generation of automated tools will assist and could replace the controller's decision-making and planning activities.

Although these technological developments have been impressive, there is also little doubt that automation is far from being able to do the whole job of air traffic control, especially to detect when the system itself is failing and what to do in the case of such failure. The technologies themselves are limited in their capabilities, in part because the underlying models of the decision-making processes are oversimplified. As we have noted, it is unlikely that technical components of any complex system can be developed in such a way as to ensure that the system, including both hardware and software components, will never fail. The human is seen as an important element in the system for this reason to monitor the automation, to act as supervisory controller over the subordinate subsystems, and to be able to step in when the automation fails. Humans are thought to be more flexible, adaptable, and creative than automation and thus better able to respond to changing or unpredictable circumstances. Given that no automation technology (or its human designer) can foresee all possibilities in a complex environment, the human operator's experience and judgment will be needed to cope with such conditions.

The implementation of automation in complex human-machine systems can follow a number of design philosophies. One that has received considerable recent interest in the human factors community is the concept of *human-centered automation*. As we mentioned earlier in this report, human-centered automation is defined as "automation designed to work cooperatively with human operators in pursuit of stated objectives" (Billings, 1991:7). This design approach is discussed in more detail toward the end of this chapter.

Over a decade of human-factors research on cockpit automation has shown that automation can have subtle and sometimes unanticipated effects on human performance (Wiener, 1988). In a recent report (Federal Aviation Administration, 1996a) the impact of cockpit automation on flight deck crew performance has been documented in some detail. Similar effects have been noted in other domains in which advanced automation has been introduced, including medical systems, ground transportation, process control, and maritime systems (Parasuraman and Mouloua, 1996). Understanding these effects is important for ensuring the successful implementation of new forms of automation, although not all such influences of automation on human performance will apply to the air traffic control environment. The nature of the relationships between controllers and ground control systems on one hand, and pilots and aircraft systems on the other, will also change in as yet unknown ways. For example, at one extreme, control of the flight path and maintenance of separation could be achieved by automated systems on the ground, data-linked to flight deck computers. At the other extreme, as in some of the concepts involved in free flight, all responsibility for maintaining separation could rest with the pilot and on-board traffic display and collision avoidance systems (Planzer, 1995). Whether or not the most advanced automated tools are implemented, however, it is likely that the nature of the controller's tasks will change dramatically. At the same time, future air traffic control will require much greater levels of communication and integration between ground and airborne resources.

In this chapter, we focus on four aspects of automation in air traffic control. We first describe the different forms and levels of automation that can be implemented in human-machine systems in general. Second, we describe the functional characteristics of several examples of air traffic control automation, covering past, current, and new systems slated for implementation in the immediate future. We do not discuss automation concepts that are still in the research and development stage, such as free flight, or the national route program, which will provide indications of air traffic control capabilities and requirements relevant for free flight considerations. Third, we discuss a variety of important human factors issues related to automation in general, with a view to drawing implications for air traffic control (see also Hopkin, 1995). Recent empirical investigations and human factors analyses of automation have been predicated on the view that general principles of human operator interaction with automation apply across domains of application (Mouloua and Parasuraman, 1994; Parasuraman and Mouloua, 1996). Thus, many of the ways in which automation changes the nature of human work patterns may apply to air traffic control as well. At the same time, there may also be some characteristics of human interaction with automation that are specific to air traffic control. This caveat should be kept in mind because most of what has been learned in this area has come from studies of cockpit automation and, to a lesser extent, automation in manufacturing and process control. Finally, we discuss the attributes of human-centered automation as they apply to air traffic control.

FORMS OF AUTOMATION

The term *automation* has been so widely used as to have taken on a variety of meanings. Several authors have discussed the concept of automation and tried to define its essence (Billings, 1991, 1996; Edwards, 1977; Sheridan, 1980, 1992, 1996; Wiener and Curry, 1980). The American Heritage Dictionary (1976) definition is quite general and therefore not very illuminating: "automatic operation or control of a process, equipment, or a system." Other definitions of automation as applied to human-machine systems are quite diverse, ranging from, at one extreme, a tendency to consider any technology addition as automation, and, at the other extreme, to include only devices incorporating "intelligent" expert systems with some autonomous decision-making capability as automation.

Definition

A middle ground between the two extreme views of automation would be to define automation as: *a device or system that accomplishes (partially or fully) a function that was previously carried out (partially or fully) by a human operator.*

Because this definition emphasizes a change in the control of a function from a human to a machine (as opposed to the machine control of a function never before carried out by humans), what is considered automation will change over time with technological development and with human usage. Once a function is allocated to a machine in totality, then after a period of time the function will tend to be seen simply as a machine operation, not as automation. The reallocation of function is permanent. According to this reasoning, the electric starter motor of an automobile, which serves the function of turning over the engine, is no longer considered automation, although in the era when this function was carried out manually with a crank (or when both options existed), it would have been so characterized. Other examples of devices that do not meet this definition of automation are the automatic elevator and the fly-by-wire flight controls on many modern aircraft. By the same token, cruise controls in automobiles and autopilots in aircraft represent current automation. In air traffic control, electronic flight strips represent a first step toward automation, whereas such decision-aiding automation as the final approach spacing tool (FAST)(Erzberger, 1992) represents higher levels of automation that could be implemented in the future. Similar examples can be found in other systems in aviation, process control, manufacturing, and other domains.

From a control engineering perspective, automation can be categorized into (and historically automation has progressed through) several forms:

- *Open-loop mechanical or electronic control.* This was the only automation at first, as epitomized by the elegant water pumping and clockworks of the Middle Ages: gravity or spring motors driving gears and cams to perform continuous or repetitive tasks. Positioning, forcing, and timing were dictated by the mechanism and whatever environmental disturbances (friction, wind, etc.) that happened to be present. The automation of factories in the early parts of the Industrial Revolution was also of this form. In this form of automation, there is no self-correction by the task variables themselves. Much automation remains open loop, and precision mechanical parts or electronic timing circuits ensure sufficient constancy.
- *Classic linear feedback control.* In this form of automation, the difference between a reference setting of the desired output and a measurement of the actual output is used to drive the system into conformance. The flyball governor on the steam engine was probably the first such device. The gun-aiming servomechanisms of World War II enlarged the scope of such automation tremendously. What engineers call conventional proportional-integral-derivative (PID) control also falls into this category.
- *Optimal control.* In this type of control, a computer-based model of the controlled process is driven by the same control input as that used to control the actual process. The model output is used to predict system state and thence to determine the next control input. The measured discrepancy between model and actual output is then used to refine the model. This "Kalman filtering" approach to estimating (observing) the system state determines the best control input, under conditions of noisy state measurement and time delay, "best" being defined in terms of a specified trade-off between control error, resources used, and other key variables. Such control is inherently more complex than PID control but, when computer resources are available, it has been widely adopted.
- *Adaptive control.* This is catchall term for a variety of techniques in which the structure of the controller is changed depending on circumstances. This category includes the use of rule-based controllers (either "crisp" or "fuzzy" rules or some combination), neural nets, and many other nonlinear methods.

Levels of Automation

It is useful to think of automation as a continuum rather than as an all-or-

TABLE 12.1 Levels of Automation

1.	The computer offers no assistance, the human must do it all.
2.	The computer offers a complete set of action alternatives, and
3.	narrows the selection down to a few, or
4.	suggests one, and
5.	executes that suggestion if the human approves, or
6.	allows the human a restricted time to veto before automatic execution, or
7.	executes automatically, then necessarily informs the human, or
8.	informs the human after execution only if he asks, or
9.	informs the human after execution if it, the computer, decides to.
10.	The computer decides everything and acts autonomously, ignoring the human.

Source: Sheridan (1987).

nothing concept. The notion of levels of automation has been discussed by a number of authors (Billings, 1991, 1996; Hopkin, 1995; McDaniel, 1988; National Research Council, 1982; Parasuraman et al., 1990; Sheridan, 1980; Wickens, 1992). At the extreme of total manual control, a particular function is continuously controlled by the human operator, with no machine control. At the other extreme of total automation, all aspects of the function (including its monitoring) are delegated to a machine, so that only the end product and not its operation is made available to the human operator. In between these two extremes lie different degrees of participation in the function by the human and by automation (Table 12.1). At the seventh level, for example, the automation carries out a function and is programmed to inform the operator to that effect, but the operator cannot influence the decision. McDaniel (1988) similarly described the level of *monitored automation* as one at which the automation carries out a series of operations autonomously that the human operator is able to monitor but cannot change or override. Despite the relatively high level of automation autonomy, the human operator may still monitor the automation because of implications elsewhere for the system, should the automation change its state.

What is the appropriate level of automation? There is no easy or single answer to this question. Choosing the appropriate level of automation can be relatively straightforward in some cases. For example, most people would probably prefer to have an alarm clock or a washing machine operate at a fairly high level of automation (level 7 or higher), and a baby-sitting robot set at a fairly low level of automation, or not at all (level 2 or 1). In most complex systems, however, the choice of level of automation may not be so simple. Furthermore, the level of automation may not be fixed but context dependent; for example, in dynamic systems such as aircraft, the pilot will select whatever level he or she considers appropriate for the circumstances of the maneuver.

The concept of levels of automation is also useful in understanding distinctions in terminology with respect to automation. Some authors prefer to use the term *automation* to refer to functions that do not require, and often do not permit, any direct participation or intervention in them by the human operator; they use the term *computer assistance* to refer to cases in which such human involvement is possible (Hopkin, 1995). According to this definition of automation, only technical or maintenance staff could intervene in the automated process. The human operator who applies the products of automation has no influence over the processes that lead to those products. For example, most previous applications of automation to air traffic control have affected simple, routine, continuous functions such as data gathering and storage, data compilation and correlation, data synthesis, and the retrieval and updating of information. These applications are universal and unselective. When some selectivity or adaptability in response to individual needs has been achieved or some controller intervention is permitted, automation then becomes computer assistance.

Because the notion of multiple levels of automation includes both the concepts of automation and computer assistance, only the term *automation* is used in the remainder of this chapter. However, our use of this term does not imply, at this stage in our analysis, adoption of a general position on the appropriate level for the air traffic control system, whether full automation or computer assistance.

Authority and Autonomy

Higher levels of automation are associated with greater machine autonomy, with a corresponding decrease in the human operator's ability to control or influence the automation. The level of authority may also be characterized by the degree of emergency or risk involved. Figure 12.1 shows a two-dimensional characterization of where authority might reside (human versus computer) in this respect.

Sarter and Woods (1994a, 1995a, 1995b) have suggested that automation, particularly high-level automation of the type found in advanced cockpits, needs to be decomposed with respect to critical properties such as autonomy, authority, and observability. Although these terms can be defined in many instances, there are cases in which they are independent of one another and cases in which they are not. Autonomous automation, once engaged, carries out many operations with only early initiating input from the operator. The operations respond to inputs other than those provided by the operator (from sensors, other computer systems, etc.). As a result, autonomous automation may also be less "observable" than other forms of automation, though machine functions can be programmed to keep the human informed of machine activities. Automation authority refers to the power to carry out functions that cannot be overridden by the human operator. For example, the flight envelope protection function of the Airbus 320 cannot be overridden except by turning off certain flight control computers. The envelope protection systems in other aircraft (e.g., MD-11 and B-777), however, have "soft" limits that can be overridden by the flight crew.

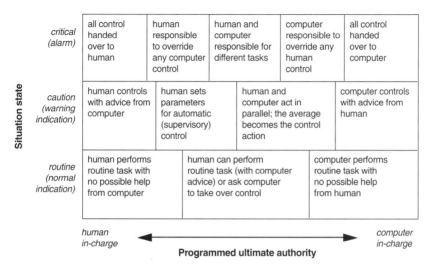

FIGURE 12.1 Alternatives for ultimate authority as related to risk.

Autonomy and authority are also interdependent automation properties (Sarter and Woods, 1994a).

Sarter and Woods (1995b) propose that the combination of these properties of high-level automation creates multiple "agents" in the workplace who must work together for effective system performance. Although the electronic copilot or pilot's associate (Chambers and Nagel, 1985; Rouse et al., 1990) is still a future concept for the cockpit, current cockpit automation possesses many qualities consistent with autonomous, agent-like behavior. Unfortunately, because the properties of the automation can create strong but silent partners to the human operator, mutual understanding between machine and human agents can be compromised (Sarter and Woods, 1995b). Future air traffic management systems may incorporate multiple versions of such agents, both human and machine, and both on the ground and in the air.

FUNCTIONAL CHARACTERISTICS

Despite the high-tech scenarios that are being contemplated for the 21st century (e.g., free flight), current air traffic control systems, while including some automation (e.g., automated handoffs, conflict warnings), remain largely manual. The influence on air traffic control operations of existing and proposed automation, both near-term and long-term, has been discussed extensively in recent years as the prospects for implementing various forms of advanced automation have become clear (Harwood, 1993; Hopkin, 1991, 1994, 1995; Hopkin and Wise, 1996; Vortac, 1993). In this section we describe the functional characteristics of air traffic control automation, emphasizing currently implemented technologies or immediate near-term proposals for automation, rather than "blue-sky" concepts. Before describing specific examples, however, we discuss the rationale for implementing automation, both generally and as represented in the strategic planning of the FAA.

The Need for Automation

More people wish to use air transportation, and more planes are needed to get them to their destinations. All current forecasts foresee substantial future increases in air traffic and a continuing mix of aircraft types. The FAA must face and accommodate increasing demands for air traffic control services. Current systems were never intended to handle the quantities of air traffic now envisaged for the future. Many of them are already functioning at or near their planned maximum capacity for traffic handling much of the time. Present practices, procedures, and equipment are often not capable of adaptation to cope with a lot more traffic. In addition, traffic patterns, predictability, and laws of control may change. New systems therefore have to be designed not merely to function differently initially but also to evolve differently while they are in service. The apparent option of leaving air traffic control systems unchanged and expecting them to cope with the predicted large increases in traffic is therefore not a practical option at all. Air traffic control must change and evolve (Wise et al., 1991) to meet the foreseen changes in demand.

The limited available airspace in regions of high traffic density constrains the kinds of solutions to the problem of increased air traffic. The only way to handle still more air traffic within regions that are already congested is to permit each aircraft to occupy less airspace. This means that aircraft will be closer together. Their safety must not be compromised in any way, they must not be subjected to more delays, and each flight should be efficient in its timing, costs, and use of resources. To further these objectives, flight plans, navigational data, on-board sensors, prediction aids, and computations can provide information about the state of each flight, the flight objectives, its progress, and its relationships to other flights, and about any potential hazards such as adverse weather, terrain proximity, and airspace restrictions or boundaries. All this information, combined with high-quality information about the position of each aircraft on its route and about the route itself, could allow the minimum separation standards between aircraft to be reduced safely.

Since changes in air traffic control tend to be evolutionary rather than revolutionary, current systems have to be designed so that they can evolve to integrate and make effective use of such developments in technology during the lifetime of the system. The appearance and functionality of some current systems have therefore been influenced, sometimes considerably, by the ways in which they are expected to be improved during their operational lifetime. Common current examples are the replacement of paper flight progress strips by electronic ones and the advent of data links. Many of the remaining practical limitations on the availability of data for air traffic control purposes are expected eventually to disappear altogether.

One apparent response to the problem of increased air traffic can be ruled out: that is simply to recruit more controllers and to make each controller or each small team of controllers responsible for the air traffic in a smaller region of airspace. Unfortunately, in many of the most congested airspace regions, this process has already been taken about as far as is practicable, because the smaller the region of airspace of each air traffic control sector, the greater the amount of communication, coordination, liaison, and handoff required in the air traffic control system itself and in the cockpit, whenever an aircraft flies out of one controller's jurisdiction and into another's. Eventually any gains from having smaller regions are more than canceled by the associated additional activities incurred by the further partitioning of the airspace.

Automation has therefore been seen by some as the best alternative to the problem of increased traffic demand. Automation may include technologies for such functions as information display, communication, decision making, and cooperative problem solving. Free flight is an alternate remedy that has been recently proposed (RTCA, 1995; Planzer and Jenny, 1995). However, increased traffic is only one factor in the drive to automate the air traffic control system. Another factor is the increasing tendency of air traffic control providers to serve rather than to control the aviation community in its use of airspace resources. Automation is seen as one of the ways in which service providers can meet the needs of airspace customers, both now and in the future (Federal Aviation Administration, 1995).

The advanced automation system (AAS) was a major program of technology enhancement that was initiated by the FAA in the 1980s. Despite its title, the program was largely concerned with the *modernization* of equipment and, although some parts of the program did deal with automation, the major thrust was not with automation per se. In 1994, following delays and other problems in meeting the goals of the program, the AAS was divided into smaller projects, each concerned with replacement of aging equipment with newer, more efficient, and powerful capabilities. In contrast to these efforts at improving the air traffic control infrastructure, other planning efforts have focused more specifically on automation. The FAA's plans for automation are contained in its *Automation Strategic Plan* (Federal Aviation Administration, 1994) and the *Aviation System Capital Investment Plan* (Federal Aviation Administration, 1996b). Two major goals for automation are identified: the improvement of system safety and an increase in system efficiency. In the context of safety automation is proposed because of its potential to:

- Reduce human error through better human-computer interfaces and improved data communications (e.g., datalink),
- Improve surveillance (radar and satellite-based),
- Improve weather data,
- Improve reliability of equipment, and
- Prevent system overload.

Automation is proposed to improve system efficiency because of its potential to:

- Reduce delays,
- Accommodate user-preferred trajectories,
- Provide fuel-efficient profiles,
- Minimize maintenance costs, and
- Improve workforce efficiency.

Air Traffic Control Automation Systems: General Characteristics

There is no easy way to classify or to even list comprehensively all of the various automation systems that have been deployed in air traffic control. In this section, we describe the major systems that have been fielded in the United States and discuss the principal functional characteristics of the automation to date, making reference to specific systems as far as possible. Although there is considerable diversity in the technical features, functionality, and operational experience with different automation systems, some generalizations are possible.

1. On the whole, automation to date has been successful,[1] in the sense that the new technologies that have been fielded over the past 30 years have been fairly well integrated into the existing air traffic control system and have generally been found useful by controllers. This is clearly a positive feature, and efforts should be made to ensure its continuity in the future as new systems are introduced.

2. There has been a steady increase in the complexity and scope of automated systems that have been introduced into the air traffic control environment. In terms of functionality, automation to date has largely been concerned with improving the quality of the information provided to the controller (e.g., automated data synthesis and information presentation) and with freeing the controller from simple but necessary routine activities (e.g., automated handoffs to other sectors) and less so with the automation of decision-making and planning functions. Historically, following the development of automation of data processing,

---

[1] With the exception of the AAS, which, as noted before, was not primarily an automation program.

aspects of communications were automated, followed by conflict alerts and conflict prediction.

3. Automation systems that are currently under field testing or about to be deployed in the near future (e.g., the center-TRACON automation system—CTAS), will have a greater impact on the controller's complex cognitive activities, the role of which was discussed extensively in Chapter 5. Higher levels of automation, with possibly greater authority and autonomy, may be the next to follow. Some of these systems are briefly mentioned in this chapter, but a more detailed discussion is expected in Phase 2 of the panel's work.

4. Although processing of flight data and radar data automatically distributes common data to team members, and although automated handoff supports pairs of controllers, automation to date, as well as most future projected systems, has been designed to support individual controllers rather than teams of controllers. Given the importance of team activities to system performance (discussed in Chapter 7), this fact will need to be taken into account in evaluating the impact of future automation.

5. Different forces have led to the development and deployment of air traffic control automation systems. In some cases, technology has become available, or there has been technology transfer from other domains, as in the case of the global positioning system (GPS). Other contributing sources to the development of particular automation systems include controllers, users of air traffic control services, the FAA, and human factors research efforts.

6. Finally, cockpit automation has generally been more pervasive than air traffic control automation. Some of these systems, such as the FMS and TCAS, have implications for air traffic control and hence must be considered in any discussion of air traffic control automation.

Table 12.2 provides an overview of the automated systems that have been implemented in the air traffic control environment. Automation technologies introduced over time are represented along the columns, with the rows representing the four major types of environments. The distinction among tower, TRACON, en route, and oceanic air traffic control is just one of many ways in which the system could be subdivided, but it is convenient for the purpose of identifying specific systems that have been implemented, from the initial automation of data-gathering functions to the more advanced decision-aiding technologies that are currently under production. Because some cockpit automation systems (such as TCAS) have significant implications for control of the airspace, relevant cockpit systems are also shown in the table.

TABLE 12.2 Introduction of Automation Systems in Air Traffic Control, 1960s-1990s

ENVIRONMENT	AUTOMATION SYSTEMS			
	1960s	1970s	1980s	1990s
Tower		ASDE RVR	LLWAS	AMASS TDWR HABTCC
Terminal	ARTS		CA MSAW	CTAS (DA/TMA) ERM	CTAS (FAST) VSCS
En Route	RDP	FDP	CA MSAW HOST	CTAS (DA/TMA)
Oceanic			ODAPS	OSDS
Cockpit Systems with air traffic control implications			GPWS FMS	TCAS GPS

AMASS	Airport Movement Area Safety System	GPS	Global Positioning System
ARTS	Automated Radar Terminal System	GPWS	Ground Proximity Warning System
ASDE	Airport Surface Detection Equipment	HOST	Host Computer Upgrade
HABTCCC	High Availability Basic Tower Control Computer Complex	LLWAS	Low Level Windshear Advisory System
CA	Conflict Alert	MSAW	Minimum Safe Altitude Warning
CTAS	Center-TRACON Automation System	ODAPS	Oceanic Display and Planning System
DA	Descent Advisor	OSDS	Oceanic System Development and Support
ERM	En-Route Metering System	RDP	Radar Data Processmg
FAST	Final Approach Spacing Tool	RVR	Runway Visual Range System
FDP	Flight Data Processing	TCAS	Traffic Alert Collision Avoidance System
FMS	Flight Management System	TDWR	Terminal Doppler Weather Radar
		TMA	Traffic Management Advisor
		VSCS	Voice Switching Control System

Flight Data Gathering and Processing

Data Gathering

Originally the gathering of data about forthcoming and current flights was entirely a human activity, prior to the introduction of the radar data processing system to the en route environment. The controller or an assistant wrote on a strip of paper, called a flight progress strip, all the essential information about a forthcoming flight, amending and updating it by hand as further information was gathered about the progress of the flight and as instructions were issued by the controller and agreed to by the pilot. When the responsibility for a flight passed to another controller, either its flight strip was physically handed over or parallel strips were prepared in advance and marked appropriately when responsibility was transferred. Flight strips plus speech communication channels between the controller and pilots were the basic procedural tools of air traffic control—and they still are. They are found wherever a minimally equipped air traffic control service is provided, which is often in regions with low levels of air traffic, in regions in which minimally equipped aircraft are the main traffic, and at locations in which the air traffic control service is not provided on a permanent basis. Controllers still rely on these nonautomated tools wherever there is no radar, either because the flight is beyond radar coverage, which is a common event and applies to much oceanic air traffic control, or because the radar has failed, which is a rare event.

The automation of data gathering has progressed so far that the controller may now spend very little time gathering data. The data are now presented automatically in forms that satisfy the task requirements, with appropriate data formats and levels of detail and with appropriate timing so that some data are not presented all the time but only when they are needed. Behind this revolution for the controller are many technical advances in the sensing, processing, storage, compilation, and synthesis of data. Much time formerly spent by the controller in gathering data is saved.

Data Storage

A significant area of automation is data storage, beginning with the automated radar terminal system (ARTS) for terminals and the flight data processing (FDP) for en route centers (see Table 12.2). The automation of data storage, associated with the widespread use of computers, seems so obvious that it is taken for granted, but it has a big effect on the controller. Originally the paper flight progress strips were the data storage, and there was no other. It is still quite common for the paper flight progress strip to constitute the official record of a flight for purposes such as the levying of charges for air traffic control services (in Europe) and subsequent incident investigation. Now a great deal of information is stored about each flight and applied for data synthesis, for data presentation, and for computations. The controller relies on this data storage and makes extensive use of it, albeit often indirectly through the aspects of it represented on visual displays or aurally. This stored database has to be matched with the controller's tasks. Sometimes the controller has extra tasks because of it. For example, the written amendment of a paper flight progress strip to record a control action agreed to by a pilot is not sufficient to enter that agreement into the database of the system, and the change may have to be keyed in as well in order to enter it into the computer and store it automatically. This has prompted a quest for forms of automation that could avoid such duplication of functions, so that the same executive controller action both informs the pilot and updates the data storage.

Data Smoothing

Automation is widely applied to smooth data for human use. The original unprocessed data, very frequently sampled, contain numerous generally small anomalies concerning the aircraft's position, altitude, speed, heading, and other quantitative flight parameters. These are too small to be relevant to air traffic control but not too small to be distracting. They are removed by smoothing so that aircraft appear to turn, climb, descend, accelerate, and decelerate consistently along flight paths depicted as smoothed. In earlier times, aircraft were seen to perform more raggedly and with more vacillations, as an artifact of sensing methods, sampling rates, and anomalies in the processing and depiction of information, and the controller had to do data smoothing by extrapolation. This is now done automatically, striking a balance so that genuine and operationally significant aberrations are not smoothed away.

Data Compilation and Synthesis

Not only are the sensed data summarized and smoothed automatically, but sometimes they are also combined, represented in different forms, or otherwise transformed to make them suitable for human use. Such functions are automated. Much of the information depicted on radar displays has undergone processing of this kind.

The ARTS system, which was extensively discussed in Chapter 2, fulfills many of these data gathering and processing functions for the TRACON controller.

Information About Aircraft

Identification

There has always been a need for controllers to identify aircraft positively so that instructions can be given the correct aircraft. This function has now been widely automated, but it was not always so. For a time after radar was introduced, a controller might request a change of heading of an aircraft and identify that aircraft positively by noting which aircraft blip on the radar display changed heading appropriately in response to the instruction. Identification has now been largely automated by recognizable signals transmitted between air and ground. This function is contained within the ARTS system. A data block on the radar display beside the symbol or blip denoting the current position of an aircraft also gives its identity. This saves a lot of controller work and removes a potential source of confusion, although call signs that are visually or acoustically similar can be confused (Conrad and Hull, 1964), irrespective of automation.

Flight Level

Following departure from an airfield, aircraft are cleared by the controller to climb to a flight level. In light traffic, this may be their cruising altitude, but in heavier traffic they are more likely to require reclearance through one or more intermediate levels. Formerly aircraft could be at any level between their last reported one and their cleared one, but transponded altitude information (for equipped aircraft), appearing as numerals in the data block for the aircraft on the controller's radar display, automatically updates its actual altitude continuously. This removes uncertainty, facilitating conflict prediction, and curtails the need for spoken messages between controller and pilot. Other displays showing altitude may also be updated automatically. Similarly, aircraft altitude during its final approach is also continuously updated automatically by ARTS. Associated automated warnings of various kinds are provided if it strays from the glide slope, particularly if it is too low.

The aircraft data blocks on the controller's radar display often also include symbols to show if the aircraft is climbing or descending, the absence of a symbol denoting that it is in level flight. These symbols are derived and maintained automatically. They replace the controller's memory and annotations on the flight strips. Aircraft in climb or descent may also be coded differently on tabular displays of the traffic, without controller intervention. Checks for future potential conflicts have then to cover a band of flight levels rather than a single one.

Aircraft Heading and Speed

Aircraft heading is computationally quite complex from the point of view of air traffic control, since it is derived from angles and distances between the aircraft and fixed points on the ground. The heading flown has to take account of wind speed and direction, but it may be presented digitally on the aircraft's data block or on a tabular display. Aircraft speeds may also be depicted digitally within the air traffic control workspace, and speeds and headings may each be represented relatively by lines attached to the aircraft's positional symbol and derived automatically from computations. The controller's knowledge of the type of aircraft and of its route provide additional information about expected speeds and headings. Therefore these information categories tend to be the subject of relatively few spoken messages.

Reporting Times and Handoffs

A manual method of updating the progress of a flight is for the pilot to report to air traffic control on reaching a predesignated position on a route, or on leaving a particular flight level during climb or descent. In particular, spoken communications were an essential aspect of the handoff of control responsibility from one controller to another as the aircraft left one flight sector and entered the adjacent one. A change of frequency in communications would also be involved. The practice of silent handoffs is widely adopted in some air traffic control regions, in which the air traffic control handoff of responsibility occurs routinely at the sector boundary without any spoken interchange with the pilot until there is a need to make verbal contact for other reasons (e.g., the need to update communications frequencies). In these instances, more automation renders some human procedures redundant by replacing them with some form of continuous updating of information.

Automated systems for handoffs, introduced in the 1970s, simply replaced the requirement for air traffic controllers to verbally communicate when an aircraft moves from one sector to another, with the automated system to accomplish the same goals. It represented a reliable system that had a clear and unambiguously favorable impact on controller workload and communications load. With continuously and automatically updated information about the state and position of the aircraft, such regular reporting points become redundant.

Overlap of Aircraft Data Blocks

In dense traffic, aircraft data blocks may overlap and become difficult to read. A practical form of automation prevents data blocks from overlapping so that all the information on them remains clear, no matter how closely aircraft that are safely separated vertically overlap in the plan view. An alternative solution is to alternate automatically data blocks that overlap. There may be penalties in the form of new sources of human error: without such automation, controllers may expect a given data block to maintain a fixed positional relationship with its associated aircraft symbol and may then compare the relative positions of data blocks to assess the relative positions of aircraft. However, if the automation keeps data blocks separated, it achieves this by interfering with the relative position of each aircraft and its data block, so that the data blocks themselves no longer depict the relative positions of the aircraft.

Communications: Datalink

Applications of automation to communications have concentrated on the quantitative information exchanged between controller and pilot, since this can be readily identified and digitized for efficient automated transmission (in comparison to other aspects of speech communication, such as voice quality, formality, degree of stereotyping, pronunciation, accent, pace, pauses, level of detail, redundancy, courtesy, acknowledgment). The datalink system is designed to replace the traditional audio-voice link between pilot and controller with an automated transfer of digital information (Kerns, 1991; Corwin, 1991). Datalink is already currently in use for transmission of information between aircraft and the airline dispatch center. In the future, it will also provide to the pilot a visual (or synthesized speech) display of clearance, runway, wind, and other key information needed for landing.

Implementation of datalink systems has a number of generic implications. First, at the interface, a datalink system relies on a manual-visual interface both on the ground and in the air. This shift from audio-voice channels has potential workload implications (Wickens et al., 1996). Second, and potentially more profound, datalink has the potential to make available to both controller and pilot digital information from automated agents at either location, of which one or the other human participant may be unaware. For the pilot, datalink may also increase "head down" time (i.e., visual attention is diverted from primary displays) in terminal areas (Groce and Boucek, 1987). Datalink also reduces both the need for human speech and the reliance on speech for controller-pilot communication. In principle, time is saved, although not always in practice (Cardosi, 1993). Pilots also glean further party-line information that can be very useful to them from overhearing messages between controllers and other pilots on the same frequency to which they are tuned (Pritchett and Hansman, 1993).

Two contrary challenges should be applied to the automation of communications channels. One challenge, which would favor the retention of more human speech, is to demonstrate that the qualitative information contained in human speech cannot be safely eliminated because it is actually essential in terms of either efficiency or safety. The contrary challenge, which could favor the introduction of further automation of speech, is to prove that the judgments based on the qualitative attributes of human speech have no benefit at all, for it is possible that human judgments based on such impressions are sufficiently false that the system is better without them. It is important also to consider the implications of a mixed audio-voice and datalink environment.

Navigation Aids

Ground-based aids that can be sensed or interrogated from aircraft mark standard routes between major urban areas. Navigation aids that show the direction from the aircraft of sensors at known positions on the ground permit computations about current aircraft track and heading. Data from navigation aids can be used to make comparisons between aircraft and hence help the controller to maintain safe separations between them. The information available to the controller depends considerably on the navigation aids in use. From the point of view of the controller, these functions are highly automated, and the controller has access only to the limited product from them necessary for the control of the aircraft as traffic, showing where they are, how they relate to each other, and whether there are discrepancies between their actual and intended routes, the latter being derived from flight plans and updates of flight progress, some of which are themselves automated.

Implementation of the GPS will enhance greatly the amount and quantity of navigational data derived from satellites. GPS can provide much more accurate positional information than is needed for most air traffic control purposes (e.g., on the order of 10 meters), but its greatest impact will be on traffic currently beyond radar coverage.

Information Displays

Because all the information needed by the controller about the aircraft cannot be presented within the data blocks on a radar display without their becoming too large and cluttered and inviting data block overlap, and because much of the information is not readily adaptable to such forms of presentation, there have always been additional types of information displays in air traffic control, such as maps and tables. In the latter, aircraft can be listed or categorized according to time, flight level, direction, route, destination, or other criteria appropriate for the controller's tasks. Tabular displays of automatically compiled and updated information can be suitable for presentation as windows in other air traffic control displays. Some of these additional displays are normally semipermanent and are maintained automatically. Others containing more dynamic information that is tabulated or in the form of windows are often updated automatically, wholly or in part. Examples include displays of weather information, of serviceability states, of communications channels and their availability, and of the categories of updated information referred to above. CTAS, the suite of automation tools introduced to assist in terminal area traffic management, incorporates extensive reliance on such information displays (Erzberger et al., 1993). As automation progresses, controllers have to spend less time in maintaining these displays, as more of these functions are done automatically for them.

Alerting Devices

Various visual or auditory alerting signals are provided for the controller as automated aids. They may serve as memory aids, prompts, or instructions and may signify a state of progress or a change of state. Their intention is to draw the controller's attention to particular information or to initiate a response. They are triggered automatically when a predefined set of circumstances actually arises in the course of routine recurring computations within the system. The automation is normally confined to their detection and identification and does not extend to instructing the controller in the human actions that would be expected or appropriate, although this is an intended future development. Such alerting devices aid monitoring, searching, and attending rather than more complex cognitive functions. The human factors problems in their usage tend to concern timing of presentation and methods of display, since this kind of automated aid has the potential to be distracting and should not interrupt more important activities. An additional human factors consideration is the adequacy of computational technology underlying alerting devices.

The minimum safe altitude warning (MSAW) system is a specific example of alerting automation. This system provides an alerting function that is parallel to the ground proximity warning system in the cockpit. MSAW alerts the controller that a conflict with terrain or obstructions (e.g., radar towers) is projected to take place if an aircraft continues along its current trajectory. The projection is based on the aircraft's transgression of a specified minimum altitude during descent segments.

Track Deviation

Controllers spend considerable cognitive effort in searching their displays for track deviations. To reduce the need for searching, computations can be made automatically by comparing the intended and the actual track of an aircraft and by signaling to the controller whenever an aircraft deviates from its planned track by more than a predetermined permissible margin. In principle this can be applied to all four dimensions of the flight: the lateral, longitudinal, and altitude dimensions of the track and its time of arrival at a given location. The controller is then expected to contact the pilot to ascertain the reason for the deviation and to correct it when appropriate and when possible and to issue an immediate instruction if the deviation could be potentially hazardous. The significance and degree of urgency of a track deviation depend on the phase of flight and on traffic densities: it can be very urgent if it occurs during the final approach.

Conflict Detection

One of the principal duties of the controller is to ensure that aircraft do not conflict with each other. This requires the estimation of future aircraft positions. Conflict detection automation is designed to support the controller in this function. Comparisons between aircraft are made frequently and automatically, and the controller's attention is drawn by changing the coding of any displayed aircraft that are predicted to infringe the separation standards between them within a given time or distance. Introduction of the ARTS system in the 1960s provided a feature that could detect and predict conflicts between aircraft, thus automating what had previously been a vulnerable, effort-intensive human function of trying to visualize future traffic situations. This capability has been extended in order to alert the controller with a visual and auditory alarm whenever two trajectories are predicted to initiate a loss of separation in the future—the conflict resolution advisor. Although occasional false alarms in this system may be a source of irritation (e.g., if the controller was aware of the pending conflict and was planning to resolve it before a loss of separation occurred), controller acceptance of these devices has been generally favorable. Indeed, controllers have expressed specific concern at the loss of these predictive functions when equipment failures have temporarily disabled them (Wald, 1995).

Depending on the quality of the data about the aircraft, a balance is struck in the design of conflict detection systems to give as much forewarning as possible without incurring too many false alarms. The practical value of the aid relies on computational correctness and on getting this balance right. (The necessary procedures for ensuring an optimal balance between false alarms and missed warnings are discussed further in a later part of this chapter.) Sometimes the position or the time of occurrence of the anticipated conflict are depicted, but a conflict detection aid provides no further information about it. Unlike TCAS resolution advisories (described below), which mandate specific pilot maneuvers, no specific action to relieve the alert condition is recommended. The controller is left to resolve the problem in whatever way he or she sees fit.

Computer-Assisted Approach Sequencing and Ghosting

Approach sequencing and ghosting are applied to flows of air traffic approaching an airfield from diverse directions, amalgamating them into a single flow approaching one runway or into parallel flows approaching two or more parallel runways. Computer-assisted approach sequencing depicts on a display the predicted flight paths of arriving aircraft. It shows directly, or permits extrapolation of, their expected order of arrival at the amalgamation position and the expected gaps between consecutive aircraft when they arrive. The controller can issue instructions for minor changes in flight path or speed in order to adjust and smooth gap sizes, and particularly to ensure that the minimum separation standards applicable during final approach are met. Ghosting fulfills a similar function in a different way: the data block of each aircraft appears normally on the radar at its position on its actual route, but it also appears as a ghost with much reduced brightness contrast in the positions on the other converging routes that it would occupy to arrive at the amalgamation position at the same time (Mundra, 1989). In effect, each route depicts all the traffic before route convergence as if convergence had already occurred, showing what the gaps between consecutive aircraft will be and permitting adjustment and smoothing of any gaps before the traffic streams actually merge. One of the main ways in which these kinds of aids may differ is in the length of the concluding phase of the flight to which they apply. Some deal only with traffic already in the terminal maneuvering area and preparing to turn onto final approach to the runway, and others extend into en route flight and apply to the aircraft from before the start of its descent from cruising until its point of touchdown on the runway. This kind of aid may thus cut across some of the traditional divisions of responsibility within air traffic control, between the en route sector controller and the terminal area controller. It relies on extensive automated computation facilities.

Flows and Slots

Various schemes have evolved that treat aircraft as items in traffic flows and that exercise control by assigning slots and slot times to each aircraft in the flow. Separations can then be dealt with by reference to the slots. The maximum traffic-handling capacities of flows can be defined in terms of slots, and the air traffic control is planned to ensure that all the available slots are actually utilized. One complication is that slots are not all the same size: because of wake turbulence from heavy aircraft, slot sizes vary, and slot adjustment can be problematic at peak traffic times. Tactical adjustments can be minimized by allowing for the intersection or amalgamation of traffic flows in the initial slot allocation. Slots are being widely adopted in European air traffic control, where they have generally been beneficial and have spread delays attributable to insufficient capacity more evenly among users.

Other Automation Systems

In this section we briefly consider some other systems that are either being considered for deployment or are being field tested. Because this is more properly a topic for Phase 2 of this report, only two examples are discussed.

Electronic Flight Progress Strips

Electronic flight progress strips are a particular kind of tabular display intended to replace paper flight progress strips. Being electronic, they can be generated and updated automatically in ways that paper strips cannot be, and the controller must use a keyboard instead of handwritten annotations to amend them (Manning, 1995). Provision is often made to expand visually an electronic flight progress strip while it is being amended in order to see the amendments clearly and to place them correctly on the strip. The approach to the automation of flight strips has essentially been to capture electronically the functions of paper flight strips. This has proved quite straightforward to accomplish in part and impossible to accomplish entirely, although complete automation of human functionality may not be feasible or necessary. For example, a paper strip is sometimes cocked sideways on the display board as a reminder of an outstanding action to be taken with it, and electronic equivalents of this memory aid could be devised.

There has been concern that some more complex cognitive functions that controllers engage in do not lend themselves so readily to conversion into electronic form. Some human factors problems of design have been posed by difficulties in capturing electronically the full functionality of paper flight strips, which are more complex than they seem (Hughes et al., 1993). However the results of a series of extensive studies examining the use of both paper and electronic flight progress strips do not support the view that electronic strips impair the controller's memory or ability to build up a picture of the traffic (Vortac and Gettys, 1990; Manning, 1995). Further research is needed to resolve these issues.

Center-TRACON Automation System

The center-TRACON automation system (Erzberger et al., 1993) is the newest ground-based automation system and is only now being evaluated at certain operational centers and terminals. Like conflict alerting automation, CTAS is not designed to replace controller actions, but rather to provide an automated means for integrating and displaying information, to assist the TRACON controller with optimally scheduling and spacing aircraft in three-dimensional space on arrival to and departure from the runway. Direct displayed representations that are designed to replace (or augment) cognitive visualizations is a key element. CTAS currently consists of three subsystems, the traffic management advisor (TMA), the descent advisor (DA), and the final approach spacing tool (FAST). CTAS development has involved close collaboration with users and attention to issues of the user interface. Recent human factors evaluations of CTAS subsystems have validated the benefits of this close collaboration with users during the development of the system (Harwood, 1994; Hilburn et al., 1995).

Cockpit Automation

Aviation automation has generally been more extensive and far-reaching in the air than it has been on the ground. Several of the many automated systems that have been introduced in the cockpit have had major implications for air traffic control operations. Accordingly, a discussion of air traffic control automation would not be complete without examining the role of these automated systems.

Aircraft Flight Guidance Automation

Automation of aircraft guidance began with devices for flight stabilization, progressed through simple autopilots, to the high-level flight management system and flight protection systems found in the modern glass cockpit, in which multimodal electronic displays and multifunction keyboards are used extensively (see Billings, 1991, 1996, for an overview). The introduction of these systems has undoubtedly improved aircraft performance. The safety record of modern aircraft is also good, aircraft accident rates having remained relatively stable over the past two decades. The mean time between accidents involving hull loss is substantially higher on glass cockpit aircraft than on the previous generation of aircraft, although the supporting data do not include two recent B-757 accidents (Wiener, 1995).

Although the most mature level of flight deck guidance automation, the FMS (Sarter and Woods, 1994b), is a pure airborne system, it has two important implications for air traffic control automation. First, several important lessons have been learned regarding the human factors of automation introduction, both in design of the interface and in implementing the philosophy of human-centered automation. Second, the tremendous potential power of the FMS to define optimal routes has revealed the limitations of the current air traffic control technology to enable these routes to be flown. Hence, the FMS has served as an impetus for greater authority for route planning and adjustment to be shifted to the cockpit and the air carrier—the issue of free flight. Whether there is pressure to shift authority from the human controller to other system users (as in free flight) or to an automated agent, the human factors implications for the controller—of being removed from the flight control loop—are similar.

Ground Proximity Warning System

The ground proximity warning system (GPWS) was widely implemented in commercial airlines following a series of incidents involving controlled flight into terrain (CFIT) in the early 1970s (Wiener, 1977). The device within the cockpit alerts the pilot when combinations of parameters (altitude, vertical velocity) suggest that the aircraft may be on a collision path with the terrain or not configured for landing. As such, it provides the pilot with clear and unambiguous information to supplement whatever terrain information the controller may be able to provide. Early renditions of the system provided pilots with an excessive number of irritating false alarms; subsequent refinement has reduced their number. The record of the GPWS has generally been a successful one. In the 5 years following its widespread installation in 1975, CFIT incidents decreased dramatically compared with the previous 5 years, for nations that required GPWS installation (Diehl, 1991). Nonetheless, the CFIT problem remains one of the most dangerous facing transport aviation.

Traffic Alert Collision Avoidance System

The traffic alert collision avoidance system (TCAS) is one of the more recent cockpit systems with significant implications for air traffic control. It is a cockpit-based device that allows an aircraft fitted with it to detect other aircraft in close proximity on potential collision courses with it, in time to make emergency avoidance maneuvers, the nature of which is recommended by the device itself. GPWS provides information regarding separation between aircraft and ground. TCAS, however, although also an airborne system, has considerably more direct implications for air traffic control.

Because conflicts between two moving aircraft are far more complex to predict than between a moving aircraft and a static ground object, TCAS algorithms are far more complex, have taken much longer to evolve than those for the GPWS, and are still undergoing refinement (Chappell, 1990). Billings (1996) summarized the impacts of inadequacies of TCAS algorithms, including nuisance warnings in high-density traffic areas. Hence, TCAS has only recently been introduced in force into the commercial airline cockpit.

A TCAS alert should be rare. If it is ever needed, the whole of the normal air traffic control system has in a sense failed to function as planned, since one of its objectives is to ensure that there is no need for emergency maneuvers of the kind that TCAS involves, although TCAS still protects equipped aircraft from aircraft flying under visual flight rules that are not in the air traffic control system. TCAS is also difficult to reconcile with normal air traffic control practices, because all the planning of air traffic control is done on the assumption that aircraft will fly as planned and will not make sudden unexpected maneuvers. A TCAS maneuver therefore always has the potential to generate longer-term problems in solving a short-term and very urgent problem.

Nevertheless, it is a final safeguard, which air traffic control must accommodate in an essentially retrospective way, since TCAS is a very short-term solution from an air traffic control perspective. It therefore has to be handled tactically, and its circumstances are exceptional. Normal conflict detection procedures, and conflict resolution ones if they are available, would resolve a TCAS situation long before it was needed, provided that the basic information had been in the air traffic control system, the system had functioned as planned, and the aircraft concerned had behaved as predicted. If TCAS occurrences become more common, controllers may have considerable difficulty in accommodating their consequences; this would reveal a fundamental flaw in the air traffic control system itself, which would have to be tackled as a matter of urgency.

Not surprisingly, like GPWS, TCAS was initially plagued by many false alarms, which have triggered further efforts to refine and optimize the detection algorithms, in order to preserve pilot trust in the system. However, unlike GPWS, by alerting the pilot to vertical spacing issues and recommending avoidance maneuvers, TCAS in effect shifts authority to an automated agent (and the pilot) and hence away from the controller, whose primary responsibility is to maintain lateral and vertical separation between the aircraft in his or her purview.

In fact, in two respects this shift has already occurred. First, on some occasions, pilots responding to a TCAS alert may change their altitude radically without informing controllers concurrently (Mellone and Frank, 1993). Second, on some occasions, pilots may use TCAS to adjust their spacing in such a way as to maintain separation just beyond the TCAS alert zone. In both cases, the controller's perception of an erosion of authority, the result of this cockpit-based automated device, is quite accurate.

HUMAN FACTORS ASPECTS OF AUTOMATION

Technology-Centered Versus Human-Centered Design

The design and implementation of automation in general has followed what has been termed a *technology-centered* approach. Typically, a particular incident or accident identifies circumstances in which human error was seen to be a major contributing factor. Technology is designed in an attempt to remove the source of error and improve system performance by automating functions carried out by the human operator. The design questions revolve around the hardware and software capabilities required to achieve machine control of the function. There is not much concern for how the human operator will subsequently use the automation in the new system, or to how the human operator's tasks will be changed by the automation—only the assumption that automation will simplify the operator's job and reduce errors and costs.

The available evidence suggests that this assumption is often supported: automation shrinks costs and reduces or even eliminates certain types of human error. However, the limitation of a purely technology-centered approach to automation design is that some potential human performance costs (discussed later in this chapter) can become manifest—that is, entirely new human error forms can surface (Wiener, 1988), and "automation surprises" can puzzle the operator (Sarter and Woods, 1995a). This can reduce system efficiency or compromise safety, negating the other benefits that automation provides. These costs and benefits have been noted most prominently in the case of cockpit automation (Wiener, 1988), but they may also occur in other domains of transportation.

For example, automated solutions have been proposed for virtually every error that automobile drivers can make: automated navigation systems for route-finding errors, collision avoidance systems for braking too late behind a stopped vehicle, and alertness indicators for drowsy drivers. No doubt some of the new proposed systems will reduce certain accident types. But there has been insufficient consideration of how well drivers will be able to use these systems and whether new error forms will emerge (Hancock and Parasuraman, 1992; Hancock et al., 1993, 1996). There is a real danger that the automated systems being marketed under the rubric of intelligent transportation systems, although beneficial in some respects, will also be subject to some of the same problems as cockpit automation. After two decades of human factors experience with and research on modern aviation automation, the benefits and costs of automation are beginning to become well understood.

The technology-centered approach has dominated aviation for a number of years, only to be tempered somewhat in recent years by growing acceptance of human factors in aviation design.[2] Several alternative philosophies for automation design, based on the concept of *user-centered design*, have also been proposed (Norman, 1993; Rouse, 1991). The human-centered approach to automation (Billings, 1991, 1996) has much in common with the concept of user-centered design. This approach has been applied to the design of personal computers, consumer products, and other systems (Norman, 1993; Rouse, 1991). The success of this design philosophy, particularly as reflected in the marketplace, testifies to the viability of the human-centered approach. Its applicability to air traffic control automation is considered later in this chapter; at this point, however, it should be noted that these alternatives to strictly technology-centered design approaches do not mandate human involvement in a system, at any level. Rather, a user-centered approach to the design of a system is a process, the outcome of which could be function allocation either to a machine or to a human. In this sense, the user-centered approach is broader than a technology-centered approach and in principle can anticipate problems that strict adherence to the latter design approach can bring.

Benefits and Costs of Automation

In a seminal paper, Wiener and Curry (1980) pointed out the promises and problems of the technology-centered approach to automation in aviation. They questioned the premise that automation could improve aviation safety by removing the major source of error in aviation operations—the human. In another early analysis of the impact of automation on human performance, Bainbridge (1983) also pointed out several ironies of automation, principal among them being that automation designed to reduce operator workload sometimes increased it. These early papers described some of the potential problems associated with automation, including manual skill deterioration, alteration of workload patterns, poor monitoring, inappropriate responses to alarms, and reduction in job satisfaction. Although Edwards (1976) had earlier raised some of the same concerns regarding cockpit automation, the Wiener and Curry (1980) paper was notable for proposing the beginnings of an automation design philosophy that would minimize

---

[2]Another recent indicator of this trend is the appointment of human factors experts to key positions in several major airlines and aircraft manufacturers (Hughes, 1995).

some of these problems and would support the human operator of complex systems—an approach later developed further by Billings (1991, 1996) as "human-centered automation."

Billings (1991, 1996) has traced the existence of incidents in modern aviation to problems in the interaction of humans and advanced cockpit automation. Many of these problems derive from the complexity of cockpit systems and from the difficulties pilots have in understanding the dynamic behavior of these systems, which in turn is related to the relative lack of feedback that they provide (Norman, 1990).

Since the pioneering Wiener and Curry (1980) paper, a number of conceptual analyses, laboratory experiments, simulator studies, and field surveys have enlarged our understanding of the human factors of automation. For reviews of this work, see Boehm-Davis et al. (1983), Billings (1991, 1996), Hopkin (1994, 1995), Mouloua and Parasuraman (1994), National Research Council (1982), Parasuraman and Mouloua (1996), Rouse and Morris (1986), Wiener (1988), and Wickens (1994). In most of this research, it has become common to point out, as did Wiener and Curry (1980), both the benefits and the pitfalls of automation. Two things should be kept in mind, however, when discussing the problems associated with automation. First, some of the problems can be attributed not to the automation per se but to the way the automated device is implemented in practice. Problems of false alarms from automated alerting systems (e.g., the TCAS), automated systems that provide inadequate feedback to the human operator (e.g., the FMS; Norman, 1990), and automation that fails "silently" without salient indications, fall into this category. Many of this class of problems can be alleviated to some extent by more effective training of users of the automated system (Wickens, 1994).

Second, and perhaps more frequent, a class of problems can arise from unanticipated interactions between the automated system, the human operator, and other systems in the workplace. These are not problems inherent to the technological aspects of the automated device itself, but to its behavior in the larger, more complex, distributed human-machine system into which the device is introduced (Woods, 1993). This is particularly true in the cockpit environment, in which the introduction of high-level automation with considerable autonomy and authority has produced a situation in which system performance is determined by qualitative aspects of the interaction of multiple agents (Sarter, 1996).

The benefits of aircraft automation include more precise navigation and flight control, fuel efficiency, all-weather operations, elimination of some error types, and reduced pilot workload during certain flight phases. Air traffic control automation, which has been more modest to date in comparison to cockpit automation, has provided benefits in the form of improved awareness of hazardous conditions (conflict alerts) and elimination of certain routine actions that allow the controller to concentrate on other tasks (e.g., automated sector handoffs).

TABLE 12.3 Potential Human Performance Costs of Automation

Automation Cost	Source
New error forms	Sarter and Woods (1995b); Wiener (1988)
Increased mental workload	Wiener (1988)
Increased monitoring demands	Parasuraman et al. (1993); Wiener (1985)
Unbalanced trust, including mistrust	Lee and Moray (1992); Parasuraman et al. (1996)
Overtrust	Parasuraman et al. (1996); Riley (1994)
Decision biases	Mosier and Skitka (1996)
Skill degradation	Hopkin (1994); Wiener (1988)
Reduced situation awareness	Sarter and Woods (1992, 1994a, 1994b)
Cognitive overload	Kirlik (1993)
Masking of incompetence	Hopkin (1991, 1994)
Loss of team cooperation	Foushee and Helmreich (1988); Parsons (1985)

The benefits of aviation automation, whether in the air or on the ground, are not guaranteed but represent possible outcomes. Nevertheless, the economic arguments that initially stimulate investment in automation are clearly reinforced by the financial return on that investment. In some instances the prospective benefits of automation have accrued independently of any costs, for example, more fuel-efficient flight. Other benefits may be mitigated or even eliminated by the costs. For example, although automation has reduced operator workload in some work phases, the overall benefit of automation on workload has been countered by some costs. Sometimes automation decreases workload only during a short work phase, but not otherwise. At other times, automation increases workload because of increased demands on monitoring or because of the extensive reprogramming that is required. In either case, the anticipated workload reduction benefit of automation is not realized.

Table 12.3 lists the kinds of human performance costs that have been associated with automation. The list is not comprehensive and is not meant to indicate that automation is inevitably associated with these problems (for a more detailed listing of automation-related human factors concerns in the cockpit, see Funk et al., 1996). Rather, the table gives an indication of the kinds of problems that can potentially arise with automated systems. These effects are now quite well documented in the literature, at least with respect to cockpit automation. There is empirical support for each of the effects noted in the table, although the quality, quantity, and generalizability to air traffic control operations of the empirical evidence varies from effect to effect. Given that it is by now almost axiomatic that automation does not always work as planned (or as advertised) by designers, a better understanding of these effects of automation on human performance is vital for designing new automated systems that are safe and efficient. The effects that have been the most well studied are discussed here.

New Error Forms

Wiener and Curry (1980) first pointed out that, although automation can reduce or eliminate certain kinds of human error, it can also produce new error forms. Such cases do not necessarily represent a failure of the automation per se. On the contrary, the automation may work exactly as designed. However, if incorrect inputs are provided and the automation proceeds to act on these inputs in a manner that is not monitored by the human operator, or if the automation behavior is unexpected, errors can result.[3] Although it is likely that such errors arise in many domains in which automation has been implemented, they have been most widely studied in the aircraft cockpit, and particularly well documented for one automated system, the flight management system.

The FMS is a high-level automation system that represents a significant increase in complexity, authority, and autonomy over previous systems introduced into the cockpit. The FMS includes such elements as the flight management computer, the mode control panel, the autothrottle, and the flight director. The FMS has several modes of operation pertaining to flight planning, navigation, monitoring of the flight path, thrust control, and other functions. Modes can be selected for vertical and lateral navigation, heading, level change, altitude capture, etc. The FMS represents a more complex, high-level automated system than previously available in the cockpit. The added complexity has increased the number of intervening subsystems between the pilot and the aircraft control surfaces, decreasing the direct control functions of the pilot and increasing the "peripheralization" of the pilot (Norman et al., 1988; Satchell, 1993).

FMS mode errors represent a direct consequence of increased complexity. Several studies have shown that even experienced pilots do not have a complete understanding of all FMS modes or their interactions with each other, particularly in unusual circumstances (Sarter and Woods, 1992, 1994a, 1994b, 1995a, 1995b). These studies have shown, for example, that pilots of the Boeing 737-300 had some gaps in their knowledge of FMS modes and mode behavior in unusual conditions, such as an aborted takeoff. Similar mode awareness and confusion problems were noted in a subsequent study of Airbus A320 pilots (Sarter and Woods, 1995b). The difficulties were attributed to the pilots having an imperfect mental model of the various functions of the FMS (Sarter and Woods, 1992). Moreover, some pilots were not aware of the gaps in their knowledge (Sarter and Woods, 1994b). This led to automation surprises, or automation behavior that was unexpected. The resulting confusion led the pilots to ask questions such as:

---

[3]Early examples of this kind of error include numerous aircraft incidents involving the inertial navigation system that were attributed to incorrect loading of way point data (Wiener, 1988). More recent examples include incidents arising from incorrect data entry into the control and display unit of the FMS (Vakil, Midkiff, et al., 1995).

What is the automation doing now? What will it do next? and How in the world did we ever get into that mode? (Sarter and Woods, 1995a; Wiener, 1988).

Mode awareness problems were also identified in a recent analysis of ASRS reports by Vakil, Midkiff, et al. (1995). Of 184 incidents reported during the period 1990-1994, 74 percent involved FMS mode confusion or errors associated with vertical navigation. Although aircraft type was not explicitly identified (because ASRS reports previously did not allow it, for confidentiality reasons), the researchers suggested that the incidents were not unique to a single aircraft type.

The various new error forms provide considerable challenges to human factors professionals and designers of existing and proposed automation systems. Any single approach to the problem—e.g., integrated displays for vertical navigation (Vakil, Hansman, et al., 1995), training for developing improved mental models of the FMS—is unlikely to provide all the answers. It is widely believed in the human factors community that the multidimensional approach that has been characterized as human-centered automation (Billings, 1991, 1996) is necessary, although there is less agreement on the details of this approach.

Near-term proposals for automation in air traffic control do not approach the complexity, authority, and autonomy of the automated systems described previously. Nevertheless, it is worthwhile to keep in mind the lessons learned from studies of cockpit automation. Furthermore, some long-term automation concepts (e.g., the automated en route advisor and free flight) approach and even exceed the highest levels of current cockpit automation. Hence the knowledge gained from human factors studies of other automated systems is likely to be helpful no matter which automation concepts become reality in air traffic control.

Workload

Although automation can reduce workload, it does not inevitably do so. Wiener (1988) surveyed commercial glass-cockpit pilots and found that, although one-third of the sample agreed with the statement "automation reduces workload," an equal number disagreed. For example, although the FMS is meant to reduce the pilot's workload in flight management and planning and generally does so, when the FMS must be reprogrammed (e.g., because of a runway change), pilot workload is increased, particularly if reprogramming occurs during a time-critical phase, such as final approach.

Workload is also an important factor in voluntary use of automation. Riley (1996) gave examples indicating that the decision to rely (or not to rely) on automation can be one of the most important decisions a human operator can make, particularly in time-critical situations. One of the fundamental reasons for introducing automation in complex systems is to reduce workload, and thereby to reduce human error. Indeed, human operators often cite excessive workload as a factor in their choice of automation. Riley et al. (1993) tested pilots on several flight scenarios in an A320 simulator and examined the pilot's use of several automated devices (autothrottle, autopilot, FMS, flight director). Workload was cited as one of two most important factors (the other was the urgency of the situation) in the pilots' choice of automation. It should be noted that, in practice, pilots of the A320 are forced to use automated functions whose override requires disabling of substantial portions of the flight control or flight management functions normally used.

Despite the logical and intuitive rationale that operators will choose automation under heavy workload, experimental evidence is mixed. Studies both of nonpilots performing laboratory (Riley, 1994) and aviation-like tasks (Harris et al., 1993) and of pilots (Riley, 1994) have revealed little, if any, tendency to choose automation more often at higher levels of task demand. Hence this issue invites further investigation. It is possible that the influence of workload on automation use may emerge only when the workload is experienced for a sustained period of time and not transiently. Subjects in both the Riley (1994) and the Harris et al. (1993) studies viewed fatigue as another factor in choosing automation over manual control. Another possibility is that more complex attributes of workload in real environments, such as workload management and trade-offs, need to be modeled in the laboratory in order to more fully understand the impact of workload on the use of automation.

Workload may also influence the use of automation when it is difficult to engage or turn off. Such decisions can be relatively straightforward if the advantages of using the automation are clear-cut. When the benefit offered by automation is not readily apparent, however, or if the benefit becomes clear only after much thought and evaluation, then the cognitive "overhead" involved may persuade the operator not to use the automation (Kirlik, 1993). Such overhead may be particularly important when considering high-level automation aimed at providing the human operator with a solution to a complex problem. Because these aids are generally used in uncertain, probabilistic environments, it is not clear that the automated solution is better than a manual one. As a result, the human operator may expend considerable cognitive resources in generating a manual solution to the problem, comparing it to the automated solution, and then picking one of the solutions. If the operator perceives that the advantage offered by the automation is not sufficient to overcome the cognitive overhead involved, then he or she may simply choose not to use the automation and to do the task manually. This will be particularly true if the operator has unwarranted faith in his or her own capabilities.

Trust

Trust is an important factor in the use of automated systems by human operators. Sheridan (1988) discussed a number of meanings of the term *trust*, examining how trust affects the operator's use or nonuse of automation features when the occasion arises. For example, an automated tool that is reliable, accurate, and useful may nevertheless not be used if the operator believes that it is untrustworthy. Drawing on social psychological research on trust between individuals, Muir (1988) argued that similar factors influence trust between individuals and machines. She tested subjects on a simulated process control task (simulating a soft drink manufacturing plant) and found that use of an automated aid was correlated with a simple subjective measure of trust in that aid. Lee and Moray (1992) tested subjects on a similar process control task in which automation could be used to control one of the subprocesses. They also found automation reliance and subjective trust to be generally correlated, although subjects tended to show a bias toward manual control and inertia in their allocation policy. In a subsequent study, Lee and Moray (1994) found that subjective estimates of trust in the automation, in combination with subjects' self-confidence in their own skill, were jointly related to automation use. Subjects chose manual control if their confidence in their ability exceeded their trust of the automation, and they chose automation otherwise.

Trust itself is likely to be multiply determined and to vary over time. Clearly, one factor influencing trust is automation reliability. Automation that is unreliable is unlikely to be trusted by the operator and therefore will not be used, if an option is available. Automated alerting systems that emit frequent false alarms are unlikely to be trusted or even tolerated. When corporate policy or federal regulations mandate the use of an automated tool that is not trusted, then operators are likely to resort to what Satchell (1993) referred to as "creative disablement" of the automated device. Since disablement of devices is often prohibited by federal air regulations or company procedures, pilots may find themselves in a double bind.

Several studies have shown that operators choose to use automation when it works reliably and is accurate. Interestingly, occasional failures of automation seem not to be a deterrent to future use of the automation. Riley (1994) found that both college students and pilots did not delay turning on automation after a failure and in fact continued to rely on failed automation. In a study examining monitoring of automation failures (described later), Parasuraman et al. (1993) found that, even after the simulated catastrophic failure of an automated engine-monitoring system, subjects continued to rely on the automation for a period of time, although to a lesser extent than when the automation was more reliable. These findings are surprising, in view of earlier studies suggesting that operator trust in automation is slow to recover following a failure of the automation (Lee and Moray, 1992; Muir, 1988). One mitigating factor may be the overall level of automation reliability. When this is relatively high, then operators may come to rely on the automation, so that its occasional failures do not substantially reduce trust or reliance on it unless they are sustained for a period of time. Other contributing factors may be the ease with which automation behaviors can be detected and the automation enabled and disabled and overall task complexity.

Additional work is needed to identify which of these factors is important in regulating the temporal characteristics of trust following automation failure.

Mistrust

Human operators of systems tend to be conservative in their work habits. New technology, when first introduced, tends to be looked at suspiciously and perhaps mistrusted. Few technologies are an instant hit when first introduced.[4] As experience is gained with the new system, however, and given that it works reliably and accurately, most operators will tend to like and come to trust the new device. In air traffic control, many controllers were initially distrustful of automatic handoffs, but as their workload-reducing benefits were better appreciated over time, the new automation was accepted. This has not always been the case with new technology. McClumpha and James (1994) reported that pilots of advanced automation aircraft (e.g., Airbus 320) were less trusting of the automatics than they were of less advanced aircraft (e.g., Airbus 310).

Early designs of some cockpit alerting systems, such as the GPWS and TCAS, tended not to be trusted by pilots because of the frequency of false alarms. The issue of false alarms and mistrust is critical to many complex automated systems. Because of the importance of alarms and alerts in both current and future air traffic control systems, we discuss this issue in some detail.

Unfortunately, mistrust of alerting and alarm systems is widespread in many work settings because of the false alarm problem. Technologies exist for system engineers to design sensitive warning systems that are accurate in detecting hazardous conditions (ground proximity, wind shear, collision course, etc.). These systems are set with a decision threshold that minimizes the chance of a missed warning while keeping the device's false alarm rate below some low value. Two important factors that influence the false alarm rate, and hence the operator's trust in an automated alerting system, are the values of the decision threshold and the base rate of the hazardous condition.

Setting the decision threshold requires careful evaluation. For example, embedded software that decreased system sensitivity during flap movements achieved the goal of reducing nuisance alarms in a DC-9 but prevented the aircraft's wind shear advisory system from sounding during a severe microburst at Charlotte, North Carolina, in 1994. The initial consideration for setting the decision threshold of an automated warning system is the cost of a miss versus that of a false alarm. Missed signals (e.g., collisions, total engine failure) are phenomenally costly, yet their potential frequency is undoubtedly very low. Indeed, pilots may fly for decades without taking the sort of evasive action man-

---

[4]The electronic map display in the aircraft cockpit and the first Macintosh personal computer are notable exceptions to this rule. Not surprisingly, both these systems were designed with attention to human factors and with considerable input from the users of the systems.

dated by a system such as TCAS. However, if a system is designed to minimize misses at all costs, then frequent false alarms may result. A low false alarm rate (and arguably a zero false alarm rate) would appear to be critical for acceptance of warning systems by human operators. Accordingly, setting a strict or conservative decision threshold to obtain a low false alarm rate would appear to be good design practice. Should the decision threshold be set as strictly as possible? Perhaps not, because the failure to supply sufficiently advance warning could be equally problematic (but see Sorkin et al., 1988).

In an analysis of collision-warning systems for automobiles, Farber and Paley (1993) suggested that too low a false alarm rate may also be undesirable, given that police-reportable rear-end collisions are very rare events (perhaps occurring once or twice in the lifetime of a driver). If the system never emits a false alarm, then the first time the warning sounds would be just before a crash, and it is possible that the driver may not respond promptly to such an infrequent event. Farber and Paley (1993) speculated that an ideal detection algorithm might be one that gives an alarm in collision-possible conditions, even though the driver would be likely to avoid a crash. Although technically a false alarm, this type of information might be construed as a warning aid in allowing improved response to an alarm in a collision-likely situation. Thus all false alarms need not necessarily be harmful. This idea is similar to that of graded warnings and to the concept of "likelihood alarms" as espoused by Sorkin et al. (1988) and is incorporated in the current TCAS philosophy.

Setting the decision threshold for a particular device's false alarm rate may be insufficient by itself to ensure high alarm reliability. Despite the best intentions of designers, the availability of the most advanced sensor technology, and the development of very sensitive detection algorithms, one fact may conspire to limit the effectiveness of automated alarms: the low a priori probability or base rate of most hazardous events. If the base rate is low, as it often is for many real events, then the posterior odds of a true alarm can be quite low even for very sensitive warning systems (see Parasuraman et al., in press, for a signal detection theory/Bayesian analysis of the posterior probability problem as applied to collision warnings).

Figure 12.2 shows the dependence of the posterior probability on the base rate for a system with a given detection sensitivity, as predicted by Bayes's theorem. The family of curves represents the different posterior probabilities that can be achieved with different decision thresholds. For example, the decision threshold can be set so that this warning system (with a sensitivity of $d' = 4.65$) can detect hazardous conditions with a near-perfect hit rate of .999, while having a false alarm rate of .0594. Despite these impressive detection statistics, application of Bayes's theorem shows that the human operator could find that the posterior probability (or odds) of a true alarm with such a system can be quite low. When the a priori probability (base rate) is low, say .001, only 1 in 59 alarms that the system emits represents a true hazardous condition (posterior probability =

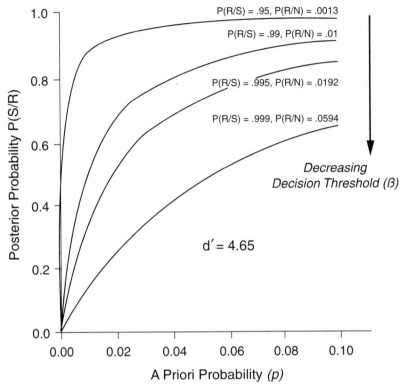

FIGURE 12.2 Posterior probability (*P(S/R)*) of a hazardous condition (*S*) given an alarm response (*R*) for an automated warning system with a fixed sensitivity $d' = 4.65$, plotted as a function of the a priori probability *p* (base rate) of *S*. Source: Parasuraman et al. (1996).

.0168; see Parasuraman et al., in press). Little wonder, then, that many human operators tend to ignore, mistrust, and turn off alarms—they have cried wolf once too often (Sorkin, 1988). As Figure 12.2 indicates, reliably high posterior alarm probabilities are guaranteed for only certain combinations of detection sensitivity and base rate. Even if operators do not ignore an alarm, they may be sluggish to respond when the posterior probability is low. This was confirmed in a laboratory study by Getty et al. (1995), who had subjects respond to a visual alert while performing a tracking task. Accurate tracking and rapid response to a true alarm were rewarded with a bonus; poor performance on either task was penalized. For a given set of bonuses and penalties, Getty et al. (1995) found that subjects became progressively slower to respond to a true alarm as the posterior probability of the alarm was reduced from .75 to .25.

Given these results, how should the parameters for an automated alerting system be set? Consistently true alarm response occurs only when the a priori probability of the hazardous event is relatively high. There is no guarantee that this will be the case in many real systems. Thus, designers of automated alerting systems must take into account not only the decision threshold at which these systems are set (Kuchar and Hansman, 1995; Swets, 1992) but also the a priori probabilities of the condition to be detected (Parasuraman et al., in press). Only then will operators tend to trust and use the system. In addition, a possible effective strategy to avoid operator mistrust would seem to be to inform users of the inevitable occurrence of false alarms when base rates are low.

Finally, in addition to mistrust related to systems that give frequent false alarms, human operators may also express leeriness of automated systems that they do not understand well. As discussed earlier, pilots have been found to have an incomplete knowledge of the various modes and mode behaviors of the FMS (Sarter and Woods, 1994b). Despite the designer's intent to produce a useful software product, an incomplete or underspecified mental model of the automation on the part of the operator can undermine the benefit of the automation because of operator mistrust. As automated systems become more complex and their behaviors less predictable, efforts must be made to make automation more transparent, so as not to generate mistrust.

Overtrust (Complacency)

If some automated alerting systems are ignored because of operator mistrust, then others may be "overtrusted," in the sense that operators may come to rely uncritically on the automation without recognizing its limitations or may fail to monitor the inputs to the automation. High trust in automation could lead operators to not carry out vigilant monitoring of their displays and instruments. In numerous aviation incidents over the past two decades, problems of monitoring of automated systems have been involved as one, if not *the* major cause of the incident. An early example is the crash of Eastern Flight 401 in the Florida Everglades, in which the crew, preoccupied with diagnosing a possible problem with the landing gear, did not notice the disengagement of the autopilot and did not monitor their altitude, even though the descent was apparent from the instruments and despite a query (although ambiguous) from a controller who noticed the loss of altitude (National Transportation Safety Board, 1973).

Although poor monitoring in the cockpit can have multiple determinants, pilot overreliance on automation is thought to be a contributing factor. Analyses of pilots' subjective impressions of cockpit automation have revealed that overreliance on automation is common among pilots. Empirical studies have also shown that skilled, subject-matter experts sometimes have misplaced trust in diagnostic expert systems (Will, 1991) and other forms of computer technology that give wrong advice (Weick, 1988). Analyses of ASRS reports have provided evidence of monitoring failures linked to excessive trust in, or overreliance on, automated systems such as the autopilot and flight management system (Mosier et al., 1994; Singh et al., 1993). Mosier et al. (1994) examined a number of similar reports in the ASRS database. They found that 77 percent of the incidents in which overreliance on automation was suspected involved a probable failure in monitoring. Such incidents are not restricted to aviation. In a recent maritime accident involving a cruise ship that ran aground off Nantucket Island, a GPS-based navigational system failed "silently." The ship's crew did not monitor other sources of position information that would have indicated that they had drifted off course (Phillips, 1995).

The factors influencing the monitoring of automation have been studied by Parasuraman and colleagues (Parasuraman, 1987; Parasuraman et al., 1993, 1994, 1996). One factor is the overall task load imposed on the operator. When monitoring an automated system is the only task given to humans, then, despite the extensive evidence of the fragility of human vigilance, human operators are very efficient at detecting any failures in the automation. However, when they have to perform other manual tasks simultaneously, then monitoring of automation failures is degraded (Parasuraman et al., 1993). This complacency effect seems to reflect an overreliance on automated systems when task load is high. The consistency of automation reliability is also a relevant factor. If the automation is inconsistent, being sometimes highly reliable and sometimes not, then subjects are better at monitoring the automation, presumably because their low trust precludes excessive reliance on the automation (Parasuraman et al., 1993).

Skill Degradation

Overtrust of automation can be a particular problem if it occurs early in the implementation of an automated system that is given high authority and autonomy. The human operator who believes that the automation is 100 percent reliable will be unlikely to monitor the inputs to the automation or to second-guess its outputs. In Langer's (1989) terms, the human operator makes a "premature cognitive commitment," which affects his or her subsequent attitude toward the automation. The autonomy of the automation could be such that the operator has little opportunity to practice the skills involved in performing the automated task manually. If that is the case, then the loss in the operator's own skills relative to that of the automation will tend to lead to an even greater reliance on the automation (Lee and Moray, 1992), in a sort of vicious cycle (Mosier et al., 1994; Satchell, 1993).

Overreliance on automated solutions may thus lead to another cost that has been associated with automation, namely skill degradation. Temporary loss of manual skills following extensive use of an automated aid have been reported by both pilots (Wiener, 1988) and controllers (Hopkin, 1995). There are also several ASRS reports that indicate that pilots resort to disengaging automation in order to maintain their manual skill proficiency. During the late 1960s, United Airlines training personnel observed a higher-than-expected failure rate among first officers of DC-10 aircraft converting to captain positions in the less-automated Boeing 727. Degradation of manual flying skills was posited as a cause of the failures. Officers scheduled for conversion were thereafter advised to obtain increased manual flying practice, and the training failures were reduced. However, there is currently no objective evidence, either from laboratory or field studies, that prolonged use of automation is associated with decreases in manual performance, either for pilots or for controllers. It is also worthwhile noting that the issue of manual skill degradation may not always be moot. With the introduction of certain forms of automation, reversion to manual performance may no longer be an option, so the question of skill degradation does not arise.

Situation Awareness

In Chapters 2 and 5, we described the importance of the air traffic controller's maintenance of situation awareness—maintaining the big picture of the airspace within and around the sector. As several have argued, situation awareness of the state of automated control devices is equally important (Sarter and Woods, 1995a; Wickens, 1996), so that operators can respond in a timely and appropriate fashion if the system encounters a fault or circumstances make the programmed behavior inappropriate. Automation can influence the controller's situation awareness (Garland and Hopkin, 1994); in general, high levels of automation may endanger situation awareness in four respects.

First, if automation accomplishes operations previously under human control and fails to inform the operator of those operations (e.g., changing modes), or does so with very subtle signals (e.g., a change in the value of an alphanumeric character on a cluttered display), situation awareness is obviously degraded (Sarter and Woods, 1995a). Second, even if such state changes are more evident, as we have noted, reduced levels of vigilance can cause the operator to fail to notice state changes. Third, even if designers have provided salient alerts to automation, triggering state changes, behavioral research has established that people are more likely to remember events (i.e., state changes) if they have been the active agents in initiating those changes, rather than the passive witnesses of other agents making the same changes (Hopkin, 1995; Vortac, 1993). Finally, effective situation awareness depends not only on available and well-processed information, but also on an accurate mental model of the system under supervision.

If automation is designed to carry out procedures in a different and more complex way than humans normally carry out the task, they will again be less able to encode and remember the state changes that are signaled (Sarter and Woods, 1995a). In air traffic control, the conflict resolution advisor system will provide resolution advisories to controllers on potential conflicts. Controllers may come to accept the proposed solutions as a matter of routine, and this may lead to a loss of situation awareness compared with the case when the solution is generated manually. In an early study of decision aiding in air traffic control, Whitfield et al. (1980) reported such a loss of the mental picture in controllers, who tended to use the automated resolutions under conditions of high workload and time pressure.

Interactions Between Factors Affecting Use of Automation

We have discussed several factors that can influence a human operator's decision to use automation. Several other factors are presumably also important in influencing automation use. Some of these factors may have a direct influence, whereas others may interact with the factors already discussed. For example, the influence of cognitive overhead may be particularly evident if the operator's workload is already high. Under such circumstances, many operators may be reluctant to use automation even if it is reliable, accurate, and generally trustworthy.

The studies by Lee and Moray (1992) and Riley (1994) also identified self-confidence in one's manual skills as an important moderator of the influence of trust in automation. If trust in automation is greater than self-confidence, automation will be engaged, but not if trust is lower than self-confidence. Riley (1996) suggested that this interaction could itself be moderated by other factors, such as the risk associated with the decision to use or not use automation. On the basis of his experimental results, he outlined a model of automation use based on a number of factors (Figure 12.3). Solid lines indicate the influences of factors on automation use decisions that are supported by experimental evidence; dashed lines indicate factors that may also influence automation use, but for which

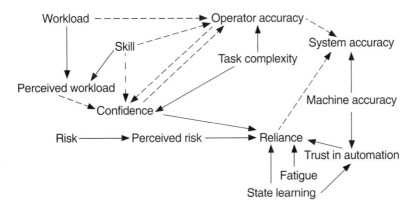

FIGURE 12.3 Factors influencing automation usage. Source: Riley (1994). Reprinted by permission.

empirical evidence is lacking. Studies in more realistic task environments are needed to validate the model shown in the figure, which provides only a general overview of the factors that potentially influence the use of automation.

HUMAN-CENTERED AUTOMATION

High workload, mistrust, overtrust (complacency), high cognitive overhead, impaired situation awareness—these represent some of the potential human performance costs of certain forms of automation. Factors such as trust, workload, and cognitive overhead also influence an operator's choice to use or not use automation in order to perform a particular task, when that choice is available. As noted previously, however, these performance costs are not inevitable consequences of automation, but rather represent outcomes associated with poorly designed automation. Can such negative outcomes be eliminated, while promoting more effective use of automation? Human-centered automation (Billings, 1991, 1996) has been proposed as a design approach that may accomplish these objectives.

Human-centered automation is a philosophy that guides the design of automated systems in a way that both enhances system safety and efficiency and optimizes the contribution of human operators. In a general sense, it requires that the benefits of automation be preserved while minimizing the human performance costs described earlier in this chapter. However, although human-centered automation is currently a fashionable idea in aviation and other contexts, its precise meaning is not well or commonly understood. It evokes many associations, some good and some not so good. The many faces of human-centered automation need to be considered. At various times and in various contexts, it can mean:

- Allocating to the human the tasks best suited to the human and allocating to the automation the tasks best suited to it.
- Maintaining the human operator as the final authority over the automation, or keeping the human in command.
- Keeping the human operator in the decision and control loop.
- Keeping the human operator involved in the system.
- Keeping the human operator informed.
- Making the human operator's job easier, more enjoyable, or more satisfying through automation.
- Empowering or enhancing the human operator to the greatest extent possible through automation.
- Generating trust in the automation by the human operator.
- Giving the operator computer-based advice about everything he or she might want to know.

- Engineering the automation to reduce human error and keep response variability to the minimum.
- Casting the operator in the role of supervisor of subordinate automatic control system(s).
- Achieving the best combination of human and automatic control, best being defined by explicit system objectives.
- Making it easy to train operators to use automation effectively, minimizing training time and costs.
- Creating similarity and commonality in various models and derivatives that may by operated by same person.
- Allowing the human operator to monitor the automation.
- Making the automated systems predictable.
- Allowing the automated systems to monitor the human operator.
- Designing each element of the system to have knowledge of the other's intent.

These seemingly innocuous objectives can often be undesirable and/or in conflict with one another. Their problems and inconsistencies are apparent when the several meanings of human-centered automation are considered further.

Allocation of Tasks to Humans and to Automation

Appropriate allocation of tasks to humans and to automation is easy to say, but not so easy to do. The Fitts (1951) approach to function allocation specifies which tasks are performed better by machines and which by humans. This approach, developed some 45 years ago, has not been able to provide an effective procedure for task allocation. Furthermore, there is the question: If designers automate those tasks that machines are better at and also require the operator to monitor the automation and maintain situation awareness of all those variables, will there be a gain in system efficiency and safety?

Human In or Out of the Loop

Keeping the human operator in the decision and control loop can mean full manual control, or it can mean tolerance of human intervention into the automatic control. Sometimes it may be best to get the operator out of the loop altogether, not letting him or her touch anything, including overriding or adjusting the automatic control. Early in the development of nuclear reactors, it was agreed that certain safety-related operations that must be performed in the case of loss of coolant must be fully automatic—the human operator was too slow and undependable in a stressful situation. The industry later evolved a standard practice that any safety-related action that must be performed within 10 minutes of a major event must be automatic, with the human operator observing, hands off.

The Human Operator as the Final Authority

Maintaining the human operator as the final authority over the automation is another meaning of human-centered automation. Many people may feel more comfortable if they know that, in the end, some human is in charge of a complex automated system. But is this belief justified? Humans are not known for their reliability; machines are. Although the human can always pull the plug or jam the machine if necessary, that may take more time than the process normally allows. In any case, the appropriateness of human or machine as final authority is likely to be context dependent. Sometimes it may be safest to require an extra confirmation action by the human (e.g., to an "are you sure?" query by the machine) or covers or guards to certain switches that must first be removed (like a fire alarm box), or tests of human capability before the human is allowed to override an automatic system. Sometimes specific action of two or more independent humans may be required before an automatic process is initiated (e.g., keys inserted and turned on by two designated officers to enable the firing of a ballistic missile).

Job Satisfaction

Some researchers claim that human-centered automation involves making the human operator's job easier, more enjoyable, or more satisfying through friendly automation, although this is clearly not the most prevalent view of its major characteristic. The operator's job may be easiest when he or she is doing nothing (or doing poorly). Designing for greater ease makes sense only if all other factors stay the same, including the tendency of the operator to become bored or drowsy—tendencies that can be enhanced by "easy" jobs. And what most satisfies the system operator may not be what most satisfies the management, the customer, or the public. Reducing the operator's mental workload, at least to a comfortable or acceptable level, is an admirable goal. But the same semiautomatic system that results in a comfortable workload under normal conditions can be quite uncomfortable under abnormal conditions. As noted previously, automation intended to decrease operator workload can end up increasing workload at the most critical times (Bainbridge, 1983; Wiener, 1988).

Empowering the Human Operator

Empowering an operator who is misguided or lacking in certain capacities can be dangerous. Empowerment may be doubly problematic if, as Hopkin (1994) has suggested in the context of future air traffic control, operator incompetence is masked because of the routine acceptance of automated solutions. The problem of empowerment was the theme of Norbert Wiener's (1964) Pulitzer prize-winning book, *God and Golem, Incorporated*. Wiener's main theme was that the computer does what it is programmed to do, whether that is what its programmer intended or not. Although stand-alone computers may not be dangerous in this respect, computers hooked up to hardware, especially rapidly responding hardware, can do significant damage before they can be stopped.

Generating Trust in Automation by the Human Operator

This view of human-centered automation can be broken down into several subgoals: making the automation more reliable and predictable, better able to adapt to a variety of circumstances, more familiar in the sense of its operation being understandable, and more open and communicative about what it is doing or about to do. These are all properties of a trustworthy friend or helper—and that is fine if trust is deserved. A system must not give the impression that it is operating normally when it is not. In some cases, operators are taught *not* to trust the computer or the machine. As noted previously, there can be negative consequences of both mistrust and overtrust. Trust must be calibrated appropriately to the system. Normally trust is built up over a period of time, but failure can give no warning. An operator once burned will have difficulty regaining trust. Trust requires in addition some capability to fail safe or fail soft.

Keeping the Human Operator Informed

Humans can absorb and make use of only very limited quantities of information. It is well established that displaying all the information that might be useful means there is too much information to be able to find what is needed when it is needed. The control panel at the nuclear power plant at Three Mile Island and the Boeing 707 cockpit are early examples of this problem. Modern control rooms and cockpits, at any given time, actually display less information than before, but make it available for the asking. But then a problem must be faced: How should the operator ask for it and, in an emergency, is there a danger that the operator will forget how to ask, or inadvertently request and act on the wrong information, information believed to be characterizing a different variable than what was actually requested?

As discussed previously, this type of mode error has been noted to occur in many highly automated cockpits and is directly responsible for at least one fatal airline accident. The computer can always be designed to second-guess the operator when the computer thinks it knows what the operator should be interested in (for example, to generate displays on the computer screen as a function of what operating mode the system is in, or to give unsolicited advice or to bring up the appropriate screen in the event of a system failure, as in the MD-11 aircraft's automated systems displays). But for some reason not known to the computer at that point, the operator may really want to see some other information. Even given that old-fashioned display clutter has been cleared up in modern systems, there remains the hazard of bombarding the operator with advice in other forms—what some pilots have referred to as "killing us with kindness."

Engineering Automation to Reduce Human Error

Human resourcefulness in case of automation failure may require taking liberties that are normally seen as human error, for they may circumvent standard emergency procedures. Such procedures may be appropriate in most cases but inappropriate in some specific case that had not been considered by the procedure writers. In any case, for the human to experiment and learn about the system, some tolerance for nonstandard behavior (variability and what would normally be called inappropriate response or error) is necessary. For example, on several occasions in the Apollo lunar spacecraft expeditions, ground controllers at Cape Canaveral had to "fool the computer" by giving it nonstandard instructions in order to cope with certain bugs and failures that were encountered. The human operator can exercise his or her marvelous capability to learn and adapt only if allowed some freedom to experiment. In the Darwinian sense, there must be some "requisite variety." Fooling the computer is commonplace in glass cockpits (Wiener, 1988).

Human Supervisory Control

Being a supervisor takes the operator out of the inner control loop for short periods or even for significantly longer periods, depending on the level at which the supervisor chooses to operate. In a human organization, the boss may not know in any detail what the subordinate employees are doing, and the more layers of middle management there are, the less the supervisor may know.

Optimal Combination of Human and Automatic Control

To explicate system objectives in a quantitative way that allows for mathematical optimization is usually not possible, or at least is very difficult. This is especially true when there are multiple conflicting objectives. It is often the case that large system planners seize on one or two easily quantifiable criteria and optimize those, totally ignoring what admittedly might be more important criteria that are not easily quantifiable. Furthermore, when the optimal combination is defined as a flexible one, which may adaptively change over time (Hancock and Chignell, 1989), such flexibility can lead to inconsistency and, worse yet, ambiguity regarding who's in charge at a given point in time. Billings (1991, 1996) emphasizes the importance of the need for both the human and automated agent to share knowledge about the other's operations and functioning, intent, and plans.

We have provided several examples to illustrate that these principles are not always upheld in current automated systems. For example, the work on FMS mode awareness indicates that the multiple agents in the cockpit do not always know each other's intent; studies on monitoring, overtrust, and silent automation failures show that human operators are not always able to monitor the automation effectively; and studies on automation surprises indicate that automated systems are sometimes unpredictable.

Billings' view of human-centered automation (1991, 1996) provides some general guidelines for the design of future automated systems and sets some boundary conditions on the types and levels of automation (see Table 12.1) that are appropriate. For example, human-centered automation would seem to rule out certain very high-level automation with complete autonomy for more complex cognitive functions (e.g., level 9 or 10), on the grounds that this would subvert the principle of the human operator's being in command of basic decision-making processes. Accordingly, concepts in which the controller is removed from responsibility for maintaining separation between aircraft, would seem to violate the principles of human-centered automation. By the same token, data-linking of information from the controller to the FMS can communicate the intent of the controller to the pilot regarding the flight path to be followed, but for the mutual intent principle of human-centered automation to be met, pilot intent, particularly if different from that programmed into the FMS, must also be communicated back to the controller. Mutual knowledge of intent (air-to-air and air-to-ground) is also likely to be an important factor in the efficient implementation of such future automation concepts as free flight.

The Architectural Framework of Supervisory Control

Human supervisory control may provide an appropriate architecture for human-centered automation in general (Sheridan, 1992) and for air traffic control in particular (see Figure 12.4). The human roles of planning what is to be done, particularly what automatic actions, teaching (programming) the computer, monitoring the automatic action while looking for abnormalities, intervening when necessary, and learning from experience, still seem appropriate to human roles, although the computer is learning to help (or encroach, depending on one's viewpoint) even here.

It is fashionable to assert that today's complex supervisory systems require more cognition than before and less motor skill. One might contend, however, that the cognitive skills have always been there, and that earlier it was just easier to integrate them with the required manual skills, since the body had learned to do this naturally over thousands of years of evolution, with much of the communication going on internally and subconsciously. Now the operator must behave more like a mother, trying to think ahead and anticipate problems for the child (the computer, the automated system). The mother must communicate quite explicitly but let the child do the acting, meanwhile monitoring the child's behavior.

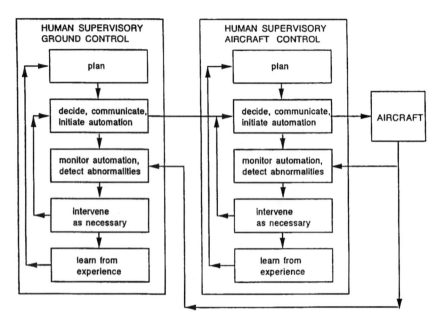

FIGURE 12.4   Architecture for supervisory control.

Supervisory control has many reasonable manifestations, but currently there is no predictive model for supervisory control that is acceptable and robust. One can say that current TRACON controllers are doing tactical supervisory control through their elaborate and high-level display systems, and also in the sense that they are giving commands to the pilots, who in turn are closing the aircraft position control loop locally with their own respective aircraft. (Currently the human pilot serves as the lower-level task-interactive "computer" in the traditional supervisory paradigm.) Current tower controllers perform supervisory control in the same sense, especially in conditions of instrument flight rules in which GCA (ground controlled approach) or AUTOLAND (automated landing) systems are used to partially automate the real-time control itself. The CTAS traffic management advisor would extend the controller's supervisory control into the strategic arena.

The supervisory functions of the controller operator are clustered under the following categories, outlined in the figure.

• *Plan*, which is performed off-line, in the form of training. It includes (a) understanding of the physical system well enough to have a working mental model of the characteristics of different aircraft (required speeds, separations, etc.). It also includes (b) knowledge of objectives (relative importance, urgency, and good-bad evaluations of events). Coming to understand both of these is augmented by (computer-based) training aids. In an ideal case in which these two functions are completely specified in mathematical form, a simultaneous solution determines optimal performance. In the real world, the supervisor must further take into account (c) procedures and guidelines specified by higher authority (FAA) in order to set strategy.

- *Monitor*, which is the controller's afferent function performed on-line. It includes (a) allocation of attention (what to look at and listen to in order to get the needed information) and is driven largely by the operator's mental model of expectations as well as by the current displays and voice communications. Next it includes (b) estimation of process state (defined as the lateral and vertical positions of all aircraft under surveillance), which can be augmented by TCAS predictor lines and other visualization aids. A final step is to evaluate the state as estimated, to determine whether there is some abnormality that requires special attention, as further aided by TCAS alarms and similar advice from the computers.
- *Decide and Communicate*, the on-line efferent function of the controller. This is broken down into steps. Step (a) involves deciding what are the proper actions to take, based on the operator's knowledge of where the errant aircraft is (are) positioned and headed, what are the options available, and the expected results of taking those options. In this case, the controller must be guided by FAA rules and procedures as well as whatever CTAS, TCAS, etc., aiding exists. Step (b) involves communication in the normal case, which must be brief and in the proper format, and in the near future will be aided by datalink. Step (c) involves communication in the abnormal case, wherein instructing the selected one or several aircraft takes priority over the other aircraft. The loop is closed to ensure that proper actions are being taken, which is important for normal as well as abnormal situations.
- *Learn*, a supervisory function that is (a) partially an on-line memory task and (b) partially a matter of later off-line reflection and study of recorded events.

In the future, supervisory control in air traffic control will probably move toward further automation, which means there will be more aids to support tasks identified in Figure 12.4, and communication will be more by datalink and less by voice. More of the functions performed now by the pilot will also be automated or at least aided by computer means.

Implementation Prospects

Although the concept of human-centered automation provides a general framework for the design of automated systems, as currently formulated it cannot provide specific details on particular automation components. It may not always be clear whether particular automated subsystems meet a particular principle of human-centered automation, and, if not, how they can be redesigned to do so.

Furthermore, some principles may require solutions to conflicting problems. For example, literal adherence to the principle of keeping the controller informed would lead to an information explosion and added workload. How much information should be provided to keep the controller sufficiently informed under normal as well as contingency conditions, and how it is displayed, are key issues in meeting this requirement. Other principles may also be difficult to achieve. For example, it is not clear how the high-level decision-making activities of the controller can be monitored covertly by the automation. Overt monitoring is possible, e.g., by query, but this can be cumbersome and aversive to the controller, who may not like to have his or her actions continually questioned (although in some cases it can be helpful). Further research and conceptual analyses are needed to address these issues of implementation of human-centered automation. The information gained from studies of human use of automation can also be added to the knowledge base of the automation. If an intelligent system can predict when the controller is likely to choose or not to choose a particular automated subsystem, then communication of intent can be facilitated.

CONCLUSIONS

Automation refers to devices or systems that execute functions that could be carried out by a human operator. Various levels of automation can be identified between the extremes of direct manual control and full automation, with higher levels being associated with greater complexity, autonomy, and authority.

A number of components of automation have been introduced in air traffic control over the past decades in the areas of sensing, warning, prediction, and information exchange. These automated systems have provided a number of system benefits, and acceptance by controllers has generally been positive. Several higher-level automated systems targeting decision-making and planning functions are being contemplated, both in the near term and in the long term. Advanced automation is to be introduced because of anticipated increases in traffic over the next decade, which threaten to outstrip the handling capabilities of the current system. It is hoped that automation will not only increase capacity, but also improve safety, increase efficiency, reduce personnel, operational, and maintenance costs, and reduce the workload levels of controllers. Achieving these outcomes will require consideration of human factors involved in operator interaction with automation.

Past and most current automated systems have been designed using a technology-centered approach. These systems have led to numerous benefits, including more efficient performance, elimination of some error types, and reduced operator workload in some cases. At the same time, several costs have been noted, including increased workload, increased monitoring demands, reduced situation awareness, unbalanced trust (mistrust and overtrust), new error forms, the masking of incompetence, and loss of team cooperation. These and related factors also influence the controller's choice to use or not to use automation, although the relative importance of the factors and their interactions are not fully understood.

There is considerable interest in the human factors community in the design philosophy known as human-centered automation, although there is less agreement on its specific characteristics. Several meanings can be discerned; however, the objectives of each view can be undesirable and/or in conflict with one another. Effective human-centered automation requires that these inconsistencies be resolved. The specific means by which the principles are realized in design also remain to be fully articulated.

References

CHAPTER 1: OVERVIEW

Billings, C.E.
- 1996a *Aviation Automation: The Search for a Human-Centered Approach.* Mahwah, NJ: Lawrence Erlbaum.
- 1996b Toward a human-centered aircraft automation philosophy. *The International Journal of Aviation Psychology* 1(4):261-270.

Cooper, G.E., M.D. White, and J.K. Lauber, eds.
- 1980 Resource Management on the Flight Deck: Proceedings of a NASA/Industry Workshop held at San Francisco, California, June 26-28, 1979 (NASA Conference Publication No. 2120).

Diehl, A.E.
- 1991 The effectiveness of training programs in preventing aircrew "error." Pp. 640-655 in *Proceedings of the Sixth International Symposium on Aviation Psychology*, R. Jensen, ed. Columbus, OH: Ohio State University.

Eckhardt, D.E., W.H. Bryant, R.M. Hueschen, and D.C. Foyle
- 1996 Low-Visibility Landing and Surface Operations for NASA's Terminal Area Productivity Program. Information Paper 4. International Civil Aviation Organization (ICAO) All Weather Operations Panel (AWOP) Advanced Surface Movement Guidance and Control Systems (A-SMGCS) Subgroup Meeting, Atlanta, GA, January.

Federal Aviation Administration
- 1996 *Aviation System Capital Investment Plan.* Washington, DC: Federal Aviation Administration.

Fischhoff, B.
- 1982 Debiasing. In *Judgment Under Uncertainty: Heuristics and Biases*, D. Kahneman, P. Slovic, and A. Tversky, eds. New York: Cambridge University Press.

Foushee, H.C.
- 1984 Dyads and triads at 35,000 ft.: Factors affecting group processes and aircrew performance. *American Psychologist* 39:885-893.

290

Luffsey, W.S.
    1990   *How to Become an FAA Air Traffic Controller.* New York: Random House.

Mellone, V.J., and S.M. Frank
    1993   Behavioral impact of TCAS II on the National Air Traffic Control System. In *Proceedings of the Seventh International Symposium on Aviation Psychology*, R. Jensen, ed. Columbus, OH: Ohio State University.

Monan, W.P.
    1987   *Cleared for the Visual Approach: Human Factors Problems in Air Carrier Operations.* NASA Contractor Report 166573 (September).

Murphy, M.
    1980   Analysis of eighty-four commercial aviation incidents: Implications for a resource management approach to crew training. *Proceedings of the Annual Reliability and Maintainability Symposium.* IEEE.

Planzer, N., and M. Jenny
    1995   Managing the evolution to free flight. *Journal of Air Traffic Control* March:18-20.

Sarter, N., and D.D. Woods
    1995   *"Strong, Silent, and Out-of-the-Loop": Properties of Advanced (Cockpit) Automation and Their Impact on Human-Automation Interaction.* Technical Report CSEL 95-TR-01. Columbus, OH: Cognitive Systems Engineering Laboratory, Ohio State University.

Stix, G.
    1994   Aging airways. *Scientific American* May:96-104.

U.S. Congress Office of Technology Assessment
    1988   *Federal Research and Technology for Aviation.* OTA-SET-610. Washington, DC: U.S. Government Printing Office.

Wiener, E.L.
    1977   Controlled flight into terrain accidents: system-induced errors. *Human Factors* 19:171-181.
    1989   *Human Factors of Advanced Technology (Glass Cockpit) Aircraft.* Washington, DC: National Aeronautics and Space Administration.
    1995   Opening remarks on debate on automation and safety. Presented at the Annual meeting of Human Factors and Ergonomics Society, San Diego, CA.

Wiener. E.L., and R.E. Curry
    1980   Flight deck automation: promises and problems. *Ergonomics* 23(10), 995-1011.

CHAPTER 2: TASKS IN AIR TRAFFIC CONTROL

Cheaney, E.S., and C.E. Billings
    1981   Application of the epidemiological model in studying human error in aviation. *1980 Aircraft Safety and Operating Problems Conference.* NASA CP 2170.

Federal Aviation Administration
    1988   *FAA Organization Manual.* Order 1100.148B (December 6). Washington, DC: U.S. Department of Transportation.
    1989   *FAA Organization—Field.* Order 1100.5C (February 6). Washington, DC: U.S. Department of Transportation.
    1990   *Standard Organization of Air Traffic Control Terminal Facilities.* Order 1100.126F (April 13). Washington, DC: U.S. Department of Transportation.
    1994   FAA reorganizes along its key lines of business. *Headquarters Intercom* (December 13). Washington, DC: U.S. Department of Transportation.

Huey, B.M., and C.D. Wickens, eds.
  1993 *Workload Transition: Implications for Individual and Team Performance.* Committee on Human Factors, National Research Council. Washington, DC: National Academy Press.
Luffsey, W.S.
  1990 *How to Become an Air Traffic Controller.* New York: Random House.
Monan, W.P., Cpt.
  1986 *Human Factors in Aviation Operations: The Hearback Problem* (NASA Contractor Report 177398). Moffett Field, CA: NASA Ames Research Center.
Pritchett, A., and R.J. Hansman
  1993 Preliminary analysis of pilot ratings of "party line" information importance. In *Proceedings of the Seventh International Symposium on Aviation Psychology*, R. Jensen, ed. Columbus: Ohio State University

CHAPTER 3: PERFORMANCE ASSESSMENT, SELECTION, AND TRAINING

Ackerman, P.L., R. Kanfer, and M. Goff
  1995 Cognitive and noncognitive determinants and consequences of complex skill acquisition. *Journal of Experimental Psychology: Applied* 4:270-304.
Barrick, M.R., and M.K. Mount
  1991 The big five personality dimensions and job performance: A meta-analysis. *Personnel Psychology* 44:1-26.
Borman, W.C., J.W. Hedge, and M.A. Hanson
  1992 Criterion Development in the SACHA Project: Toward Accurate Measurement of ATCS Performance. Institute Report #222. Prepared by Personnel Decisions Research Institutes, Inc.
Broach, D., and J. Brecht-Clark
  1994 *Validation of the Federal Aviation Administration Air Traffic Control Specialist Pre-Training Screen.* DOT/FAA/AM-94/4. Prepared for the Federal Aviation Administration, Office of Aviation Medicine. Available through the National Technical Information Service, Springfield, VA 22161. Washington, DC: U.S. Department of Transportation.
Buckley, E.P., W.F. O'Connor, and T. Beebe
  1969 *A Comparative Analysis of Individual and System Performance Indices for the Air Traffic Control System.* (NA-69-40; RD-69-50; Government accession #710795). Atlantic City, NJ: Department of Transportation, Federal Aviation Administration, National Aviation Facilities Experimental Center, Systems Research and Development Service.
Buckley, E.P., B.D. DeBaryshe, N. Hitchner, and P. Kohn
  1983 *Methods and Measurements in Real-Time Air Traffic Control System Simulation.* Prepared for the Federal Aviation Administration. Washington, DC: U.S. Department of Transportation.
Cattell, R.B., H.W. Eber, and M.M. Tatsuoka
  1970 *Handbook for the 16PF.* Champaign, IL: Institute for Personality and Aptitude Testing (IPAT).
Collins, W.E., L.G. Nye, and C.A. Manning
  1990 *Studies of Poststrike Air Traffic Control Specialist Trainees: III. Changes in Demographic Characteristics of Academy Entrants and Biodemographic Predictors of Success in Air Traffic Controller Selection and Academy Screening.* DOT/FAA/AM- 90/4. Prepared for the Federal Aviation Administration, Office of Aviation Medicine. Washington, DC: U.S. Department of Transportation.

Della Rocca, P., C.A. Manning, and H. Wing
    1990 *Selection of Air Traffic Controllers for Automated Systems: Applications from Current Research.* DOT/FAA/AM-90/13. Prepared for the Federal Aviation Administration, Office of Aviation Medicine. Available through the National Technical Information Service, Springfield, VA 22161. Washington, DC: U.S. Department of Transportation.

Druckman, D., and R. Bjork, eds.
    1994 *Learning, Remembering, Believing: Enhancing Human Performance.* Committee on Techniques for the Enhancement of Human Performance, National Research Council. Washington, DC: National Academy Press.

Federal Aviation Administration
    1995 3120.4H Air Traffic Technical Training. Prepared by Air Traffic Program Management.

Greeno, J.G., D.R. Smith, and J.L. Moore
    1993 Transfer of situated learning. Pp. 99-167 in *Transfer on Trial: Intelligence, Cognition, and Instruction,* D.K. Detterman and R.J. Sternberg, eds. Norwood, NJ: Ablex.

Hedge, J.W., W.C. Borman, M.A. Hanson, G.W. Carter, and L.C. Nelson
    1993 Progress Toward Development of ATCS Performance Criterion Measures. Institute Report #235. Prepared for Personnel Decisions Research Institutes, Inc.

Knerr, C.M., J.E. Morrison, R.J. Mumaw, D.J. Stein, P.J. Sticha, R.G. Hoffman, D.M. Buede, and D.M. Holding
    1987 Simulation-Based Research in Part-Task Training. AF HRL-TR-86-12, AD-B107 293. Air Force Human Resources Laboratory, Brooks Air Force Base, TX.

Lave, J., and E. Wenger
    1991 *Situated Learning: Legitimate Peripheral Participation.* Cambridge, England: Cambridge University Press.

Lewis, M.
    1978 Use of the Occupational Knowledge Test to Assign Extra Credit in Selection of Air Traffic Controllers. FAA Office of Aviation Report No. FAA-AM-78-7, 1978.

Manning, C.A., and D. Broach
    1992 *Identifying Ability Requirements for Operators of Future Automated Air Traffic Control Systems.* DOT/FAA/AM-92/26. Prepared for the Federal Aviation Administration, Office of Aviation Medicine. Available through the National Technical Information Service, Springfield, VA 22161. Washington, DC: U.S. Department of Transportation.

Manning, C.A., P.S. Della Rocco, and K.D. Bryant
    1989 *Prediction of Success in FAA Air Traffic Control Field Training as a Function of Selection and Screening Test Performance.* DOT/FAA/AM-89/6. Prepared for the Federal Aviation Administration, Office of Aviation Medicine. Available through the National Technical Information Service, Springfield, VA 22161. Washington, DC: U.S. Department of Transportation.

Manning, C.A., P.S. Kegg, and W.E. Collins
    1988 *Studies of Poststrike Air Traffic Control Specialist Trainees: II. Selection and Screening Programs.* DOT/FAA/AM-88/ 3. Prepared for the Federal Aviation Administration, Office of Aviation Medicine. Available through the National Technical Information Service, Springfield, VA 22161. Washington, DC: U.S. Department of Transportation.

Milkovich, G.T., and A.K. Wigdor, eds., with R.F. Broderick and A.S. Mavor
    1991 *Pay for Performance: Evaluating Performance Appraisal and Merit Pay.* Committee on Performance Appraisal for Merit Pay, National Research Council. Washington, DC: National Academy Press.

Miller, R.B.
    1954 Psychological Considerations in the Design of Training Equipment. Report No. WADC-TR-54-563, AD 71202. Wright Air Development Center, Wright-Patterson Air Force Base, Ohio.

Naylor, J.C.
- 1962 Parameters Affecting the Relative Efficiency of Part and Whole Practice Methods: A Review of the Literature. NAVTRADEVCEN 950-1, AD-275 921. U.S. Naval Training Device Center, Port Washington, N.Y.

Nickels, B.J., P. Bobko, M.D. Blair, and E.L. Tartak
- 1995 *Separation and Control Hiring Assessment: Final Job Report.* Bethesda, MD: University Research Corporation.

Nye, L.G., and W.E. Collins
- 1991 *Some Personality Characteristics of Air Traffic Control Specialist Trainees: Interactions of Personality and Aptitude Test Scores with FAA Academy Success and Career Expectations.* DOT/FAA/AM-91/8. Prepared for the Federal Aviation Administration, Office of Aviation Medicine. Available through the National Technical Information Service, Springfield, VA 22161. Washington, DC: U.S. Department of Transportation.

O'Hare, D., and S.N. Roscoe
- 1990 *Flight Deck Performance: The Human Factor.* Ames, IA: Iowa State University Press.

Patrick, J.
- 1992 *Training: Research and Practice.* San Diego, CA: Academic Press.

Redding, R.E., J.R. Cannon, and T.L. Seamster
- 1992 Expertise in air traffic control (ATC): What is it, and how can we train for it? Pp. 1326-1330 in *Proceedings of the 36th Annual Meeting of the Human Factors Society.* Santa Monica, CA: Human Factors Society.

Rodgers, M.D., and D.A. Duke
- 1994 SATORI: situation assessment through the re-creation of incidents. Pp. 217-225 in *Situational Awareness in Complex Systems*, R.D. Gilson, D.J. Garland, and J.M. Koonee, eds. Florida: Embry-Riddle University Press.

Schroeder, D.J., D. Broach, and W.C. Young
- 1993 *Contribution of Personality to the Prediction of Success in Initial Air Traffic Control Specialist Training.* DOT/ FAA/AM-93/4. Prepared for the Federal Aviation Administration, Office of Aviation Medicine. Available through the National Technical Information Service, Springfield, VA 22161. Washington, DC: U.S. Department of Transportation.

Sells, S.B., J.T. Dailey, and E.W. Pickrel, eds.
- 1984 *Selection of Air Traffic Controllers.* FAA-AM-84-2. Prepared for the Federal Aviation Administration, Office of Aviation Medicine. Available through the National Technical Information Service, Springfield, VA 22161. Washington, DC: U.S. Department of Transportation.

Spielberger, D.C.
- 1979 *Preliminary Manual for the State-Trait Personality Inventory.* Tampa, FL: University of South Florida, Human Resources Institute.

Stein, E., and R.L. Sollenberger
- 1996 Another look at air traffic controller performance. In *Proceedings of the 40th Annual Meeting of the Human Factors Society.*

VanDeventer, A.D., W.E. Collins, C.A. Manning, D.K. Taylor, and N.E. Baxter
- 1984 *Studies of Poststrike Air Traffic Control Specialist Trainees: I. Age, Biographic Factors, and Selection Test Performance Related to Academy Training Success.* FAA-AM-84-6. Prepared for the Federal Aviation Administration, Office of Aviation Medicine. Available through the National Technical Information Service, Springfield, VA 22161. Washington, DC: U.S. Department of Transportation.

VanDeventer, A.D., D.K. Taylor, W.E. Collins and J.O. Boone
- 1983 *Three Studies of Biographical Factors Associated with Success in Air Traffic Control Specialist Screening/Training at the FAA Academy.* FAA-AM-83-6. Prepared for the

Federal Aviation Administration, Office of Aviation Medicine. Available through the National Technical Information Service, Springfield, VA 22161. Washington, DC: U.S. Department of Transportation.

Weltin, M., D. Broach, K. Goldback, and R. O'Donnel
- 1992 *Concurrent Criterion-Related Validation of Air Traffic Control Specialist Pre Training Screen*. Fairfax, VA: Aerospace Sciences, Inc.

Wigdor, A.K., and Bert F. Green, Jr., eds.
- 1991 *Performance Assessment for the Workplace*. Committee on Performance of Military Personnel, National Research Council. Washington, DC: National Academy Press.

Wing, H., and C.A. Manning, eds.
- 1991 *Selection of Air Traffic Controllers: Complexity, Requirements, and Public Interest*. DOT/FAA/AM-91/9. Prepared for the Federal Aviation Administration, Office of Aviation Medicine. Available through the National Technical Information Service, Springfield, VA 22161. Washington, DC: U.S. Department of Transportation.

CHAPTER 4: AIRWAY FACILITIES

Blanchard, R.E., and J.J. Vardaman
- 1994 *Human Factors in Airway Facilities Maintenance: Development of a Prototype Outage Assessment Inventory*. Document number DOT/FAA/AM-94/5. Office of Aviation Medicine, Civil Aeromedical Institute, Federal Aviation Administration. Washington, DC: U.S. Department of Transportation.

Booz, Allen and Hamilton, Inc.
- 1993 U.S. Department of Transportation Final Report for Airway Transportation Systems Specialists. GS-2101. Washington, DC: U.S. Department of Transportation.
- 1994 U.S. Department of Transportation Test Phase Data Analysis Report for Airway Transportation Systems Specialist. GS-2101. Washington, DC: U.S. Department of Transportation.

Federal Aviation Administration
- 1970 Air Traffic and Airway Facilities Responsibilities at NAS Computer-Equipped ARTCCs. Order number 1100.124, July 20. Washington, DC: U.S. Department of Transportation.
- 1974 Air Traffic and Airway Facilities Responsibilities at NAS Computer-Equipped Terminal Facilities. Order number 1100.139, May 21.
- 1976 Airway Facilities Career Planning Program. Order number 3410.12, April 7. Washington, DC: U.S. Department of Transportation.
- 1985 Airway Facilities Maintenance Technical Training Program. Order number 3000.10A, May 8. Washington, DC: U.S. Department of Transportation.
- 1991a Airway Facilities Sector Configuration (RIS: AF 1100-1). Order number 1100.127C, May 22. Washington, DC: U.S. Department of Transportation.
- 1991b Airway Facilities Technical Inspection Program. Order number 6040.6D, August 8. Washington, DC: U.S. Department of Transportation.
- 1991c Maintenance Control Center (MCC) Operations Concept. Order number 6000.39, August 8. Washington, DC: U.S. Department of Transportation.
- 1991d General Maintenance Handbook - Airway Facilities. Order number 6000.15B, August 15. Washington, DC: U.S. Department of Transportation.
- 1992a Airway Facilities Maintenance Personnel Certification Program. Order number 3400.3F, August 6. Washington, DC: U.S. Department of Transportation.
- 1992b Airway Facilities Sector Level Staffing Standard System. Order number 1380.40C, December 21. Washington, DC: U.S. Department of Transportation.
- 1993a Demographic Profiles of the Airway Facilities Work Force. Washington, DC: U.S. Department of Transportation.

1993b General Maintenance Handbook for Automated Logging. Order number 6000.48, December 1. Washington, DC: U.S. Department of Transportation.
1993c The GS-2101 Airway Transportation Systems Specialist. Videotape. Washington, DC: U.S. Department of Transportation.
1993d GS-2101 Implementation. Videotape. Washington, DC: U.S. Department of Transportation.
1994a Airway Facilities Operational Plan for Training. Washington, DC: U.S. Department of Transportation.
1994b Airway Transportation System Specialists Implementation/Conversion Document. Washington, DC: U.S. Department of Transportation.
1994c Remote Maintenance Monitoring System (RMMS), Remote Monitoring Subsystem (RMS) Requirements. Document number NAS-MD-793A. Washington, DC: U.S. Department of Transportation.
1994d National Maintenance Coordination Center: Monitoring the Heartbeat of the NAS. Produced by ASM-100. Washington, DC: Federal Aviation Administration, U.S. Department of Transportation.
1995 Classification Guide and Qualification Standard. GS-2101.

CHAPTER 5: COGNITIVE TASK ANALYSIS OF AIR TRAFFIC CONTROL

Adams, M.J., Y.J. Tenney, and R.W. Pew
  1995 Situation awareness and the cognitive management of complex systems. *Human Factors* 37(1):85-104.
Ammerman, H.L., E.S. Becker, G.W. Jones, and W.K. Tobey
  1987 *FAA Air Traffic Control Operations Concepts.* Volume I: ATC Background and Analysis Methodology (Technical Report DOT/FAA/AP-87-01). Colorado Springs, CO: Computer Technology Associates, Inc.
Baddeley, A.D.
  1986 *Working Memory.* Oxford: Oxford University Press.
Burke-Cohen, J.
  1995 Traffic control clearances. Pp. 29-38 in *How To Say It and How Much: The Effect of Format and Complexity on Pilot Recall of Air Traffic Control Clearances*, B.G. Kanki and O.V. Prinzo, eds. (DOT/FAA/AM-96/10). Prepared for Office of Aviation Medicine, Federal Aviation Administration. Washington, DC: U.S. Department of Transportation.
Chi, M.T., P.J. Feltovich, and R. Glaser
  1981 Categorization and representation of physics problems by experts and novices. *Cognitive Science* 5:121-152.
Cohen, M.S., J.T. Freeman, and B.B. Thompson
  1996a Training the naturalistic decision maker. In *Naturalistic Decision Making*, C. Zsambok et al., eds. Hillsdale, NJ: Lawrence Erlbaum.
Cohen, M.S., J.T. Freeman, and S. Wolf
  1996b Meta-recognition in time-stressed decision making: Recognizing, critiquing, and correcting. *Human Factors* Special Issue on Decision Making in Complex Environments.
Corwin, W.H.
  1991 Data link integration in commercial transport operations. In *Proceedings of the Sixth International Aviation Psychology Symposium.* Columbus, OH: Department of Aviation, Ohio State University.

Endsley, M.R.
　1994　Automation and situation awareness. In *Automation and Human Performance: Theory and Application*, R. Parasuraman and M. Mouloua, eds. Hillsdale, NJ: Lawrence Erlbaum.

Endsley, M.R.
　1995　Toward a theory of situation awareness in dynamic systems. *Human Factors* 37(1):32-64.

Fischer, U., and J. Orasanu
　1993　Effective decision strategies on the flight deck. In *Proceedings of the 7th Biannual Symposium on Aviation Psychology*, R. Jensen, ed. Columbus, OH: Ohio State University.

Fowler, F.D.
　1980　Air traffic control problems: A pilot's view. *Human Factors* 22:645-653.

Gopher, D.
　1993　The skill of attention control: Acquisition and execution of attention strategies. Pp. 299-322 in *Attention and Performance XIV: Synergies in Experimental Psychology, Artificial Intelligence, and Cognitive Neuroscience—A Silver Jubilee*, D.E. Meyer and S. Kornblum, eds. Cambridge, MA: Massachusetts Institute of Technology Press.

Gopher, D., M. Weil, and T. Bareket
　1994　Transfer of skill from a computer game trainer to flight. *Human Factors* 36(4):387-405.

Harris, J.E., and A.J. Wilkins
　1982　Remembering to do things: A theoretical framework and an illustrative experiment. *Human Learning* 1:123-136.

Hart, S.G., and C.D. Wickens
　1990　Workload assessment and prediction. Pp. 257-300 in *MANPRINT: An Emerging Technology. Advanced Concepts for Integrating People, Machines and Organizations*, H.R. Booher, ed. New York: Van Nostrand Reinhold.

Harwood, K., R. Roske-Hofstrand, and E. Murphy
　1991　Exploring conceptual structures in air traffic control (ATC). In *Proceedings of the Sixth International Symposium on Aviation Psychology*. Columbus, OH: Department of Aviation, Ohio State University.

Hawkins, F.H.
　1987　*Human Factors in Flight*. UK: Gower Technical Press.

Holding, D.
　1987　Concepts in training. In *Handbook of Human Factors*, G. Salvendy, ed. New York: Wiley & Sons.

Hopkin, V.D.
　1988a　Air traffic control. Pp. 639-662 in *Human Factors in Aviation*, E.L. Wiener and D.C. Nagel, eds. New York: Academic Press.
　1988b　Training Implications of Technological Advances in Air Traffic Control. Keynote address at the Federal Aviation Symposium on Air Traffic Control Training for Tomorrow's Technology, Oklahoma City, OK. Federal Aviation Administration.
　1995　*Human Factors of Air Traffic Control*. London: Taylor & Francis.

Huey, B.M., and C.D. Wickens, eds.
　1993　*Workload Transition: Implications for Individual and Team Performance*. Committee on Human Factors, National Research Council. Washington, DC: National Academy Press.

Kanki, B.G., and O.V. Prinzo, eds.
　1995　*Methods and Metrics of Voice Communications* (DOT/FAA/AM/96-10). Washington, DC: Office of Aviation Medicine, Federal Aviation Administration.

Kerns, K.
    1991   Data-link communication between controllers and pilots: A review and synthesis of the simulation literature. *International Journal of Aviation Psychology* 1(3):181-204.

Klein, G.A., J. Orasanu, R. Calderwood, and C.E. Zsambok, eds.
    1993   *Decision Making in Action: Models and Methods*. Norwood, NJ: Ablex.

Klein, G., and B.W. Crandall
    1995   The role simulation in problem solving and decision making. In *Local Applications of the Ecological Approach to Human-Machine Systems*, P. Hancock, J. Flach, J. Caird, and K. Vicente, eds. Hillsdale, NJ: Lawrence Erlbaum.

Liu, Y.Y., and C.D. Wickens
    1992   Visual scanning with or without spatial uncertainty and divided and selective attention. *Acta Psychologica* 79:131-153.

Logie, R.
    1995   *Spatial Working Memory*. Hillsdale, NJ: Lawrence Erlbaum.

May, P., M. Campbell, and C.D. Wickens
    1995   Perspective displays for air traffic control: Display of terrain and weather. *Air Traffic Control Quarterly* 3:1-17.

Means, B., E. Salas, B. Crandall, and T.O. Jacobs
    1993   Training decision makers for the real world. Pp. 306-326 in *Decision Making in Action: Models and Methods*, G.A. Klein, J. Orasanu, R. Calderwood, and C.E. Zsambok, eds. Norwood, NJ: Ablex.

Monan, W.P.
    1986   *Human Factors in Aviation Operations: The Hearback Problem* (NASA Contractor Report 177398). Moffett Field, CA: NASA Ames Research Center.

Moray, N.
    1986   Monitoring behavior and supervisory control. Pp. 40-51 in *Handbook of Perception and Performance, Vol. II*, K.R. Boff, L. Kaufman, and J.P. Thomas, eds. New York: Wiley & Sons.

Morrow, D., A. Lee, and J.A. Purcell
    1993   Analysis of problems in routine controller-pilot communications. *International Journal of Aviation Psychology* 3:285-302.

Murphy, E.D.
    1989   *Methodology for Correlation of ATC Controller Workload and Errors Phase II* (DOT SBIR Final Technical Report). Rockville, MD: Computer Technology Associates, Inc.

Nagel, D.C.
    1988   Human error in aviation operations. Pp. 263-303 in *Human Factors in Aviation*, E.L. Wiener and D.C. Nagel, eds. New York: Academic Press.

National Transportation Safety Board
    1992   *Aircraft Accident Report* (PB91-910409 - NTSB/AAR-91/08). Washington, DC: Department of Transportation.

Norman, D.A.
    1981   Categorization of action slips. *Psychological Review* 88:1-15.

Norman, D.
    1988   *The Psychology of Everyday Things*. New York: Basic Books.

Orasanu, J.
    1993   Decision-making in the cockpit. Pp. 137-173 in *Cockpit Resource Management*, E.L. Wiener, B.G. Kanki, and R.L. Helmreich, eds. San Diego: Academic Press.

Pennington, N., and R. Hastie
    1993   A theory of explanation-based decision making. Pp. 188-201 in *Decision Making in Action: Models and Methods*, G.A. Klein, J. Orasanu, R. Calderwood, and C.E. Zsambok, eds. Norwood, NJ: Ablex.

Pritchett, A., and R.J. Hansman
- 1993 Preliminary analysis of pilot ratings of "party line" information importance. In *Proceedings of the Seventh International Symposium on Aviation Psychology*. Columbus, OH: Department of Aviation, Ohio State University.

Raby, M., and C.D. Wickens
- 1994 Strategic workload management and decision biases in aviation. *International Journal of Aviation Psychology* 4(3):211-240.

Rasmussen, J., A.-M. Pejtersen, and L. Goodstein
- 1995 *Cognitive Engineering: Concepts and Applications.* New York: Wiley & Sons.

Rasmussen, J.
- 1986 *Information Processing and Human-Machine Interaction: An Approach to Cognitive Engineering.* New York: North Holland.

Reason, J.T.
- 1990 *Human Error.* Cambridge, Eng.: Cambridge University Press.

Redding, R.E., J.R. Cannon, and T.L. Seamster
- 1992 Expertise in air traffic control (ATC): What is it, and how can we train for it? Pp. 1326-1330 in *Proceedings of the 36th Annual Meeting of the Human Factors Society*. Santa Monica, CA: Human Factors Society.

Rodgers, M.
- 1993 An Examination of the Operational Error Database for En Route Traffic Control Centers. Federal Aviation Administration Civil Aeromedical Institute report.

Rouse, W.B., and N.M. Morris
- 1987 Conceptual design of a human error tolerant interface for complex engineering systems. *Automatica* 23(2):231-235.

Salas, E., C. Prince, D.P. Baker, and L. Shrestha
- 1995 Situation awareness in team performance: Implications for measurement and training. *Human Factors* 37(1):123-136.

Sarter, N.B., and D.D. Woods
- 1995 How in the world did we ever get into that mode? Mode error and awareness in supervisory control. *Human Factors* 37(1):5-19.

Schank, R.C., and R. Abelson
- 1977 *Scripts, Plans, Goals, and Understanding.* New York: John Wiley & Sons.

Seamster, T.L., R.E. Redding, J.R. Cannon, J.M. Ryder, and J.A. Purcell
- 1993 Cognitive task analysis of expertise in air traffic control. *International Journal of Aviation Psychology* 3(4):257-283.

Segal, L.
- 1995 Desiging team workstations: In *Local Applications of the Ecological Approach to Human Machine Systems*, P. Hancock, J. Flach, J. Caird, and K. Vicente, eds. Hillsdale, NJ: Laurence Erlbaum.

Singley, M.K., and J.R. Anderson
- 1989 *The Transfer of Cognitive Skill.* Cambridge, MA: Harvard University Press.

Slamecka, N.J., and P. Graf
- 1978 The generation effect: Delineation of a phenomenon. *Journal of Experimental Psychology: Human Learning and Memory* 4:592-604.

Sperandio, J.C.
- 1976 From the plane space to the air mobile space: Experimental comparison between two displays of spatio temporal information. *Travail Humain* 30:130-154.

Stager, P., and D. Hameluck
- 1990 Ergonomics in air traffic control. *Ergonomics* 33:93-499.

Steenblik, J.W.
- 1996 Runway incursion. *Air Line Pilot* January:20-24.

Stein, E.S.
  1992 Air Traffic Control Visual Scanning. Report Number DOT/FAA/CT-TN92/16. Atlantic City, NJ: Federal Aviation Administration.
Stein, W.
  1993 Visual Scanning of Air Traffic Controllers. In *OSO Aviation Psychology Symposium*, R. Jensen ed.
Vortac, O.U., and C.F. Gettys
  1990 *Cognitive Factors in the Use of Flight Progress Strips: Implications for Automation.* Norman, OK: Cognitive Processes Laboratory, University of Oklahoma.
Vortac, O.U., M.B. Edwards, D.K. Fuller, and C.A. Manning
  1993 Automation and cognition in air traffic control: An empirical investigation. *Applied Cognitive Psychology* 7:731-751.
Waag, W.L., and M.R. Houck
  1994 Tools for assessing situational awareness in an operational fighter environment. *Aviation, Space, and Environmental Medicine* (May):A13-A19
Wickens, C.D.
  1992 *Engineering Psychology and Human Performance*, 2nd ed. New York: Harper Collins.
  1996 Situation awareness: Impact of automation and display technology. In *Situation Awareness: Limitations and Enhancement in the Aviation Environment.* NATO AGARD Report CP-575. Neuilly-sur-Seine, France: AGARD.
Wickens, C.D., and J. Flach
  1988 Human information processing. Pp. 111-155 in *Human Factors in Aviation*, E. Wiener and D. Nagel, eds. New York: Academic Press.
Wickens, C.D., and C.M. Carswell
  1995 The proximity compatibility principle: Its psychological foundation and relevance to display design. *Human Factors* 37(3):473-494.

CHAPTER 6: WORKLOAD AND VIGILANCE

Andre, A.D., and P.A. Hancock
  1995 Special issue on pilot workload. *The International Journal of Aviation Psychology* 5:1-4.
Arad, B.A.
  1964 The control load and sector design. *Journal of Air Traffic Control* May:12-31.
Bainbridge, L.
  1983 Ironies of automation. *Automatica* 19:775-779.
Baker, C.H.
  1962 *Man and Radar Displays.* New York: Macmillan.
Barton, J., and S. Folkard
  1993 Advanced versus delaying shift systems. *Ergonomics* 36:59-64.
Bisseret, A.
  1981 Application of signal detection theory to decision making in supervisory control. *Ergonomics* 24:81-94.
Brown, I.D., and E.C. Poulton
  1961 Measuring the spare "mental capacity" of car drivers by a subsidiary task. *Ergonomics* 4:35-40.
Bruce, D.S., N.E. Freeberg, and D.A. Rock
  1993 An explanatory model for influences of air traffic control task parameters on controller work pressure. Pp. 108-112 in *Proceedings of the Human Factors and Ergonomics Society 37th Annual Meeting.*

Bulloch, C.
   1982   Cockpit automation and workload reduction. Too much of a good thing? *Interavia* 3:263-264.

Burbank, N.S.
   1994   *The Development of a Task Network Model of Operator Performance in a Simulated Air Traffic Control Task.* Technical Report DCIEM No 94-05. North York, Ontario, Canada: Defence and Civil Institute of Environmental Medicine.

Byrne, E.A.
   1993   A Psychophysiological Investigation of Individual Differences Affecting Performance During a Complex Monitoring Task. Ph.D. Dissertation. University of Maryland, College Park.

Canadian Aviation Safety Board
   1990   Report on a Special Investigation Into Air Traffic Control Services in Canada. Report No. 90-SO001. Supply and Services Canada Catalog No. TU4-5/1990E.

Cardosi, K.
   1993   Time required for transmission of time-critical ATC messages in an en-route environment. *International Journal of Aviation* Psychology 3(4):303-313.

Costa, G.
   1993   Evaluation of workload in air traffic controllers. *Ergonomics* 36:1111-1120.

Craig, A.
   1985   Vigilance: Theories and laboratory studies. Pp. 107-121 in *Hours of Work*, S. Folkard and T.H. Monk, eds. Chichester, UK: Wiley.

Damos, D.
   1991   *Multiple Task Performance.* London: Taylor and Francis.

Danaher, J.
   1980   Human error in ATC systems operations. *Human Factors* 22:535-545.

Davies, D.R., and R. Parasuraman
   1982   *The Psychology of Vigilance.* London: Academic Press.

Edwards, E.
   1976   Some aspects of automation in civil transport aircraft. In *Monitoring Behavior and Supervisory Control*, T.B. Sheridan and G. Johannsen, eds. New York: Plenum.

Endsley, M., and M. Rodgers
   1996   Attention distribution and situation awareness in air traffic control. In *Proceedings of the 40th Annual Meeting of the Human Factors Society.*

Federal Aviation Administration
   1994   *Minimum Flight Crew Certification* (FAR 25.1523). Washington DC: U.S. Department of Transportation.

Flach, J.
   1994   Situation awareness: The emperor's new clothes. Pp. 242-248 in *Human Performance in Automated Systems: Recent Research and Trends*, M. Mouloua and R. Parasuraman, eds. Hillsdale, NJ: Erlbaum.

Folkard, S.
   1980   Shiftwork and its effects on performance. Pp. 293-306 in *Chronobiology: Principles and Applications to Shift in Schedules*, L. Scheving and F. Halberg, eds. Alphen aan den Rijn, The Netherlands, and Rockville, MD: Sijthoff and Noordhoff.

Gilson, R., D. Garland, and J. Koonce
   1994   *Situational Awareness in Complex Systems.* Daytona Beach, FL: Embry-Riddle Aeronautical University Press.

Graeber, C.
   1988   Aircrew fatigue and circadian rhythmicity. In *Human Factors in Aviation*, E. Wiener and D. Nagel, eds. New York: Academic Press.

Green, D.M., and J.A. Swets
- 1966 *Signal Detection Theory and Psychophysics.* New York: Holt.

Hancock, P.A., and M.H. Chignell
- 1988 Mental workload dynamics in adaptive interface design. *IEEE Transactions on Systems, Man, and Cybernetics* 18:647-658.

Hancock, P.A., and N. Meshkati
- 1988 *Human Mental Workload.* Amsterdam: North-Holland.

Hancock, P.A., M. Rahimi, T. Mihaly, and N. Meshkati
- 1988 A bibliographic listing of mental workload research. Pp. 329-382 in *Human Mental Workload*, P.A. Hancock and N. Meshkati, eds. Amsterdam: North-Holland.

Hancock, P.A., and J.S. Warm
- 1989 A dynamic theory of stress and sustained attention. *Human Factors* 31:519-537.

Hart, S.G., and C.D. Wickens
- 1990 Workload assessment and predictions. In *MANPRINT: An Emerging Technology*, H.R. Booher, ed. New York: Van Rostrand Reinhold.

Hopkin, V.D.
- 1971 Conflicting criteria in evaluating air traffic control systems. *Ergonomics* 14:557-564.
- 1988 Air traffic control. Pp. 639-663 in *Human Factors in Aviation*, E. Wiener and D. Nagel, eds. New York: Academic Press.
- 1991 The impact of automation on air traffic control systems. Pp. 3-20 in *Automation and Systems Issues in Air Traffic Control*, J.A. Wise, V.D. Hopkin, and M.L. Smith, eds. Berlin: Springer Verlag.
- 1992 Human factors issues in air traffic control. *Human Factors Society Bulletin* 35(6):1-4.
- 1995 *Human Factors in Air-Traffic Control.* London: Taylor and Francis.

Horne, J.
- 1988 *Why We Sleep.* New York: Oxford University Press.

Huey, B.M., and C.D. Wickens
- 1993 *Workload Transition: Implications for Individual and Team Performance.* Committee on Human Factors, National Research Council Washington, DC: National Academy Press.

Hurst, M.W., and R.M. Rose
- 1978 Objective job difficulty, behavioral response, and sector characteristics in air route traffic control centres. *Ergonomics* 21:697-708.

Jorna, P.G.A.M.
- 1991 Operator workload as a limiting factor in complex systems. Pp. 281-292 in *Automation and Systems Issues in Air Traffic Control*, J.A. Wise, V.D. Hopkin, and M.L. Smith, eds. Berlin: Springer Verlag.

Kahneman, D.
- 1973 *Attention and Effort.* Englewood-Cliffs, NJ: Prentice Hall.

Kalsbeek, J.W.H.
- 1965 Mésure objective de la surcharge mentale: Nouvelle applications de la méthode des doubles tâches. *Le Travail Humain* 28:121-132.

Kanki, B.G., and O.V. Prinzo, eds.
- 1995 Methods and Metrics of Voice Communications. Report DOT/FAA/AM/96-10. Office of Aviation Medicine. Washington, DC: Federal Aviation Administration.

Kantowitz, B.H., and B.L. Campbell
- 1996 Pilot workload and flight-deck automation. In *Automation and Human Performance: Theory and Applications*, R. Parasuraman, and M. Mouloua, eds. Hillsdale, NJ: Erlbaum.

Kantowitz, B.H., and P.A. Casper
- 1988 Human workload in aviation. Pp. 157-185 in *Human Factors in Aviation*, E. Wiener and D. Nagel, eds. New York: Academic Press.

Laudeman, I.V., and E.A. Palmer
  1995 Quantitative measurement of observed workload in the analysis of aircrew performance. *International Journal of Aviation Psychology* 5:187-197.
Leplat, J., and R. Browaeys
  1965 Analyse et mésure de la charge du travail du controleur du trafic aérienne. *Bulletin du Centre d'Études et de Recherches Psychotechniques* 14:69-79.
Leplat, J., and J-C. Sperandio
  1967 La mésure de la charge de travail par la technique de la tâche ajoutée. *L'Année Psychologique* 67:255-277.
Lysaght, R.J., S.G. Hill, A.O. Dick, et al.
  1989 *Operator Workload: Comprehensive Review and Evaluation of Operator Workload Methodologies.* Report No. 851. Fort Bliss, TX: Army Research Institute.
Mackie, R.R., D.C. Wylie, and M.J. Smith
  1994 Countering loss of vigilance in sonar watchstanding using signal injection and performance feedback. *Ergonomics* 47:1157-1164.
Mackworth, N.H.
  1957 Some factors affecting vigilance. *Advancements in Science* 53:389-393.
Matthews, G., D.R. Davies, and J. Lees
  1990 Arousal, extroversion, and visual sustained attention: The role of resource availability. *Personality and Individual Differences* 11:1159-1173.
May, P., M. Campbell, and C.D. Wickens
  1995 Perspective displays for air traffic control: Display of terrain and weather. *Air Traffic Control Quarterly* 3:1-17.
McAdaragh, R.M.
  1995 Human circadian rhythms and the shift work practices of air traffic controllers. *Journal of Aviation/Aerospace Education and Research* 5(3):7-15.
Melton, C.E.
  1985 *Physiological Responses to Unvarying (Steady) and 2-2-1 Shifts: Miami International Flight Service Station.* Report No. FAA-AM-85, Civil Aeromedical Institute, Office of Aviation Medicine. Oklahoma City, OK: Federal Aviation Administration.
Melton, C.E., J.M. McKenzie, R.C. Smith, B.D. Polis, A.E. Higgins, S.M. Hoffmann, G.E. Funkhouser, and J.T. Saldivar
  1973 *Physiological, Biochemical, and Psychological Responses in Air Traffic Control Personnel: Comparison of the 5-Day and 2-2-1 Shift Rotation Patterns.* Technical Report No. FAA-AM-73-22, Civil Aeromedical Institute, Office of Aviation Medicine. Oklahoma City, OK: Federal Aviation Administration.
Melton, C.E., R.C. Smith, J.M. McKenzie, J.T. Saldivar, S.M. Hoffmann, and P.R. Fowler
  1975 *Stress in Air Traffic Controllers: Comparison of Two Air Route Traffic Control Centers on Different Shift Rotation Patterns.* Technical Report No. FAA-AM-75-7, Civil Aeromedical Institute, Office of Aviation Medicine. Washington, DC: Federal Aviation Administration.
Molloy, R., and R. Parasuraman
  In press Monitoring an automated system for a single failure: Vigilance and task complexity effects. *Human Factors.*
Moray, N.
  1967 Where is capacity limited? A survey and a model. *Acta Psychologica* 27:84-93.
  1979 *Mental Workload.* New York: Plenum.
  1988 Mental workload since 1979. *International Reviews of Ergonomics* 2:123-150.
Morrow, D.
  In press Mitigating age decrements in pilot performance: Implications for design. *Ergonomics in Design.*

Morrow, D., A. Lee, and M. Rodvold
1993 Analyzing problems in routine controller-pilot communication. *International Journal of Aviation Psychology* 3:285-302.

Navon, D.
1984 Resources: A theoretical stone soup? *Psychological Review* 91:216-234.

Norman, D.A., and D. Bobrow
1975 On data-limited and resource-limited processing. *Cognitive Psychology* 7:44-60.

North, R.A., and V.A. Riley
1989 A predictive model of operator workload. Pp. 81-90 in *Applications of Human Performance Models to System Design*, G.R. McMillan et al., eds. New York: Plenum Press.

National Transportation Safety Board
1991 Aircraft Accident Report. NTSB/AAR-91408. Washington, DC: National Transportation Safety Board.

Nuechterlein, K., R. Parasuraman, and Q. Jiang
1983 Visual sustained attention: Image degradation produces rapid sensitivity decrement over time. *Science* 220:327-329.

O'Donnell, R.D., and F.T. Eggermeier
1986 Workload assessment methodology. In *Handbook of Perception. Volume 2. Cognitive Processes and Performance*, K. Boff, L. Kaufman, and J. Thomas, eds. New York: Wiley.

Parasuraman, R.
1979 Memory load and event rate control sensitivity decrements in sustained attention. *Science* 205:924-927.

1984 Psychobiology of sustained attention. Pp. 61-101 in *Sustained Attention and Human Performance*, J.S. Warm, ed. London: Wiley.

1985 Detection and identification of abnormalities in chest x-rays: Effects of reader skill, disease prevalence, and reporting standards. Pp. 69-66 in *Trends in Ergonomics/Human Factors II*, R.E. Eberts and C.G. Eberts, eds. Amsterdam: North-Holland.

1986 Vigilance, monitoring, and search. Pp. 43.1-43.39 in *Handbook of Perception. Volume 2. Cognitive Processes and Performance*, K. Boff, L. Kaufman, and J. Thomas, eds. New York: Wiley.

1993 Effects of adaptive function allocation on human performance. Pp. 147-157 in *Human Factors and Advanced Aviation Technologies*, D.J. Garland and J.A. Wise, eds. Daytona Beach, FL: Embry-Riddle Aeronautical University Press.

Parasuraman, R., and D.R. Davies, D.R.
1977 A taxonomic analysis of vigilance. Pp. 559-574 in *Vigilance: Theory, Operational Performance, and Physiological Correlates*, R.R. Mackie, ed. New York: Plenum.

Parasuraman, R., R. Molloy, and I.L. Singh
1993 Performance consequences of automation-induced "complacency." *International Journal of Aviation Psychology* 3:1-23.

Parasuraman, R., M. Mouloua, and R. Molloy
1994 Monitoring automation failures in human-machine systems. Pp. 45-49 in *Human Performance in Automated Systems: Current Research and Trends*, M. Mouloua and R. Parasuraman, eds. Hillsdale, NJ: Erlbaum.

Parasuraman, R., M. Mouloua, R. Molloy, and B. Hilburn
1996 Monitoring automated systems. In *Automation and Human Performance: Theory and Applications*, R. Parasuraman and M. Mouloua eds. Hillsdale, NJ: Erlbaum.

Parasuraman, R., J. Warm, and W.N. Dember
1987 Vigilance: Taxonomy and utility. Pp. 11-32 in *Ergonomics and Human Factors: Recent Research*, L.S. Mark, J.S. Warm, and R.L. Huston, eds. New York: Springer-Verlag.

Parks, A.M., and G.P. Boucek
- 1989 Workload prediction, diagnosis and continuing challenges. Pp. 47-64 in *Applications of Human Performance Models to System Design*, G.R. McMillan et al., eds. New York: Plenum Press.

Pawlak, W.W., C.R. Brinton, L. Courch, and K.M. Lancaster
- 1996 A framework for the evaluation of air traffic control complexity. *Proceedings of the 1996 Meeting of Aeronautics and Astronautics*.

Pigeau, R.A., R.G. Angus, P. O'Neill, and I. Mack
- 1995 Vigilance latencies to aircraft detection among NORAD surveillance operators. *Human Factors* 37(3):624-637.

Prinzo, O.V., and T.W. Britton
- 1993 *ATC/Pilot Voice Communications - A Survey of the Literature*. Technical Report DOT/FAA/AM-93/20, Office of Aviation Medicine. Washington, DC: Federal Aviation Administration.

Rhodes, W., I. Szlapetis, K. Hahn, R. Heslegrave, and K.V. Ujimoto
- 1994 *A Study of the Impact of Shiftwork and Overtime on Air Traffic Controllers. Phase I: Determining Appropriate Research Tools and Issues*. Publication No. TP 12257E. Montréal: Transport Canada.

Rodgers, M.D., and D.A. Duke
- 1994 SATORI: Situation assessment through the re-creation of incidents. Pp. 217-225 in *Situational Awareness in Complex Systems*, R.D. Gilson, D.J. Garland, and J.M. Koonce, eds. Daytona Beach, FL: Embry-Riddle Aeronautical University.

Rodgers, M.D., R.H. Mogford, and L.S. Mogford
- 1995 Air traffic controller awareness of operational error development. In *Proceedings of the 2nd Annual Situation Awareness Conference*, M. Endsley and D. Garland, eds. Daytona Beach, FL: Embry-Riddle Aeronautical University.

Rodgers, M.
- 1993 An Examination of the Operational Error Database for En Route Traffic Control Centers. Federal Aviation Administration Civil Aeromedical Institute report.

Rosa, R.
- In press Extended workshifts and excessive fatigue. *Journal of Sleep Research*.

Rouse, W.B., S.L. Edwards, and J.M. Hammer
- 1993 Modeling the dynamics of mental workload and human performance in complex systems. *IEEE Transactions on Systems, Man, and Cybernetics* 23:1662-1671.

Sanders, M., and E.J. McCormick
- 1992 *Human Factors in Engineering and Design*. New York: McGraw Hill.

Sarno, K.J., and C.D. Wickens
- 1995 The role of multiple resources in predicting time-sharing efficiency: An evaluation of three workload models in a multiple task setting. *International Journal of Aviation Psychology* 5(1):107-130.

Sawin, D.A., and M.W. Scerbo
- 1994 Vigilance: How to do it and who should do it. Pp. 1312-1316 in *Proceedings of the 38th Annual Meeting of the Human Factors and Ergonomics Society*.

Sheridan, T.B.
- 1970 On how often the supervisor should sample. Pp. 140-145 in *IEEE Transactions on Systems Science and Cybernetics, SSC-6*.

Smith, R.C., C.E. Melton, and J.M. McKenzie
- 1971 *Affect Adjective Check List Assessment of Mood Variations in Air Traffic Controllers*. Technical Report No. FAA-AM-71-21, Civil Aeromedial Institute, Office of Aviation Medicine, Federal Aviation Administration. Washington, DC: U.S. Department of Transportation.

Sperandio, J.C.
- 1971 Variation of operator's strategies and regulating effects on workload. *Ergonomics* 14:571-577.

Stager, P.
- 1991 Error models for operating irregularities: Implications for automation. Pp. 321-338 in *Automation and Systems Issues in Air Traffic Control*, J.A. Wise, V.D. Hopkin, and M.L. Smith, eds. Berlin: Springer-Verlag.

Stager, P., and D. Hameluck
- 1988 *Factors Associated With Air Traffic Control Operating Irregularities: An Analysis of Fact Finding Board Reports*. Technical Report TP 9324E. Ottawa, Ont.: Transport Canada.
- 1990 Ergonomics in air traffic control. *Ergonomics* 33:493-499.

Stager, P., D. Hameluck, and R. Jubis
- 1989 Underlying factors in air traffic control incidents. In *Proceedings of the 33rd Annual Meeting of the Human Factors Society*. Santa Monica, CA: Human Factors Society.

Stokes, A., and C.D. Wickens
- 1988 Aviation displays. Pp. 387-431 in *Human Factors in Aviation*, E. Wiener and D. Nagel, eds. New York: Academic Press.

Swanson, R., L. Sauter, and A. Chapman
- 1989 The design of rest breaks for video display terminal work: A review of the relevant literature. In *Advances in Industrial Ergonomics and Safety: Vol. 1.*, A. Mital, ed. Amsterdam: Elsevier.

Teichner, W.H.
- 1974 The detection of a simple visual signal as a function of time on watch. *Human Factors* 16:339-353.

Thackray, R.I.
- 1981 The stress of boredom and monotony: A consideration of the evidence. *Psychosomatic Medicine* 43:165-176.

Thackray, R.I., J.P. Bailey, and R.M. Touchstone
- 1979 The effect of increased monitoring load on vigilance performance using a simulated radar display. *Ergonomics* 22:529-539.

Thackray, R.I., and R.M. Touchstone
- 1989a Detection efficiency on an air traffic control monitoring task with and without computer aiding. *Aviation, Space, and Environmental Medicine* 60:744-748.
- 1989b Effects of high visual taskload on the behaviors involved in complex monitoring. *Ergonomics* 32:27-38.

Tsang, P.S., and W.W. Johnson
- 1989 Cognitive demands in automation. *Aviation, Space, and Environmental Medicine* 60:130-135.

Van Cott, H., and R. Kinkade
- 1972 *Human Engineering Guide for Equipment Design*. New York: Van Nostrand.

Warm, J.S.
- 1984 *Sustained Attention in Human Performance*. London: Wiley.

Warm, J.S., W.N. Dember, and P.A. Hancock
- 1996 Vigilance and workload in automated systems. In *Automation and Human Performance: Theory and Applications*, R. Parasuraman, and M. Mouloua, eds. Hillsdale, NJ: Erlbaum.

Warm, J.S., and H.A. Jerison
- 1984 Psychophysics of vigilance. Pp. 15-59 in *Sustained Attention in Human Performance*, J.S. Warm, ed. London: Wiley.

Wickens, C.D.
　1984　Processing resources in attention. In *Varieties of Attention*, R. Parasuraman and D.R. Davies, eds. Orlando, FL: Academic Press.
　1992a　*Engineering Psychology and Human Performance*, 2nd. ed. New York: Harper Collins.
　1992b　Workload and situation awareness: An analogy of history and implications. *Insight* 14:1-3.
Wiener, E.L.
　1988　Cockpit automation. Pp. 433-461 in *Human Factors in Aviation*, E. Wiener and D. Nagel, eds. New York: Academic Press.
Wilkinson, R.T.
　1992　How fast should night shift rotate? *Ergonomics* 35:1425-1446.
Wortman, D.B., S.D. Duket, D.J. Seifert, R.L. Hann, and G.P. Chubb
　1978　*Simulation Using SAINT*. West Lafayette, IN: Pritsker.

CHAPTER 7: TEAMWORK AND COMMUNICATIONS

Butler, R.E.
　1993　LOFT: Full mission simulation as crew resource management training. Pp. 231-259 in *Cockpit Resource Management*, E. Weiner, B. Kanki, and R. Helmreich, eds. San Diego, CA: Academic Press.
Cardosi, K.M.
　1993　Time required for transmission of time-critical air traffic control messages in an en route environment. *The International Journal of Aviation Psychology* 3(4):303-314.
Cooper, G.E., M.D. White, and J.K. Lauber, eds.
　1980　Resource Management on the Flight Deck: Proceedings of a NASA/Industry Workshop Held at San Francisco, California, June 26-28, 1979 (NASA Conference Publication No. 2120).
Danaher, J.W.
　1980　Human error in ATC system operations. *Human Factors* 22:535-545.
Diehl, A.
　1993　Organizational ergonomics and aviation safety. *Proceedings of the Seventh International Symposium on Aviation Psychology*. Columbus, OH: Ohio State University.
Federal Aviation Administration
　1978　Line Oriented Flight Training (Advisory Circular AC-120-35A). Washington, DC: U.S. Department of Transportation.
　1990　*Special Federal Aviation Regulation 38-Advanced Qualification Program* (Federal Register, Vol. 55, No. 91, Rules and Regulations pp. 40262-40278). Washington, DC: National Archives and Records Administration.
　1992　Crew Resource Management (Advisory Circular 120-51A). Washington, DC: U.S. Department of Transportation.
Gregorich, S., R.L. Helmreich, and J.A. Wilhelm
　1990　The structure of cockpit management attitudes. *Journal of Applied Psychology* 75(6):682-690.
Helmreich, R.L.
　1984　Cockpit management attitudes. *Human Factors* 26:583-589.
Helmreich, Robert L., A.C. Merritt, P.J. Sherman, S.E. Gregorich, and E.L. Wiener
　1993　The Flight Management Attitudes Questionnaire (FMAQ). NASA/UT/FAA Technical Report 93-4.
Helmreich, R.L., and H.C. Foushee
　1993　Why crew resource management? Pp. 3-41 in *Cockpit Resource Management*, E.L. Wiener, B.G. Kanki, and R.L. Helmreich, eds. San Diego, CA: Academic Press.

Helmreich, R.L., and J.A. Wilhelm
　1991　Outcomes of crew resources management training. *International Journal of Aviation Psychology* 1(4):287-300.

Helmreich, R.L., E.L. Wiener, and B.G. Kanki
　1993　The future of crew resource management in the cockpit and elsewhere. Pp. 479-501 in *Cockpit Resource Management*, E.L. Wiener, B.G. Kanki, and R.L. Helmreich, eds. San Diego, CA: Academic Press.

Helmreich, R.L.
　1994　Anatomy of a system accident: The crash of Avianca Flight 052. *International Journal of Aviation Psychology* 4(3):265-284.

Helmreich, R.L., and H.G. Schaefer
　1994　Team performance in the operating room. Pp. 225-253 in *Human Error in Medicine*, M.S. Bogner ed. Hillside, NJ: Lawrence Erlbaum.

Hofstede, G.
　1980　*Culture's Consequences: International Differences in Work-related Values*. Beverly Hills, CA: Sage.

Human Technology
　1991　Analysis of Controller Communication in En Route Air Traffic Control: Report to the Federal Aviation Administration. McLean, VA: Federal Aviation Administration.

Jones, S.G.
　1993　Human factors information in incident reports. *Proceedings of the Seventh International Symposium on Aviation Psychology*. Columbus: Ohio State University.
　1995　Human Error: The Role of Group Dynamics in Error Tolerant Systems. Doctoral dissertation in progress, University of Texas at Austin.

Kinney, G.C., M.J. Spahn, and R.A. Amato
　1977　*The Human Element in Air Traffic Control: Observations and Analyses of the Performance of Controllers and Supervisors in Providing ATC Separation Services* (MTR-7655). McLean, VA: METREK Division of the MITRE Corporation.

Morrow, D.G., A.T. Lee, and M. Rodvold
　1993　Analysis of problems in routine controller-pilot communication. *International Journal of Aviation Psychology* 3(4):285-302.

Murphy, M.
　1980　Analysis of eight-four commercial aviation incidents: Implications for a resource management approach to crew training. *Proceedings of the Annual Reliability and Maintainability Symposium IEEE Spectrum*.

Nadler, E., P. Mengert, R. DiSario, D.E. Sussman, M. Grossberg, and G. Spanier
　1993　Effects of satellite- and voice-switching-equipment transmission delays on air traffic control communications. *International Journal of Aviation Psychology* 3(4):315-326.

National Transportation Safety Board
　1973　*Aircraft Accident Report: Eastern Airlines, Inc., L-1011, N310EA, Miami, Florida, December 29, 1972* (NTSB-AAR-73-14). Washington, DC: U.S. Department of Transportation.
　1990　*Aircraft Accident Report: United Airlines Flight 232, McDonnell Douglas DC-10-10, Sioux Gateway Airport, Sioux City, Iowa, July 19, 1989* (NTSB-AAR-90/05). Washington, DC: U.S. Department of Transportation.
　1991　*Aircraft Accident Report: Avianca, the Airline of Colombia, Boeing 707-321B, HK2016, Fuel Exhaustion, Cove Neck, New York, January 25, 1990* (NTSB-AAR-91/04). Washington, DC: U.S. Department of Transportation.
　1994　*Aircraft Accident Report: Controlled Flight into Terrain, Federal Aviation Administration Beech Super King Air 300/F, N82, Front Royal, Virginia, October 26, 1993* (NTSB-AAR-94/03). Washington, DC: U.S. Department of Transportation.

Predmore, S.C.
　1991　Microcoding of communications in accident analyses: Crew coordination in United 811 and United 232. *Proceedings of the Sixth International Symposium on Aviation Psychology.* Columbus: Ohio State University.

Pritchett, A., and R.J. Hansman
　1993　Preliminary analysis of pilot ratings of "party line" information importance. In *Proceedings of the Seventh International Symposium on Aviation Psychology*, R. Jensen, ed. Columbus: Ohio State University.

Seamster, T.L., R.E. Redding, J.R. Cannon, J.M. Ryder, and J.A. Purcell
　1993　Cognitive task analysis of expertise in air traffic control. *International Journal of Aviation Psychology* 3(4):257-287.

Sherman, P.J.
　1992　Factor Analysis of the Enroute Controller Resource Management Attitudes Questionnaire. Paper presented at the Annual Meeting of the Human Factors Society, Atlanta, GA.

Sherman, P.J., and R. Helmreich
　In press　Attitudes toward automation: The influence of national culture. *Proceedings of the Eighth International Symposium on Aviation Psychology.* Columbus: Ohio State University.

Stager, P., and D. Hameluck
　1990　Ergonomics in air traffic control. *Ergonomics* 33:493-499.

Taggart, W.R.
　1993　How to kill off a good CRM program. *The CRM Advocate* 93(1):11-12.
　1994　Crew resource management: Achieving enhanced flight operations. In *Aviation Psychology in Practice*, N. Johnston, R. McDonald, and R. Fuller, eds. Brookfield, VT: Ashgate.

Waller, M.
　1995　Multitasking in Work Groups: Coordination Processes in Work Groups with Multiple Tasks. Unpublished doctoral dissertation, University of Texas at Austin.

CHAPTER 8: SYSTEM MANAGEMENT

Breenlove, M.S.
　1993　*Vectors to Spare: The Life of an Air Traffic Controller.* Ames, IA: Iowa State University.

Broderick, A.
　1995　*The Challenge of Aviation Safety.* Washington, DC: Federal Aviation Administration.

Collins, W., and M. Wayda
　1994　*Index of FAA Office of Aviation Medicine Reports: 1961 through 1993.* Federal Aviation Administration report DOT/FAA/AM-94/1, January, 1994.

Daschle, L.H.
　1995　Remarks delivered at the 1995 FAA Aviation Forecast Conference, March 3, 1995.

Deci, E.L., Connell, J.P., and Ryan, R.M.
　1989　Self-determination in a work organization. *Journal of Applied Psychology* 74(4):580-593.

Federal Aviation Administration
　1989　Aviation Safety Indicators - Concept Definition. November.
　1991　Quality Through Partnership. Order 3710.16. November 15.
　1993a　The Future Traffic Flow Management System (Draft), Volume I: Operational Description. June 8, 1993.
　1993b　Human Factors Policy. FAA Order 9550.8. October.
　1993c　Employee Indemnification Policy and Procedures. FAA Order 2300.2A. October 22.

1994a *Aviation System Indicators. 1994 Annual Report.*
1994b Memorandum from the Office of the FAA Administrator, November 30.
1994c New structure: FAA reorganizes along its key lines of business. FAA Headquarters Intercom, December 13.
1995a *Strategic Plan, Volume 2.*
1995b Air Traffic Control. FAA Order 7110.65J.
1995c *Annotated Summary Briefing of the Results of the Employee Attitude Survey for 1995.*
1995d *National Plan For Civil Aviation Human Factors: An Initiative for Research and Application.* March.
1995e *Air Traffic Services Plan 1995-2000.* August.
1995f FAA vision, mission, and values. FAA Internet Service. October.
1995g Federal Aviation Administration Technical Center Internet Homepage, December 21.
1996a *Air Traffic Services FY-96 Business Plan.*
1996b Human factors research project initiatives: 1996. FAA Human Factors Internet Home Page.
1996c *Report to the Committee on the Status and Organization of Human Factors Within the FAA.* Final report of the Human Factors Subcommittee, August 5, 1996.

Harss, C., J. Lichtenfeld, M. Kastner, and J. Goodrich
    1990 Air traffic controller working conditions and organization: Suggestions for analysis and improvements from a psychological point of view. In *Automation and Systems Issues in Air Traffic Control*, J. Wise, V.D. Hopkin, and M. Smith, eds. New York: Springer-Verlag.

Hinson, D., FAA administrator
    1995a Remarks delivered at the National Airspace System (NAS) Architecture Meeting, Washington, DC, June 27.
    1995b Remarks delivered at the 1995 Aviation Safety Initiative Review Conference, New Orleans, December 6.

International Civil Aviation Organization (ICAO)
    1993 Human factors, management, and organization. *ICAO Circular: Human Factors Digest* 10 Circular 247-AN/148.

LaPorte, T.
    1987 The self-designing high-reliability organization: Aircraft carrier flight operations at sea. *Naval War College Review* Autumn.
    1988 The United States air traffic system: Increasing reliability in the midst of rapid growth. In *The Development of Large Technical Systems*, R. Mayntz and T. Hughes, eds. Boulder, CO: Westview Press.
    1996a High reliability organizations: Unlikely, demanding and at risk. *Journal of Contingency and Crisis Management* 4(2).
    1996b Technologies as systems and networks: Issues of dependence, public confidence and constancy. *FLUX: International Quarterly of Networks and Technology.*

LaPorte, T., and P. Consolini
    1991 Working in practice but not in theory: Theoretical challenges of "high-reliability organizations." *Journal of Public Administration Research and Theory* 1:19-47.

Lee, J., and N. Moray
    1992 Trust, control strategies, and allocation of function in human-machine systems. *Ergonomics* 35:1243-1270.

Moray, N., and B. Huey, eds.
    1988 *Human Factors Research and Nuclear Safety.* Committee on Human Factors, National Research Council. Washington, DC: National Academy Press.

Mundra, A.
    1996   Introduction of the Ghosting Display Aid. Presentation to the National Research Council Panel on Human Factors in Air Traffic Control Automation, April 25.

Office of Technology Assessment
    1988   *Safe Skies for Tomorrow: Aviation Safety in a Competitive Environment.* OTA-SET-381. Washington, DC: U.S. Government Printing Office, July.

Planzer, N.
    1995   Air Traffic Control Requirements. Presentation to the National Research Council Panel on Human Factors in Air Traffic Control Automation, August.

Planzer, N., and M. Jenny
    1995   Managing the evolution to free flight. *Journal of Air Traffic Control* March:18-20.

Reason, J.
    1987a   A framework for classifying errors. In *New Technology and Human Error*, J. Rasmussen, K. Duncan, and J. LePlat, eds. Chichester, U.K.: Wiley.
    1987b   Generic error-modelling system (GEMS): A cognitive framework for locating common human error forms. In *New Technology and Human Error*, J. Rasmussen, K. Duncan, and J. LePlat, eds. Chichester, U.K.: Wiley.

Schroeder, D.
    1996   CAMI Human Factors Research. Federal Aviation Administration internal memorandum, January 8.

Stager, P.
    1990   Error Models for Operating Irregularities: Implications for Automation. In *Automation and Systems Issues in Air Traffic Control*, J. Wise, V.D. Hopkin, and M. Smith, eds. New York: Springer-Verlag.

Stein, E., and E. Buckley
    1994   *Human Factors at the FAA Technical Center: Bibliography 1958-1994.* Federal Aviation Administration report DOT/FAA/CT-TN94/50.

Westrum, R.
    1990   Automation, information and consciousness in air traffic control. In *Automation and Systems Issues in Air Traffic Control*, J. Wise, V.D. Hopkin, and M. Smith, eds. New York: Springer-Verlag.

Wood, R.
    1991   *Aviation Safety Programs: A Management Handbook.* Englewood, CO: Jeppeson Sanderson, Inc.

CHAPTER 9: HUMAN FACTORS IN AIRWAY FACILITIES

Blanchard, R.E., and J.J. Vardaman
    1994   *Human Factors in Airway Facilities maintenance: Development of a Prototype Outage Assessment Inventory.* Document number DOT/FAA/AM-94/5. Office of Aviation Medicine, Civil Aeromedical Institute, Federal Aviation Administration. Washington, DC: U.S. Department of Transportation.

Collins, W., and M. Wayda
    1994   Index of FAA Office of Aviation Medicine Reports: 1961 through 1993. FAA report DOT/FAA/AM-94/1, January 1994.

Federal Aviation Administration
    1991a   Maintenance Control Center (MCC) Operations Concept. Order number 6000.39, August 8. Washington, DC: U.S. Department of Transportation.
    1991b   General Maintenance Handbook - Airway Facilities. Order number 6000.15B, August 15. Washington, DC: U.S. Department of Transportation.

1991c Policy for Maintenance of the National Airspace System (NAS) through the Year 2000. Order number 6000.30B, October 8. Washington, DC: U.S. Department of Transportation.
1993a Airway Facilities Strategic Plan. Washington, DC: U.S. Department of Transportation.
1993b Demographic Profiles of the Airway Facilities Work Force. Washington, DC: U.S. Department of Transportation.
1994a Airway Facilities Operational Plan for Training. Washington, DC: U.S. Department of Transportation.
1994b Airway Transportation System Specialists Implementation/Conversion Document. Washington, DC: U.S. Department of Transportation.
1994c Remote Maintenance Monitoring System (RMMS), Remote Monitoring Subsystem (RMS) Requirements. Document number NAS-MD-793A. Washington, DC: U.S. Department of Transportation.
1995a Classification Guide and Qualification Standard. GS-2101.
1995b Plan For Engineering, Research, and Development. Washington, DC: U.S. Department of Transportation.
1995c Airway Facilities Concept of Operations for the Future, 1994 Edition. Washington, DC: U.S. Department of Transportation.
1995d Airway Facilities Strategic Plan. Washington, DC: U.S. Department of Transportation.
1995e National Plan For Civil Aviation Human Factors: An Initiative for Research and Application. Washington, DC: U.S. Department of Transportation.
1996 Human Factors Research Project Initiatives. Internet: FAA Human Factors Home Page.

Johannssen, H.
1992 FAA Operations and Staffing. Testimony before the Subcommittee on Aviation of the Committee on Public Works and Transportation, House of Representatives, March 3, 5.

Reynolds, P., and V. Prabhu
1993 Future operations environment of Airway Facilities. McLean, VA: MITRE Center for Advanced Aviation System Development.

Schroeder, D., and J. Deloney
1983 Job attitudes toward the new maintenance concept of the Airway Facilities Service. FAA report FAA-AM-83-7.

Stein, E.S., and E. Buckley
1994 Human Factors at the FAA Technical Center: Bibliography 1958-1994. FAA report DOT/FAA/CT-TN94/50.

Theisen, C.J., Jr., A. Salvador, and W.J. Hoffman
1987 Development of a System Engineer Workstation. Pp. 1421-1424 in *Proceedings of the Human Factors Society 31st Annual Meeting*.

Wagner, D., J. Birt, M. Snyder, and J. Duncanson
1996 *Human Factors Design Guide (HFDG) for Acquisition of Commercial-off-the-Shelf (COTS) Subsystems, Nondevelopmental Items (NDI), and Developmental Systems*. Department of Transportation report DOT/FAA/CT-96/1.

CHAPTER 10: STRATEGIES FOR RESEARCH

Albright, C.A., T.R. Truitt, A.L. Barile, O.U. Vortac, and C.A. Manning
1994 Controlling traffic without flight progress strips: Compensation, workload, performance, and opinion. *Air Traffic Control Quarterly* 2(4):229-248.

Amaldi, P.
1994 Radar controller's problem solving and decision making skills. Pp. 33-57 in *Verification and Validation of Complex Systems: Additional Human Factors Issues*, J.A. Wise, V.D. Hopkin, and P. Stager, eds. Daytona Beach, FL: Embry-Riddle Aeronautical University Press.

Andre, A.D., and C.D. Wickens
  1995 When users want what's not best for them. *Ergonomics in Design* (October):10-14.
Bailey, R.W.
  1993 Performance vs. preference. Pp. 282-286 in *Proceedings of the Human Factors and Ergonomics Society 37h Annual Meeting.* Santa Monica, CA: Human Factors and Ergonomics Society.
Bainbridge, L.
  1988 Multiple representations or "good" models. Pp. 1-11 in *Training, Human Decision Making and Control,* J. Patrick and K.D. Duncan, eds. Amsterdam: North-Holland.
Baker, S.
  1993 The role of incident investigation in system validation. Pp. 239-250 in *Verification and Validation of Complex Systems: Human Factors Issues,* J.A. Wise, V.D. Hopkin, and P. Stager, eds. NATO ASI Series Vol. F110. Berlin: Springer-Verlag.
Baker, S., and E. Marshall
  1988 Evaluating the man-machine interface—The search for data. Pp. 79-92 in *Training, Human Decision Making and Control,* J. Patrick and K.D. Duncan, eds. Amsterdam: North-Holland.
Baron, S., and K. Corker
  1989 Engineering-based approaches to human performance modeling. In *Applications of Human Performance Models to System Design,* in G. McMillan, ed. New York: Plenum Press.
Beevis, D., and G. St. Denis
  1992 Rapid prototyping and the human factors engineering process. *Applied Ergonomics* 23:155-160.
Benel, R.A., and D.A. Domino
  1993 Future ATC system integration tools for developing a shared vision. Pp. 313-317 in *Proceedings of the Seventh International Symposium on Aviation Psychology,* R. Jensen, ed. Columbus, OH: Ohio State University.
Boff, K.R., L. Kaufman, and J.P. Thomas
  1986 *Handbook of Perception and Human Performance.* 2 Vols. New York: Wiley.
Boff, K.R., and J.E Lincoln
  1988 *Engineering Data Compendium. Human Perception and Performance.* 3 Vols. AAMRL. Dayton: Wright-Patterson Air Force Base.
Brown, I.D., and E.C. Poulton
  1961 Measuring the spare "mental capacity" of car drivers by a subsidiary task. *Ergonomics* 4:35-40.
Buckley, E.P., B.D. DeBaryshe, N. Hitchner, and P. Kohn
  1983 Methods and Measurements in Real-Time Air Traffic Control System Simulation, Atlantic City, NJ. Federal Aviation Administration Report No. DOT/FAA/CT-83/26.
Burbank, N.S.
  1994 The Development of a Task Network Model of Operator Performance in a Simulated Air Traffic Control Task. DCIEM Technical Report 94-05. Ontario, Canada: Defence and Civil Institute of Envorinmental Medicine.
Bussolari, S.R.
  1991 Real-time control tower simulation for evaluation of airport surface traffic automation. Pp. 502-507 in *Proceedings of the Sixth International Symposium on Aviation Psychology,* R.S. Jensen, ed. Columbus, OH: Department of Aviation, Ohio State University.
Byrne, E.A., and R. Parasuraman
  1996 Psychophysiology and adaptive automation. *Biological Psychology* 42:249-268.

Byrne, E.A., and S.W. Porges
   1994   Heart rate variability, effort, and complex monitoring: Implications for research on automation. Pp. 175-182 in *Human Performance in Automated Systems: Recent Research and Trends*, M. Mouloua and R. Parasuraman, eds. Hillsdale, NJ: Erlbaum.
Campbell, D.T.
   1957   Factors relevant to the validity of experiments in social settings. *Psychological Bulletin* 54(4):297-312.
   1969   Reforms as experiments. *American Psychologist* 24:409-429.
Campbell, D.T., and J.C. Stanley
   1966   *Experimental and Quasi-Experimental Designs For Research*. Chicago: Rand McNally.
Canadian Aviation Safety Board
   1990   Report on a Special Investigation into Air Traffic Control Services in Canada. Report No. 90-SP001. Supply and Services Canada Catalog No. TU4-5/1990E.
Cardosi, K.M., and M.S. Huntley
   1993   *Human Factors for Flight Deck Certification Personnel: Final Report, July 1993*. DOT/FAA/RD-93/5, DOT-VNTSC-FAA-93-4, Research and Development Service, Federal Aviation Administration. Washington, DC: U.S. Department of Transportation.
Cardosi, K.M.
   1993   Time required for transmission of time-critical air traffic control messages in an en route environment. *International Journal of Aviation Psychology* 3(4):303-313.
Cardosi, K.M., and E.D. Murphy, eds.
   1995   *Human Factors in the Design and Evaluation of Air Traffic Control Systems*. DOT/FAA/RD-95-3, DOT-VNTSC-FAA-95-3, Office of Aviation Research, Federal Aviation Administration. Washington, DC: U.S. Department of Transportation.
Chapanis, A.
   1967   The relevance of laboratory studies to practical situations. *Ergonomics* 10:557-577.
   1988   Some generalizations about generalization. *Human Factors* 30:253-267.
Christensen, J.M.
   1958   Trends in human factors. *Human Factors* 1:2-7.
Collins, W.E., and M.E. Wayda, eds.
   1994   *Index of FAA Office of Aviation Medicine Reports: 1961 through 1993*. Final Report (January). DOT/FAA/AM-94/1. Office of Aviation Medicine. Washington, D. C: U.S. Department of Transportation.
Corker, K.M., and B.R. Smith
   1995   An Architecture and Model for Cognitive Engineering Simulation Analysis: Application to Advanced Aviation Automation. Paper presented at the AIAA Computing in Aerospace 9 Conference, San Diego, CA, October 21.
Costa, G.
   1991   Shiftwork and circadian variations of vigilance and performance. Pp. 267-280 in *Automation and Systems Issues in Air Traffic Control*, J.A. Wise, V. Hopkin, and M.L. Smith, eds. NATO ASI Series Vol. F 73. Berlin: Springer-Verlag.
Cushing, S.
   1994   *Fatal Words: Communication Clashes and Aircraft Crashes*. Chicago: University of Chicago Press.
Danaher, J.W.
   1980   Human error in ATC system operations. *Human Factors* 22(5):535-545.
David, H.
   1993   Systems theory versus verification and validation. Pp. 465-474 in *Verification and Validation in Complex Systems: Human Factors Issues*, J.A. Wise, V.D. Hopkin, and P. Stager, eds. Berlin: Springer-Verlag.

Day, P.O.
1991 Human factors in system design. Pp. 201-208 in *Automation and Systems Issues in Air Traffic Control*, J.A. Wise, V.D. Hopkin, and M.L. Smith, eds. NATO ASI Series Vol. F73. Berlin: Springer-Verlag.

Dennison, T.W., and V.J. Gawron
1995 Tools and methods for human factors test and evaluation: Mockups, physical and electrical human models, and simulation. Pp. 1228-1232 in *Proceedings of the Human Factors and Ergonomics Society 39th Annual Meeting*. Santa Monica, CA: Human Factors and Ergonomics Society.

Diehl, A.E.
1991 Human performance and system safety considerations in aviation mishaps. *International Journal of Aviation Psychology* 1:97-106.

Dipboye, R.L.
1990 Laboratory vs. field research in industrial and organizational psychology. Pp. 1-34 in *International Review of Industrial and Organizational Psychology* Vol. 5, C.L. Cooper and I.T. Robertson, eds.

Druckman, D., and R.A. Bjork, eds.
1994 *Learning, Remember, Believing: Enhancing Human Performance*. Committee on Techniques for Enhancing Human Performance, National Research Council. Washington, DC: National Academy Press.

Dujardin, P.
1993 The inclusion of future users in the design and evaluation process. Pp. 435-441 in *Verification and Validation of Complex Systems: Human Factors Issues*, J.A. Wise, V.D. Hopkin, and P. Stager, eds. NATO ASI Series Vol. F110. Berlin: Springer-Verlag.

Durso, F.T., T.R. Truitt, C.A. Hackworth, D.D. Ohrt, J.M. Hamic, and C.A. Manning
1995 Factors characterizing en route operational errors: Do they tell us anything about situation awareness. In the *Proceedings of the International Conference on Experimental Analysis and Measurement of Situation Awareness*, Daytona Beach, FL, November 1-3.

Edwards, J.L.
1991 Intelligent dialogue in air traffic control systems. Pp. 137-151 in *Automation and System Issues In Air Traffic Control*, J.A. Wise, V.D. Hopkin, and M.L. Smith, eds. Berlin: Springer-Verlag.

Elkind, J.I., S.K. Card, J. Hochberg, and B.M. Huey, eds.
1990 *Human Performance Models for Computer-Aided Engineering*. Committee on Human Factors, National Research Council. San Diego, CA: Academic Press.

Empson, J.
1991 Cognitive failures in military air traffic control. Pp. Pp. 339-348 in *Automation and Systems Issues in Air Traffic Control*, J.A. Wise, V.D. Hopkin, and M.L. Smith, eds. NATO ASI Series, Vol. F73. Berlin: Springer-Verlag.

Endsley, M.R., and M.D. Rodgers
1994 *Situation Awareness Information Requirements for En Route Air Traffic Control*. Final Report. FAA Technical Report No. DOT/FAA/AM-94/27, Federal Aviation Administration Office of Aviation Medicine, Washington, D.C.: U.S. Department of Transportation.

Erzberger, H., T.J. Davis, and S.M. Green
1993 Design of Center-TRACON Automation System. Pp. 11-1—11-12 in the Proceedings 538 of the AGARD Guidance and Control Panel 56th Symposium on Machine Intelligence in Air Traffic Management, Berlin, Germany.

Fassert, C., and I. Pichancourt
1994 Evaluation and use of prototypes: Cases in air traffic control. Pp. 69-75 in *Verification and Validation of Complex Systems: Additional Human Factors Issues*, J.A. Wise, V.D. Hopkin, and P. Stager, eds. Daytona Beach, Fl.: Embry-Riddle Aeronautical University Press.

Federal Aviation Administration
  1996 SIMMOD: The FAA's Airport and Airspace Simulation Model. Internet On-Line Publications, 08/06/96.
Fisk, A.D., W.L. Derrick, and W. Schneider
  1986 A methodological assessment and evaluation of dual-task paradigms. *Current Psychological Research and Reviews* 5:315-327.
Fitts, P.M.
  1947 Psychological research on equipment design in the AAF. *American Psychologist* 2:93-98.
  1951 *Human Engineering for an Effective Air-Navigation and Traffic-Control System.* Washington, DC: National Research Council.
Frolow, I., and J.H. Sinnott
  1989 National airspace system demand and capacity modeling. *Proceedings of the IEEE* 77:1612.
Flanagan, J.C.
  1954 The critical incident technique. *Psychological Bulletin* 51:327-358.
Fuller, R., N. Johnston, and N. McDonald, eds.
  1995 Human factors in aviation operations. Proceedings of the 21st Conference of the European Association for Aviation Psychology (EAAP) Volume 3. Aldershot, Hants, Eng.: Avebury Aviation.
Garland, D., and M. Endsley
  1996 Experimental analysis and measurement of situation awareness. Daytona Beach, FL: Embry-Riddle Aeronautical University Press.
Garvey, W.D., and F.V. Taylor
  1959 Interactions among operator variables, system dynamics, and task-induced stress. *Journal of Applied Psychology* 43:79-85.
Green, R.J., H.C. Self, and T.S. Ellifritt, eds.
  1995 *50 Years of Human Engineering: History and Cumulative Bibliography of the Fitts Human Engineering Division.* Armstrong Laboratory Dayton, OH: Wright-Patterson Air Force Base.
Hancock, P.A.
  1987 Arousal theory, stress, and performance: Problems of incorporating energetic aspects of behavior into human-machine systems function. Pp. 170-179 in *Ergonomics and Human Factors*, L.S. Mark, J.S. Warm, and R.L. Huston, eds. New York: Springer-Verlag.
  1993 On the future of hybrid human-machine systems. Pp. 61-85 in *Verification and Validation of Complex Systems: Human Factors Issues*, J.A. Wise, V.D. Hopkin, and P. Stager, eds. NATO ASI Series Vol. F110. Berlin: Springer-Verlag.
Hart, S.G., and L.E. Staveland
  1988 Development of NASA-TLX: Results of empirical and theoretical research. Pp. 139-183 in *Human Mental Workload*, P.A. Hancock and N. Meshkati, eds. Amsterdam: North-Holland.
Hart, S.G., and C.D. Wickens
  1990 Workload assessment and predictions. In *MANPRINT: An Emerging Technology*, H.R. Booher, ed. New York: Van Rostrand and Reinhold.
Harwood, K.
  1993 Defining human-centered system issues for verifying and validating air traffic control systems. Pp. 115-129 in *Verification and Validation of Complex Systems: Human Factors Issues*, J.A. Wise, V.D. Hopkin, and P. Stager, eds. NATO ASI Series, Vol. F110. Berlin: Springer-Verlag.

1994 CTAS: An alternative approach to developing and evaluating advanced automation. Pp. 332-338 in *Human Performance in Automated Systems: Current Research and Trends*, M. Mouloua and R. Parasuraman, eds. Hillsdale, NJ: Lawrence Erlbaum.

Harwood, K., R. Roske-Hofstrand, and E. Murphy
1991 Exploring conceptual structures in air traffic control (ATC). Pp. 466-473 in *Proceedings of the Sixth International Symposium on Aviation Psychology*, R.S. Jensen, ed. Columbus, OH: Department of Aviation, Ohio State University.

Harwood, K., and B. Sanford
1994 Evaluation in context: ATC automation in the field. Pp. 247-262 in *Human Factors Certification of Advanced Aviation Technologies*, J.A. Wise, V.D. Hopkin, and D.J. Garland, eds. Daytona Beach, FL: Embry-Riddle Aeronautical University Press.

Hawkins, F.H.
1993 *Human Factors in Flight*. 2nd Edition. Brookfield, VT: Avebury Aviation.

Hilburn, B., P.G.A.M. Jorna, and R. Parasuraman
1995 The effect of advanced ATC automation on mental workload and monitoring performance: An empirical investigation in Dutch airspace. In *Proceedings of the International Symposium on Aviation Psychology*, Columbus, OH.

Hill, S.G., H.P. Iavecchia, A.C. Byers, A.L. Zaklad, and R.E. Christ
1992 Comparison of four subjective workload scales. *Human Factors* 34:429-439.

Holding, D.
1987 Concepts of training. In *Handbook of Human Factors*, G. Salvendy, ed. New York: Wiley.

Hollnagel, E.
1988 Mental models and model mentality. Pp. 261-268 in *Tasks, Error, and Mental Models*, L.P. Goodstein, H.B. Anderson, and S.E. Olson, eds. London: Taylor and Francis.
1993a The reliability of interactive systems: Simulation based assessment. Pp. 205-221 in *Verification and Validation in Complex Systems: Human Factors Issues*, J.A. Wise, V.D. Hopkin, and P. Stager, eds. Berlin: Springer-Verlag.
1993b *Human Reliability Analysis Context and Control*. New York: Academic Press.

Hopkin, V.D.
1979 Mental workload measurement in air traffic control. Pp. 381-386 in *Mental Workload: Its Theory and Measurement*, N. Moray, ed. London: Plenum Press.
1980 The measurement of the air traffic controller. *Human Factors* 22(5):547-560.
1982a *Human Factors in Air Traffic Control*. NATO AGARDograph No. 275. Paris: North Atlantic Treaty Organization.
1982b *Subjective Assessment Techniques in Air Traffic Control Evaluations*. Report No. 622. Eng.: Royal Air Force Institute of Aviation Medicine.
1994 Colour on air traffic control displays. *Information Display* 10(1):14-18.
1995 *Human Factors in Air Traffic Control*. London: Taylor and Francis.

Huey, B.M., and C.D. Wickens, eds.
1993 *Workload Transition: Implications for Individual and Team Performance*. Committee on Human Factors, National Research Council. Washington, DC: National Academy Press.

International Civil Aviation Organization
1993 Human Factors Digest No. 8. Human Factors in Air Traffic Control International Civil Aviation Organization. Circular 241-AN/145. Montreal, Canada.
1994 *Human Factors Digest No. 11. Human Factors in CNS/ATM Systems. The development of human-centred automation and advanced technology in future aviation systems*. International Civil Aviation Organization. Circular 249-AN/149. Montreal, Canada.

Jensen, R.S., ed.
1989 *Aviation Psychology*. Brookfield, VT: Gower Publishing.

Jorna, P.G.A.M.
    1993   The human component of system validation. Pp. 281-304 in *Verification and Validation in Complex Systems: Human Factors Issues*, J.A. Wise, V.D. Hopkin, and P. Stager, eds. NATO ASI Series, Vol. F110. Berlin: Springer-Verlag.

Kalsbeek, J.W.H.
    1965   Mésure objective de la surcharge mentale: Nouvelle applications de la méthode des double tâches. *Le Travail Humain* 28:121-132.

Kantowitz, B.H.
    1992   Selecting measures for human factors research. *Human Factors* 34:387-398.

Kellogg, W.A., and T.J. Breen
    1987   Evaluating user and system models: Applying scaling techniques to problems in human-computer interaction. Pp. 303-308 in *Human Factors in Computing Systems - IV and Graphics Interface: Proceedings of CHI and GI '87 Conference*, J.M. Carroll and P.P. Tanner, eds. New York: Association for Computing Machinery.

Kerns, K.
    1991   Data-link communication between controllers and pilots: A review and synthesis of the simulation literature. *International Journal of Aviation Psychology* 1(3):181-204.

Kramer, A.
    1991   Physiological measures of workload: A review of recent progress. Pp. 279-328 in *Multiple Task Performance*, D. Damos, ed. London: Taylor and Francis.

Lee, K.K., and T.J. Davis
    1995   The Development of the Final Approach Spacing Tool (FAST): A Cooperative Controller-Engineer Design Approach. NASA Technical Memorandum 110359, NASA Ames Research Center, Moffet Field, CA, August.

Leplat, J., and R. Browaeys
    1965   Analyse et mésure de la charge du travail du controleur du trafic aérienne. *Bulletin du Centre d'Études et de Recherches Psychotechniques* 14:69-79.

Leroux, M.
    1993a   The role of verification and validation in the design process of knowledge based components of air traffic control systems. Pp. 357-373 in *Verification and Validation of Complex Systems: Human Factors Issues*, J.A. Wise, V.D. Hopkin, and P. Stager, eds. NATO ASI Series Vol. F1110. Berlin: Springer-Verlag.
    1993b   The role of expert systems in future cooperative tools for air traffic controllers. Pp. 335-340 in *Proceedings of the Seventh International Symposium on Aviation Psychology*, R.S. Jensen and D. Neumeister, eds. Columbus, OH: Department of Aviation, Ohio State University.
    1995   ERATO: Cognitive engineering applied to ATC. Pp. 89-94 in *Human Factors in Aviation Operations. Proceedings of the 21st Conference of the European Association for Aviation Psychology (EAAP)*, Volume 3, R. Fuller, N. Johnson, and N. McDonald, eds. Aldershot, Hants, Eng.: Avebury Aviation.

Lind, M.
    1988   System concepts and the design of man-machine interfaces for supervisory control. Pp. 269-277 in *Tasks, Error, and Mental Models*, L.P. Goodstein, H.B. Anderson, and S.E. Olson, eds. London: Taylor & Francis.

Locke, E.A.
    1986   *Generalizing from Laboratory to Field Settings*. Lexington, MA: Lexington Books.

Loughery, K.R.
    1989   Micro SAINT—A tool for modeling human performance in systems. In *Applications of Human Performance Models to System Design*, G.R. McMillan, D. Beevis, E. Salas, M.H. Strub, R. Sutton, and L. Van Breda, eds. New York and London: Plenum Press.

Lysaght, R.J., S.G. Hill, and A.O. Dick
    1989   Operator Workload: Comprehensive Review and Evaluation of Operator Workload Methodologies. Report No. 851. Fort Bliss, TX: Army Research Institute.

Manning, C.A., and D. Broach
    1992   *Identifying Ability Requirements for Operators of Future Automated Air Traffic Control Systems: Final Report*. Federal Aviation Administration Report No. DOT/FAA/AM-92/26. Washington, DC: FAA Office of Aviation Medicine.

Maurino, D.E., J. Reason, N. Johnston, and R.B. Lee
    1995   *Beyond Aviation Human Factors*. Aldershot, Hants, Eng.: Avebury Aviation.

McCoy, W.E., and K.H. Funk
    1991   Taxonomy of ATC operator errors based on a model of human information processing. Pp. 532-537 in *Proceedings of the Sixth International Symposium on Aviation Psychology*, R.S. Jensen, ed. Columbus, OH: Department of Aviation, Ohio State University.

McDonald, N., N. Johnston, and R. Fuller, eds.
    1994   *Aviation Psychology in Practice*. Aldershot, Hants, Eng.: Avebury Aviation.

McMillan, G.R., D. Beevis, E. Salas, M.H. Strub, R. Sutton, and L. Van Breda, eds.
    1989   *Applications of Human Performance Models to System Design*. New York: Plenum Press.

McMillan, G.R., D. Beevis, W. Stein, M.H. Strub, E. Salas, R. Sutton, and K.C. Reynolds
    1992   A Directory of Human Performance Models for System Design. NATO Report AC/243 (Panel 8) TR/1.

Meister, D.
    1985   *Behavioral Analysis and Measurement Methods*. New York: John Wiley.
    1991   *Psychology of System Design*. Advances in Human Factors/Ergonomics, 17. Amsterdam: Elsevier.

Mellone, V.J., and S.M. Frank
    1993   The behavioral impact of TCAS II on the national air traffic control system. Pp. 352-359 in *Proceedings of the Seventh International Symposium on Aviation Psychology*, R.S. Jensen and D. Neumeister, eds. Columbus, OH: Department of Aviation, Ohio State University.

Melton, C.E.
    1982   *Physiological Stress in Air Traffic Controllers: A Review*. Report No. DOT/FAA/AM-82/17. Washington, DC: Federal Aviation Administration.

Miller, D.L., and G.J. Wolfman
    1993   Computer human interface design in tower air traffic control for aircraft flight data management. Pp. 372a-372e in *Proceedings of the Seventh International Symposium on Aviation Psychology*, R.S. Jensen and D. Neumeister, eds. Columbus, OH: Department of Aviation, Ohio State University.

MITRE Corp.
    1995   Free Flight Scenario Simulation. Presentation at Advancing Free Flight Through Human Factors Workshop, Ellicott City, MD, June 20.

Mogford, R.H.
    1991   Mental models in air traffic control. Pp. 235-242 in *Automation and Systems Issues in Air Traffic Control*, J.A. Wise, V.D. Hopkin, and M.L. Smith, eds. NATO ASI Series, Vol. F73. Berlin: Springer-Verlag.
    1994   Mental models and situation awareness in air traffic control. Pp. 199-207 in *Situational Awareness in Complex Systems*, R.D. Gilson, D.J. Garland, and J.M. Koonce, eds. Daytona Beach, FL: Embry-Riddle Aeronautical University Press.

Mogford, R.H., E.D. Murphy, and J.A. Guttman
    1994a   Using knowledge exploration tools to study airspace complexity in air traffic control. *International Journal of Aviation Psychology* 4(1):29-45.

Mogford, R.H., E.D. Murphy, R.J. Roske-Hofstrand, G. Yastrop, and J.A. Guttman
    1994b   *Application of Research Techniques for Documenting Cognitive Processes in Air Traffic Control: Sector Complexity and Decision Making*. Technical Report No. DOT/FAA/CT-TN94/3. Atlantic City, NJ: FAA Technical Center.

Monan, W.P.
    1983   *Cleared for the Visual Approach: Human Factor Problems in Air Carrier Operations*. NASA Contract Report 166573. Moffett Field, CA: NASA Ames Research Center.

Moody, C.
    1991   Operational evaluation of a tower workstation for clearance delivery. Pp. 538-549 in *Proceedings of the Sixth International Symposium on Aviation Psychology*, R.S. Jensen, ed. Columbus, OH: Department of Aviation, Ohio State University.

Moray, N.
    1988   Mental workload since 1979. *International Reviews of Ergonomics* 2:123-150.

Moroney, W.F., D.W. Biers, and F.T. Eggemeir
    1995   Some measurement and methodological considerations in the application of subjective workload measurement techniques. *International Journal of Aviation Psychology* 5(1):87-106.

Morrison, R., and R.H. Wright
    1989   ATC control and communications problems: An overview of recent ASRS data. Pp. 902-907 in *Proceedings of the Fifth International Symposium on Aviation Psychology*. Columbus, OH: Department of Aviation, Ohio State University.

Morrow, D., A. Lee, and M. Rodvold
    1993   Analysis of problems in routine controller-pilot communication. *International Journal of Aviation Psychology* 3(4):285-302.

Muckler, F.A., and S.A. Seven
    1992   Selecting performance measures: "Objective" versus "subjective" measurement. *Human Factors* 34:441-456.

Murphy, E.D., R.A. Reaux, L.J. Stewart, W.D. Coleman, and K. Bruce
    1989   Where's the workload in air traffic control? Pp. 908-913 in *Proceedings of the Fifth International Symposium on Aviation Psychology*. Columbus, OH: Department of Aviation, Ohio State University.

Nagel, D.C.
    1988   Human error in aviation operations. Pp. 263-303 in *Human Factors in Aviation*, E.L. Wiener and D.C. Nagel, eds. New York: Academic Press.

Narborough-Hall, C.S.
    1985   Recommendations for applying colour coding to air traffic control displays. *Displays* 6(3):131-137.

Narborough-Hall, C.S., and V.D. Hopkin
    1988   Human factors contributions to air traffic control evaluations. Pp. 142-147 in *Contemporary Ergonomics 1988*, E.D. Megaw, ed. London: Taylor and Francis.

National Transportation Safety Board
    1991   *Runway Collision of U.S. Air Flight 1493 and Skywest Flight 5569, Los Angeles International Airport, Los Angeles, California, February 1, 1991*. National Transportation Safety Board Report PB91-910409 NTSB/AAR-91/08. Washington, DC: National Transportation Safety Board.

Natsoulas, T.
    1967   What are perceptual reports all about. *Psychological Bulletin* 67:249-272.

Norman, D.A., and D. Bobrow
    1975   On data-limited and resource-limited processing. *Cognitive Psychology* 7:44-60.

Odoni, A.R.
    1991   *Transportation Modeling Needs: Airports and Airspace*. U.S. Department of Transportation Report. Cambridge, MA: Volpe Transportation Research Center.

O'Donnell, R.D., and F.T. Eggemeier
  1986 Workload assessment methodology. Chapter 42 in *Handbook of Perception. Volume 2. Cognitive Processes and Performance*, K. Boff, L. Kaufman, and J. Thomas, eds. New York: Wiley.

O'Hare, D., and S.N. Roscoe
  1990 *Flightdeck Performance: The Human Factor*. Ames, IA: Iowa State University Press.

Parasuraman, R.
  1990 Event-related brain potentials and human factors research. Pp. 279-300 in *Event-Related Brain Potentials: Basic and Applied Issues*, J.W. Rohrbaugh, R. Parasuraman, and R. Johnson, eds. New York: Oxford University Press.

Parsons, H.M.
  1972 *Man-Machine System Experiments*. Baltimore, MD: Johns Hopkins.

Pawlak, W.S., Brinton, C.R., Courch, K., and Lancaster, K.M.
  1996 A framework for the evaluation of air traffic control complexity. *Proceedings of the 1996 Meeting of the American Institute of Aeronautics and Astronautics*.

Pew, R.W., and S. Baron
  1983 Perspectives on human performance modelling. Pp. 1-14 in *Analysis, Design and Evaluation of Man-Machine Systems, Proceedings of the IFAC/IFIP/IFORS/IEA Conference*, G. Johannsen and J.E. Rijnsdorp, eds. Oxford, Eng.: Pergamon Press.

Planzer, N., and M.T. Jenny
  1995 Managing the evolution to free flight. *Journal of Air Traffic Control* January-March:18-20.

Porter, E.H.
  1964 *Manpower Development*. New York: Harper & Row.

Poulton, E.C.
  1973 The effects of fatigue upon inspection work. *Applied Ergonomics* 4:73-83.

Prinzo, O.V., and T.W. Britton
  1993 *ATC/Pilot Voice Communications: A Survey of the Literature: Final Report*. Report No. DOT/FAA/AM-93/20. Washington, DC: FAA Office of Aviation Medicine.

Rasmussen, J.
  1985 Trends in human reliability analysis. *Ergonomics* 28:1185-1195.
  1987 Cognitive control and human error mechanisms. Pp. 53-61 in *New Technology and Human Error*, J. Rasmussen, K. Duncan, and J. Leplat, eds. Chichester, U.K.: Wiley.

Rasmussen, J., and K.J. Vicente
  1989 Coping with human errors through system design: Implications for ecological interface design. *International Journal of Man-Machine Studies* 23:517-534.

Reason, J.
  1990 *Human Error*. Cambridge, Eng.: Cambridge University Press.
  1993 The identification of latent organizational failures in complex systems. Pp. 223-237 in *Verification and Validation of Complex Systems: Human Factors Issues*, J.A. Wise, V.D. Hopkin, and P. Stager, eds. Berlin: Springer-Verlag.

Reason, J.T., and D. Zapf, eds.
  1994 Errors, error detection and error recovery. *Applied Psychology: An International Review* 43(4):427-584 (Special Issue).

Redding, R.E.
  1992 Analysis of operational error and workload in air traffic control. Pp. 1321-1325 in *Proceedings of the Human Factors Society 36th Annual Meeting*. Santa Monica, CA: Human Factors and Ergonomics Society.

Reid, G.B., and T.E. Nygren
  1988 The subjective workload assessment technique. Pp. 185-218 in *Human Mental Workload*, P.A. Hancock and N. Meshkati, eds. Amsterdam: North-Holland.

Reynard, W.D., C.E. Billings, E. Cheaney, and R. Hardy
    1986   The Development of the NASA Aviation Reporting System. NASA Reference Publication No. 1114.

Rodgers, M.D.
    1993   *An Examination of the Operational Error Database for Air Route Traffic Control Centers: Final Report.* Federal Aviation Administration Civil Aeromedical Institute, Oklahoma City, Oklahoma, Report No. DOT/FAA/AM-93/22. Washington, DC: Office of Aviation Medicine.

Rodgers, M.D., and D.A. Duke
    1994   *SATORI: Situation Assessment Through the Re-Creation of Incidents: Final Report.* Federal Aviation Administration Report No. DOT/FAA/AM-93/12. Washington, DC: Office of Aviation Medicine

Rodgers, M.D., C.A. Manning, and C.S. Kerr
    1994   Demonstration of power: Performance and objective workload evaluation research. In *Proceedings of the 38th Annual Meeting of the Human Factors and Ergonomics Society.* Santa Monica, CA: Human Factors Society.

Rose, R.M., and L.F. Fogg
    1993   Modeling the dynamics of mental workload and human performance in complex systems. *IEEE Transactions on Systems, Man, and Cybernetics* 23:1662-1671.

Rosenthal, L.J., V.J. Mellone
    1989   Human factors in ATC operations: Anticipatory clearances. Pp. 884-889 in *Proceedings of the Fifth International Symposium on Aviation Psychology.* Columbus, OH: Department of Aviation, Ohio State University.

Rosenthal, L.J., and W. Reynard
    1991   Learning from incidents to avert accidents. *Aviation Safety Journal* Fall:7-10.

Rouse, W.B., and W.J. Cody
    1989   Designers' criteria for choosing human performance models. Pp. 7-14 in *Applications of Human Performance Models to System Design*, G.R. McMillan, D. Beevis, E. Salas, M. H. Strub, R. Sutton, and L. Van Breda, eds. New York: Plenum Press.

Rubel, P.
    1976   Tiger in the fault tree jungle. Pp. 1071-1082 in *Modeling and Simulation. Proceedings of the Seventh Annual Pittsburgh Conference.* Pittsburgh, PA: University of Pittsburgh.

Salvendy, G.
    1987   *Handbook of Human Factors.* Chichester, U.K.: Wiley.

Sandiford, W.K.
    1991   Meeting the ATC challenge through simulation. Pp. 181-184 in *Automation and Systems Issues in Air Traffic Control*, J.A. Wise, V.D. Hopkin, and M.L. Smith, eds. Berlin: Springer-Verlag.

Sarter, N.B., and D.D. Woods
    1995   How in the world did we ever get into that mode? *Human Factors* 36.

Scott, W.
    1996   Transport arrival times predicted within seconds. *Aviation Week and Space Technology* (March 11):40-41.

Seamster, T.L., R.E. Redding, J.R. Cannon, J.M. Ryder, and J.A. Purcell
    1993   Cognitive task analysis of expertise in air traffic control. *International Journal of Aviation Psychology* 3(4):257-283.

Senders, J.W., and N.P. Moray
    1991   *Human Error: Cause, Prediction, and Reduction.* Hillsdale, NJ: Lawrence Erlbaum.

Shattuck, L.G., and D.D. Woods
    1994   The critical incident technique: 40 years later. Pp. 1080-1084 in *Proceedings of the Human Factors and Ergonomics Society 38th Annual Meeting.* Santa Monica, CA: Human Factors and Ergonomics Society.

Singley, M.K., and J.R. Anderson
  1989 *The Transfer of Cognitive Skill*. Cambridge, MA: Harvard University Press.
Simolunas, A.A., and H.S. Bashinski
  1991 Computerization and automation: Upgrading the American air traffic control system. Pp. 31-38 in *Automation and Systems Issues in Air Traffic Control*, J.A. Wise, V.D. Hopkin, and M.L. Smith, eds. Berlin: Springer-Verlag.
Small, D.W.
  1994 *Lessons Learned: Human Factors in AAS Procurement*. MITRE/CAASD Human Factors Engineering Specialty Group Technical Report. Proj. No. F40C/F60C/F80C. Dept. F067. McLean, VA: MITRE.
Smith, R.C.
  1980 *Stress, Anxiety, and the Air Traffic Control Specialist: Some Conclusions From a Decade of Research*. Report No. OT/FAA/AM-80/14. Washington, DC: FAA Office of Aviation Medicine.
Smolensky, M.W., and L. Hitchcock
  1993 When task demand is variable: Verifying and validating mental workload in complex, "real-world" systems. Pp. 305-313 in *Verification and Validation in Complex Systems: Additional Human Factors Issues*, J.A. Wise, V.D. Hopkin, and P. Stager, eds. Daytona Beach, FL: Embry-Riddle Aeronautical University Press.
Sperandio, J.C.
  1971 Variation of operator's strategies and regulating effects on workload. *Ergonomics* 14:571-577.
Stager, P.
  1991a The Canadian Automated Air Traffic Control System (CAATS): An overview. Pp. 39-45 in *Automation and Systems Issues in Air Traffic Control*, J.A. Wise, V.D. Hopkin, and M.L. Smith, eds. Berlin: Springer-Verlag.
  1991b Error models for operating irregularities: Implications for automation. Pp. 321-338 in *Automation and Systems Issues in Air Traffic Control*, J.A. Wise, V.D. Hopkin, and M.L. Smith, eds. NATO ASI Series, Vol. F73. Berlin: Springer-Verlag.
  1993 Validation in complex systems: Behavioral issues. Pp. 99-114 in *Verification and Validation of Complex Systems: Human Factors Issues*, J.A. Wise, V.D. Hopkin, and P. Stager, eds. Daytona Beach, FL: Embry-Riddle Aeronautical University Press.
Stager, P., and D. Hameluck
  1986 Estimating detection probabilities for search and rescue. Pp. 312-316 in *Proceedings of the Human Factors Society 30th Annual Meeting*. Santa Monica, CA: Human Factors and Ergonomics Society.
  1989 Analysis of air traffic control operating irregularities. Pp. 890-895 in *Proceedings of the Fifth International Symposium on Aviation Psychology*. Columbus, OH: Department of Aviation, Ohio State University.
  1990 Ergonomics in air traffic control. *Ergonomics* 33:493-499.
Stager, P., D. Hameluck, and R. Jubis
  1989 Underlying factors in air traffic control incidents. Pp. 43-46 in *Proceedings of the Human Factors Society 33rd Annual Meeting*. Santa Monica, CA: Human Factors and Ergonomics Society.
Stager, P., and T.G. Paine
  1980 Separation discrimination in a simulated air traffic control display. *Human Factors* 22(5):631-636.
Stager, P., P. Proulx, B. Walsh, and T. Fudakowski
  1980a Bilingual air traffic control in Canada. *Human Factors* 22:655-670.
  1980b Bilingualism in Canadian air traffic control. *Canadian Journal of Psychology* 34:346-358.

Stein, E.S.
 1985 *Air Traffic Controller Workload: An Examination of Workload Probe.* Report No. DOT/FAA/CT-TN 84/24. Atlantic City, NJ: Federal Aviation Administration.
 1988 *Air Traffic Controller Scanning and Eye Movements: A Literature Review.* Report No. DOT/FAA/CT-TN 88/24. Atlantic City, NJ: Federal Aviation Administration.
 1992 *Air Traffic Control Visual Scanning.* Report No. DOT/FAA/CT-TN 92/16. Atlantic City, NJ: Federal Aviation Administration.
Stein, E.S., and E. Buckley, E., eds.
 1994 *Human Factors at the FAA Technical Center: Bibliography 1958-1994.* DOT Report DOT/FAA/CTTN94/50. Atlantic City, NJ: FAA Technical Center.
Stein, E.S., and D.J. Garland
 1993 *Air Traffic Control Working Memory: Considerations in Air Traffic Control Tactical Operations.* Report No. DOT/FAA/CTTN93/37. Washington, DC: Federal Aviation Administration
Tattersall, A.J., E.W. Farmer, and A.J. Belyavin
 1991 Stress and workload management in air traffic control. Pp. 255-266 in *Automation and Systems Issues in Air Traffic Control*, J.A. Wise, V.D. Hopkin, and M.L. Smith, eds. NATO ASI Series Vol. F 73. Berlin: Springer-Verlag.
Taylor, F.V.
 1947 Psychology at the Naval Research Laboratory. *American Psychologist* 2:87-92.
Taylor, R.M., and I.S. MacLeod
 1994 Quality assurance and risk management: Perspectives on human factors certification of advanced aviation systems. Pp. 97-118 in *Human Factors Certification of Advanced Aviation Technologies*, J.A. Wise, V.D. Hopkin, and D.J. Garland, eds. Daytona Beach, FL: Embry-Riddle Aeronautical University Press.
Transport Canada
 1979 *Report of the Bilingual IFR Communications Simulations Studies, Volume 1: Summary Report and Recommendations.* Publication TP 1844, Cat. No. T52-53/1979-1, ISBN-0-662-10300-9. Ottawa, Ont.: Transport Canada.
U.S. Department of Defense
 1979 *Human Engineering Requirements for Military Systems (MIL-H-46855B).* Washington, DC: U.S. Department of Defense.
Van der Veer, G.C.
 1987 Mental models and failures in human-machine systems. Pp. 221-230 in *Information Systems: Failure Analysis*, J.A. Wise and A. Debons, eds. NATO ASI Series F: Computer and Systems Sciences 32. Berlin: Springer-Verlag.
Völckers, U.
 1991 Application of planning aids for air traffic control: Design principles, solutions, results. Pp. 169-172 in *Automation and Systems Issues in Air Traffic Control*, J.A. Wise, V.D. Hopkin, and M.L. Smith, eds. NATO ASI Series, Vol. F73. Berlin: Springer-Verlag.
Vortac, O.U., M.B. Edwards, D.K. Fuller, and C.A. Manning
 1994 *Automation and Cognition in Air Traffic Control: An Empirical Investigation: Final Report.* Federal Aviation Administration, DOT/FAA/AM-94/3. Washington, DC: Office of Aviation Medicine.
Vortac, O.U., M.B. Edwards, J.P. Jones, C.A. Manning, and A.J. Rotter
 1993 En route air traffic controllers' use of flight progress: A graph-theoretic analysis. *International Journal of Aviation Psychology* 3(4):327-343.
Vortac, O.U., and C.A. Manning
 1994 Modular automation: Automating sub-tasks without disrupting task flow. Pp. 325-331 in *Human Performance in Automated Systems: Current Research and Trends*, M. Mouloua and R. Parasuraman, eds. Hillsdale, NJ: Lawrence Erlbaum.

Vortac, O.U., Manning, C.A., and A.J. Rotter
    1992   *En Route Air Traffic Controllers' Use of Flight Progress Strips: A Graph-Theoretic Analysis: Final Report.* Federal Aviation Administration Report No. DOT/FAA/AM-92/31. Washington, DC: Office of Aviation Medicine.

Waern, Y.
    1989   *Cognitive Aspects of Computer Supported Tasks.* New York: John Wiley.

Warm, J.S., W.N. Dember, and B. Hancock
    1996   Vigilance and workload in automated systems. In *Automation and Human Performance: Theory and Applications,* R. Parasuraman and M. Mouloua, eds. Hillsdale, NJ: Lawrence Erlbaum.

Westrum, R.
    1993   Cultures with requisite imagination. Pp. 401-416 in *Verification and Validation of Complex Systems: Human Factors Issues,* J.A. Wise, V.D. Hopkin, and P. Stager, eds. NATO ASI Series Vol. 110. Berlin: Springer-Verlag.
    1994   Is there a role for a "test controller" in the development of new ATC equipment? Pp. 221-228 in *Human Factors Certification of Advanced Aviation Technologies,* J.A. Wise, V.D. Hopkin, and D.J. Garland, eds. Daytona Beach, FL: Embry-Riddle Aeronautical University Press.

Whiteside, J., J. Bennett, and K. Holtzblatt
    1988   Usability engineering: Our experience and evolution. Pp. 791-817 in *Handbook of Human-Computer Interaction,* M. Helander, ed. New York: Elsevier (North Holland).

Whitfield, D.
    1979   A preliminary study of the air traffic controllers' picture. *CATCA Journal* 11:19-28.

Whitfield, D., and A. Jackson
    1983   The air traffic controller's picture as an example of a mental model. Pp. 37-44 in *Analysis, Design and Evaluation of Man-Machine Systems: Proceedings of the IFAC/IFIP/IFORS/IEA Conference,* G. Johannsen and J.E. Rijnsdorp, eds. Oxford: Pergamon Press.

Wickens, C.D.
    1984   Processing resources in attention. In *Varieties of Attention,* R. Parasuraman and D.R. Davies, eds. Orlando, FL: Academic Press.
    1990   Applications of event-related potential research to problems in human factors. Pp. 301-310 in *Event-Related Brain Potentials: Basic and Applied Issues,* J.W. Rohrbaugh, R. Parasuraman, and R. Johnson, eds. New York: Oxford University Press.
    1992   *Engineering Psychology and Human Performance, Second Edition.* New York: Harper Collins.
    1995a Aerospace techniques. Pp. 112-142 in *Research Techniques in Human Engineering,* J. Weimer, ed. Englewood Cliffs, NJ: Prentice-Hall.
    1995b The tradeoff of design for routine and unexpected performance: Implications for situation awareness. In *Proceedings of the the Second International Conference on Experimental Analysis and Measurement of Situation Awareness,* M. Endsley and D. Garland, eds. Daytona Beach, FL, November 1-3.

Wickens, C.D., and A.D. Andre
    1994   Performance-preference dissociations. Pp. 369-370 in *Society for Information Display Digest.* Playa del Rey, CA: Society for Information Display.

Wickens, C.D., and T. McCloy
    1993   ASRS and aviation psychology. Pp. 1028-1030 in *Proceedings of the Seventh International Symposium on Aviation Psychology,* R.S. Jensen and D. Neumeister, eds. Columbus, OH: Department of Aviation, Ohio State University.

Wiener, E.L.
    1980   Midair collisions: The accidents, the systems, and the realpolitik. *Human Factors* 22:521-533.

1987 Fallible humans and vulnerable systems: Lessons learned from aviation. Pp. 163-181 in *Information Systems: Failure Analysis. NATO ASI Series F: Computer and Systems Sciences, Vol. 32*, J.A. Wise and A. Debons, eds. Berlin: Springer-Verlag.
1989 Reflections on human error: Matters of life and death. Pp. 1-7 in *Proceedings of the Human Factors Society 33rd Annual Meeting*. Santa Monica, CA: Human Factors Society.

Wiener, E.L., and D.C. Nagel, eds.
1988 *Human Factors in Aviation*. New York: Academic Press.

Wiener, E.L., B.G. Kanki, and R.L. Helmreich, eds.
1993 *Cockpit Resource Management*. San Diego, CA: Academic Press.

Wise, J.A., and A. Debons, eds
1987 *Information Systems: Failure Analysis. NATO ASI Series F: Computer and Systems Sciences, Vol. 32*. Berlin: Springer-Verlag.

Wise. J.A., V.D. Hopkin, and D.J. Garland, eds.
1994a *Human Factors Certification of Advanced Aviation Technologies*. Daytona Beach, FL: Embry-Riddle Aeronautical University Press.

Wise, J.A., V.D. Hopkin, and M.L. Smith, eds.
1991 *Automation and Systems Issues in Air Traffic Control. NATO ASI Series, Vol. F73*. Berlin: Springer-Verlag.

Wise, J.A., V.D. Hopkin, and P. Stager, eds.
1993 *Verification and Validation in Complex Systems: Human Factors Issues. NATO ASI Series, Vol. F110*. Berlin: Springer-Verlag.
1994 *Verification and Validation of Complex Systems: Additional Human Factors Issues*. Daytona Beach, FL: Embry-Riddle Aeronautical University Press.

Woods, D.D.
1989 Modeling and predicting human error. Pp. 248-274 in *Human Performance Models for Computer-Aided Engineering*, J.I. Elkind, S.K. Card, J. Hochberg, and B.M. Huey, eds. Committee on Human Factors, National Research Council. Washington, DC: National Academy Press.

Woods, D.D., L.J. Johannesen, R.I. Cook, and N.B. Sarter, eds.
1994 *Behind Human Error: Cognitive Systems, Computers, and Hindsight*. Crew Systems Ergonomics Information Analysis Center Report No. CSERIAC SOAR 94-01. Wright-Patterson Air Force Base, OH: CSERIAC.

Woods, D.D., and N.B. Sarter
1993 Evaluating the impact of new technology on human-machine cooperation. Pp. 133-158 in *Verification and Validation of Complex Systems: Human Factors Issues*, J.A. Wise, V.D. Hopkin, and P. Stager, eds. Berlin: Springer-Verlag.

Yeh, Y.Y., and C.D. Wickens
1988 The dissociation of subjective measures of mental workload and performance. *Human Factors* 30:111-120.

CHAPTER 11: HUMAN FACTORS AND SYSTEM DEVELOPMENT

Akselsson, K.R., P. Bengtsson, C.R. Johansson, and J. Afklercker
1990 Computer aided participatory planning. In *Ergonomics of Hybrid Automated Systems II*, W. Karwowski and M. Rahimi, eds. New York: Elsevier.

Air Traffic Management
1995 Keeping the man in the loop. *Air Traffic Management* (March/April).

Bailey, R.W.
1982 *Human Performance Engineering*. Englewood Cliffs, NJ: Prentice-Hall.

Baldwin, R.A., and M.J. Chung
  1995 A formal approach to managing design processes. *Computer* (February):54-63.
Bennis, W.G., et al., eds.
  1976 *The Planning of Change*, 2nd edition. New York: Holt, Rinehart and Winston.
Bikson, T., S. Cammarata, S. Law, and T. West
  1995 *United Nations Economic and Social Information System (UNESIS) Phase I Report: Plans for the Progressive Implementation of UNESIS.* Santa Monica, CA: RAND.
Bikson, T.K., S. Law, M. Markovich, and B.T. Harder
  1995 Facilitating the implementation of research findings. Reprinted from the *Proceedings of the 74th Annual Meeting of the Transportation Research Board.* Santa Monica, CA: RAND.
Booher, H.R., ed.
  1990 *MANPRINT: An Approach to Systems Integration.* New York: Van Nostrand Reinhold.
Busey, J.B.
  1991 Human factors. *Aviation Safety Journal* (Fall):3.
Byte
  1984 The Apple story. *Byte* 9(Dec):A67-71.
  1985 The Apple story. *Byte* 10(Jan):167-8+).
Carrigan, R.E., and R.A. Kaufman
  1966 *Why System Engineering.* Belmont, CA: Fearon Press.
Chapanis, A.
  1960 *On Some Relations Between Human Engineering, Operations Research and Systems Engineering.* Report No. 8. Baltimore, MD: Psychology Laboratory, The Johns Hopkins University.
Coch, L., and J.P.R. French
  1948 Overcoming resistance to change. *Human Relations* 1:512-532.
Connell, J.L., and L.I. Shafer
  1995 *Object-Oriented Rapid Prototyping.* Englewood Cliffs, NJ: Prentice-Hall.
Del Balzo, J.
  1995 Lessons learned from the introduction of automation of ATC systems in the USA. *Transmit* 3:23-25.
Demaree, R.G., and M.R. Marks
  1962 *Development of the QQPRI.* MRL-TDR-62,4. Ohio: Wright Patterson Air Force Base.
Donahue, G.L.
  1996 FAA acquisition reform: The road ahead. *Aerospace America* (April):32-36.
Duffy, T.M., T. Post, and G. Smith
  1987 Technical manual production—An examination of five systems. *Written Communications* 4(4):370-93.
Eckstrand, G.A., et al.
  1962 Human resources engineering: A new challenge. *Human Factors* 9(6):517-520.
Erzberger, H., T.J. Davis, and S. Green
  1993 Design of Center-TRACON automation system. *Machine Intelligence in Air Traffic Management.* In *AGARD Conference Proceedings 538*, NATO, Berlin, Germany, May 11-14.
Fanwick, C.
  1967 *Caveat Emptor Informatibus.* SP-2900. Santa Monica, CA: System Development Corporation.
Federal Aviation Administration
  1994 *Remote Maintenance Monitoring System (RMMS) Remote Monitoring System (RMS) Requirements.* Document Number NAS-MD-793A. Washington, DC: U.S. Department of Transportation.

    1995   *National Plan for Civil Aviation Human Factors: An Initiative for Research and Application*. Washington, DC: U.S. Department of Transportation.

Fitts, P.M.
    1951a Human engineering. In *Handbook of Experimental Psychology*, S.S. Stevens, ed. New York: Wiley.

Fitts, P.M., ed.
    1951b *Human Engineering for an Effective Air-Navigation and Traffic Control System*. Washington DC: National Research Council/USGPO.

Fitts, P.M., L. Schipper, J.S. Kidd, M. Shelly, and C. Kraft
    1958   Some concepts and methods for the conduct of system research in a laboratory setting. In *Air Force Human Engineering, Personnel, and Training Research*, G. Finch, and F. Cameron, eds. Publication 516. Washington, DC: National Research Council.

Goode, H.H., and R.E. Machol
    1957   *System Engineering*. New York: McGraw-Hill.

Gould, J.D., and C. Lewis
    1983   Designing for usability: Key principles and what designers think. *Communications of the ACM* 28(3):300-311.

Government and Systems Technology Group
    1995   *The 21 CLW/GEN II Trade Study Report* (January 6). Scottsdale, AZ: Government and Systems Technology Group, Motorola, Inc.

Hornick, R.J.
    1962   Problems in vibration research. *Human Factors* 4(5):325-330.

Karat, C.
    1992   Cost-justifying human factors support on software development projects. *Human Factors Society Bulletin* 35(11):1-4.

Kidd, P.T.
    1990   Information technology: Design for human involvement or human intervention? In *Ergonomics of Hybrid Automated Systems II*, W. Karwowski, and M. Rahimi, eds. New York: Elsevier.

Lee, J.K.
    1994   *The Development of the Final Approach Spacing Tool (FAST): A Cooperative Controller-Engineer Design Approach*. NASA TM 1-10359. Moffett Field, CA: National Aeronautics and Space Administration

Lewin, K.
    1947   Frontiers of group dynamics. *Human Relations* 1:5-41.

Lingaard, G.
    1994   *Usability Testing and System Evaluation*. New York: Chapman and Hall.

Mankin, D., S.G. Cohen, and T.K. Bikson
    1996   *Teams and Technology*. Boston: Harvard Business School Press.

McCoy, C.E., P.J. Smith, J. Orasanu, C.E. Billings, A. VanHorn, R. Denning, M. Rodvold, and T. Gee
    1995   Airline dispatch and ATCSCC: A cooperative problem-solving success story with a future. In *Cooperative Problem-Solving in the Interactions of Airline Operations Control Centers with the National Aviation System*, P.J. Smith, E. McCoy, J. Orasanu, R. Denning, A VanHorn, and C.E. Billings, eds. Athens: Ohio University Department of Aviation.

McGregor, D.
    1960   *The Human Side of Enterprise*. New York: McGraw-Hill.

Mundra, A.D.
    1989   *A New Automation Aid to Air Traffic Controllers for Improving Airport Capacity*. #MP-89W00034. McLean, VA: The MITRE Corporation.

Mundra, A.D., and K.M. Levin
    1990 *Developing Automation for Terminal Air Traffic Control: Case Study of the Imaging Aid.* #MP-90W00029. McLean, VA: The MITRE Corporation.

Porter, E.H.
    1964 *Manpower Development.* New York: Harper and Row.

Price, H.A., M. Fiorello, J.C. Lowry, M.G. Smith, and J.S. Kidd
    1980 *The Contribution of Human Factors in Military System Development: Methodological Considerations.* TR-476. Alexandria, VA: USARI.

Rogers, E.
    1983 *Diffusion of Innovations*, 3rd edition. New York: Free Press.

Rouse, W.B., and W.J. Cody
    1988 Designers' criteria for choosing human performance models. In *Applications of Human Performance Models to System Design*, G.B. McMillan et al., eds. New York: Plenum.

Sahal, D.
    1975 Cross-impact analysis and prediction of economic development—Case study of farm tractors. *IEEE Management* EM 22(2):76-9.

Schipper, L.M., J.S. Kidd, M. Shelly, and A.F. Smode
    1957 *System Effectiveness as a Function of the Method Used by Controllers to Obtain Altitude Information.* WADC TR-57-278. Dayton, OH: Wright Air Development Center.

Simms, G.D.
    1993 *Statement of the National Association of Air Traffic Specialists in Delays, Technical Problems, and Cost Escalations in the Federal Aviation Administration's Advanced Automation System Program.* Hearings of the Subcommittee on Aviation of the House of Representatives, March 10. Washington, DC: U.S. Government Printing Office.

Sinaiko, H.W., and T.G. Belden
    1961 The indelicate experiment. In *Second Congress on the Information System Sciences*, R.M. Davis, ed. Hot Springs, VA: U.S. Department of Defense.

Singleton, W.T.
    1974 *Man-Machine Systems.* Baltimore: Penguin Books.

Thornton, J.F.
    1993 *Delays, Technical Problems, and Cost Escalation in the Federal Aviation Administration's Advanced Automation System.* Statement for the National Air Traffic Controllers Association, U.S. House of Representatives. Washington, DC: U.S. Government Printing Office.

U.S. Congress, House of Representatives
    1993 *Delays, Technical Problems, and Cost Escalation in the Federal Aviation Administration's Advanced Automation System.* Hearings 103-8. Washington, DC: U.S. Government Printing Office.

U.S. Department of the Army
    1994 *Manpower and Personnel Integration (MANPRINT) in the System Acquisition Process: Soldier Systems.* AR 602-21. Washington, DC: Department of the Army Headquarters.

Vroom, V.H., and A.G. Jago
    1987 *The New Leadership: Managing Participation in Organizations.* Englewood Cliffs, NJ: Prentice-Hall.

Williges, R.C., B.H. Williges, and S.H. Han
    1993 Sequential experimentation in human-computer interface design. In *Advances in Human-Computer Interaction, Vol. 4*, H.R. Hartson and D. Hix, eds. Norwood, NJ: Ablex.

Woodson, W.E., and D.W. Conover, eds.
    1964 *Human Engineering Guide to Equipment Design.* Berkeley: University of California Press.

CHAPTER 12: AUTOMATION

Bainbridge, L.
    1983   Ironies of automation. *Automatica* 19:775-779.

Billings, C.E.
    1991   Toward a human-centered aircraft automation philosophy. *International Journal of Aviation Psychology* 1(4):261-270.
    1996   *Aviation Automation: The Search for a Human-Centered Approach.* Mahwah, NJ: Erlbaum.

Boehm-Davis, D.A., R.E. Curry, E.L. Wiener, and R.L. Harrison
    1983   Human factors of flight-deck automation: Report on a NASA-industry workshop. *Ergonomics* 26:953-961.

Cardosi, K.
    1993   Time required for transmission of time-critical ATC messages in an en-route environment. *International Journal of Aviation Psychology* 3(4):303-313.

Chambers, A.B., and D.C. Nagel
    1985   Pilots of the future: Human or computer? *Communications of the Association for Computing Machinery* 28:1187-1199.

Chappell, S.L.
    1990   Pilot performance research for TCAS. Pp. 51-68 in *Proceedings of the Managing the Modern Cockpit: The Human Error Avoidance Techniques Conference.* Warrendale, PA: Society for Automative Engineers.

Conrad, R., A.J. Hull
    1964   Information, acoustic confusions, and memory span. *British Journal of Psychology* 55:429-432.

Corwin, W.
    1991   Data link integration in commercial transport operations. In *Proceedings of the Sixth International Symposium on Aviation Psychology,* in R. Jensen, ed. Columbus, OH: Ohio State University.

Diehl, A.
    1991   Human performance and systems safety considerations in aviation mishaps. *International Journal of Aviation Psychology* 1:97-106.

Edwards, E.
    1976   Some aspects of automation in civil transport aircraft. In *Monitoring Behavior and Supervisory Control,* T.B. Sheridan and G. Johannsen, eds. New York: Plenum.
    1977   Automation in civil transport aircraft. *Applied Ergonomics* 84:194-198.

Erzberger, H.
    1992   *CTAS: Computer intelligence for air traffic control in the terminal area.* NASA Technical Memorandum 103959. Moffett Field, CA: Ames Research Center.

Erzberger, H., T.J. Davis, and S.M. Green
    1993   Design of the Center-TRACON automation system. Pp. 11-1–11-12 in *Proceedings of the AGARD Guidance and Control Panel 56th Symposium on Machine Intelligence in Air Traffic Management,* Berlin, Germany.

Farber, E., and M. Paley
    1993   Using Freeway Traffic Data to Estimate the Effectiveness of Rear End Collision Countermeasures. Paper presented at the Third Annual IVHS America Meeting, Washington, DC, April.

Federal Aviation Administration
    1994   *Automation Strategic Plan.* Washington, DC: U.S. Department of Transportation.
    1995   *Air Traffic Service Plan 1995-2000.* Washington, DC: U.S. Department of Transportation.

1996a *The Interfaces Between Flightcrews and Modern Flight Deck Systems.* Human Factors Team Report, Transport Airplane Directorate. Washington, DC: U.S. Department of Transportation.

1996b *Aviation System Capital Investment Plan.* Washington, DC: U.S. Department of Transportation.

Fitts, P.
1951 *Human Engineering for an Effective Air Navigation and Traffic Control System.* Columbus, OH: Ohio State University.

Foushee, H.C., and R.L. Helmreich
1988 Group interaction and flight crew performance. Pp. 189-227 in *Human Factors in Aviation*, E.L. Wiener and D.C. Nagel, eds. San Diego: Academic Press.

Funk, K., B. Lyall, and V. Riley
1996 Flightdeck Automation Problems: Perception and Reality. Paper presented at the 2nd Automation Technology and Human Performance Conference, Cocoa Beach, FL, March.

Garland, D., and V.D. Hopkin
1994 Controlling automation in future air traffic control: The impact on situational awareness. Pp. 179-197 in *Situational Awareness in Complex Systems*, R.D. Gilson, D.J. Garland, and J.M. Koonce, eds. Daytona Beach, FL: Embry-Riddle Aeronautical University Press.

Getty, D.J., J.A. Swets, R.M. Pickett, and D. Gounthier
1995 System operator response to warnings of danger: A laboratory investigation of the effects of the predictive value of a warning on human response time. *Journal of Experimental Psychology: Applied* 1:19-33.

Groce, J.L., and G. Boucek
1987 Air Transport Crew Tasking in an ATC Data Link Environment. SAE Technical Paper Series 871764. Warrendale, PA: Society of Automotive Engineers.

Hancock, P.A., and M.H. Chignell
1989 *Intelligent Interfaces.* Amsterdam: Elsevier.

Hancock, P.A., W.L. Dewing, and R. Parasuraman
1993 A driver-centered system architecture for intelligent-vehicle highway systems. *Ergonomics in Design* 2:12-15, 35-39.

Hancock, P.A., and R. Parasuraman, R.
1992 Human factors and safety in the design of intelligent vehicle-highway systems. *Journal of Safety Research* 23:181-198.

Hancock, P.A., R. Parasuraman, and E.A. Byrne
1996 Driver-centered issues in advanced automation for motor vehicles. In *Automation and Human Performance: Theory and Applications*, R. Parasuraman, and M. Mouloua, eds. Mahwah, NJ: Erlbaum.

Harris, W., P.A. Hancock, and E. Arthur
1993 The effect of taskload projection on automation use, performance, and workload. In *Proceedings of the Seventh International Symposium on Aviation Psychology.* Columbus, OH.

Harwood, K.
1993 Defining human-centered systems for verifying and validating air traffic control systems. Pp. 115-129 *Verification and Validation of Complex Systems*, in J. Wise, V.D. Hopkin, and P. Stager, eds. Berlin: Springer-Verlag.

1994 CTAS: An alternative approach to developing and evaluating advanced ATC automation. Pp. 332-338 *Human Performance in Automated Systems: Recent Research and Trends*, in M. Mouloua and R. Parasuraman, eds. Hillsdale, NJ: Erlbaum.

Hilburn, B., P.G.A.M. Jorna, and R. Parasuraman
   1995   The effect of advanced ATC automation on mental workload and monitoring performance: An empirical investigation in Dutch airspace. In *Proceedings of the International Symposium on Aviation Psychology,* Columbus, OH.

Hopkin, V.D.
   1991   The impact of automation on air traffic control systems. Pp. 3-20 in *Automation and Systems Issues in Air Traffic Control*, J.A. Wise, V.D. Hopkin, and M.L. Smith, eds. Berlin: Springer Verlag.
   1994   Human performance implications of air-traffic control automation. Pp. 314-319 in *Human Performance in Automated Systems: Recent Research and Trends*, M. Mouloua and R. Parasuraman, eds. Hillsdale, NJ: Erlbaum.
   1995   *Human Factors in Air-Traffic Control.* London: Taylor and Francis.

Hopkin, V.D., and J.A. Wise
   1996   Human factors in air-traffic system automation. In *Automation and Human Performance: Theory and Applications*, R. Parasuraman, and M. Mouloua, eds. Mahwah, NJ: Erlbaum.

Hughes, J.A., D. Randall, and D. Shapiro
   1993   Faltering from ethnography to design. Pp. 77-90 in *Verification and Validation of Complex Systems*, J. Wise, V.D. Hopkin, and P. Stager, eds. Daytona Beach, FL: Embry-Riddle Aeronautical University Press.

Kerns, C.
   1991   Data link communications between controllers and pilots: A review and synthesis of the simulation literature. *International Journal of Aviation Psychology* 1:181-204.

Kirlik, A.
   1993   Modeling strategic behavior in human-automation interaction: Why an "aid" can (and should) go unused. *Human Factors* 35:221-242.

Kuchar, J.K., and R.J. Hansman
   1995   A probabilistic methodology for the evaluation of alerting system performance. In *Proceedings of the IFAC/IFIP/IFORS/IEA Symposium.* Cambridge, MA.

Langer, E.
   1989   *Mindfulness.* Reading, MA: Addison-Wesley.

Lee, J.D., and N. Moray
   1992   Trust, control strategies, and allocation of function in human-machine systems. *Ergonomics* 35:1243-1270.
   1994   Trust, self-confidence, and operators' adaptation to automation. *International Journal of Human-Computer Studies* 40:153-184.

Manning, C.
   1995   Empirical investigations of the utility of flight strips: A review of the VORTAC studies. In *Proceedings of the Eighth International Symposium on Aviation Psychology*, R. Jensen, ed. Columbus, OH: Ohio State University.

McClumpha, A.M., and M. James
   1994   Understanding automated aircraft. In *Human Performance in Automated Systems: Current Research and Trends*, M. Mouloua and R. Parasuraman, eds. Hillsdale, NJ: Erlbaum.

McDaniel, J.W.
   1988   Rules for fighter cockpit automation. Pp. 831-838 in *Proceedings of the IEEE National Aerospace and Electronics Conference.* New York: IEEE.

Mellone, V.J., and S.M. Frank
   1993   The behavioral impact of TCAS II on the national air traffic control system. Pp. 352-359 in *Proceedings of the Seventh International Symposium on Aviation Psychology*, R.S. Jensen and D. Neumeister, eds. Columbus, OH: Department of Aviation, Ohio State University.

Mosier, K., and L.J. Skitka
    1996   Human decision makers and automated decision aids: Made for each other? In *Automation and Human Performance: Theory and Applications*, R. Parasuraman, and M. Mouloua, eds. Mahwah, NJ: Erlbaum.

Mosier, K., L.J. Skitka, and K.J. Korte
    1994   Cognitive and social psychological issues in flight crew/automation interaction. Pp. 191-197 in *Human Performance in Automated Systems: Current Research and Trends*, M. Mouloua and R. Parasuraman, eds. Hillsdale, NJ: Erlbaum.

Mouloua, M., and R. Parasuraman, eds.
    1994   *Human Performance in Automated Systems: Current Research and Trends*. Hillsdale, NJ: Erlbaum.

Muir, B.M.
    1988   Trust between humans and machines, and the design of decision aids. Pp. 71-83 in *Cognitive Engineering in Complex Dynamic Worlds*, E. Hollnagel, G. Mancini, and D.D. Woods, eds. London: Academic Press.

Mundra, A.D.
    1989   *Ghosting: Potential applications of a new controller automation aid.* Technical Report MW-89W00030. McLean, VA: MITRE Corp.

National Research Council
    1982   *Automation in Combat Aircraft.* Air Force Studies Board. Washington, DC: National Academy Press.

National Transportation Safety Board
    1973   *Eastern Airlines L-1011, Miami, Florida, 20 December 1972.* Report No. NTSB-AAR-73-14. Washington, DC: National Transportation Safety Board.

Norman, D.
    1990   The "problem" with automation: Inappropriate feedback and interaction, not "over-automation." *Proceedings of the Royal Society of London* B237.
    1993   *Things That Make Us Smart.* New York: Basic Books.

Norman, S., C.E. Billings, D. Nagel, E. Palmer, E.L. Wiener, and D.D. Woods
    1988   *Aircraft Automation Philosophy: A Source Document.* NASA Technical Report. Moffett Field, CA: Ames Research Center,

Parasuraman, R.
    1987   Human-computer monitoring. *Human Factors* 29:695-706.

Parasuraman, R., T. Bahri, J. Deaton, J. Morrison, and M. Barnes
    1990   *Theory and Design of Adaptive Automation in Aviation Systems.* Technical Report No. CSL-N90-1. Washington, DC: Catholic University of America, Cognitive Science Laboratory.

Parasuraman, R., P.A. Hancock, and O. Olofinboba
    In press   Alarm Effectiveness in Driver-Centered Collision-Warning Systems. Submitted to *Ergonomics*.

Parasuraman, R., R. Molloy, and I.L. Singh
    1993   Performance consequences of automation-induced "complacency." *International Journal of Aviation Psychology* 3:1-23.

Parasuraman, R., and M. Mouloua
    1996   *Automation and Human Performance: Theory and Applications.* Mahwah, NJ: Erlbaum.

Parasuraman, R., M. Mouloua, and R. Molloy
    1994   Monitoring automation failures in human-machine systems. Pp. 45-49 in *Human Performance in Automated Systems: Current Research and Trends*, M. Mouloua and R. Parasuraman, eds. Hillsdale, NJ: Erlbaum.

Parasuraman, R., M. Mouloua, R. Molloy, and B. Hilburn
  1996   Monitoring automated systems. In *Automation and Human Performance: Theory and Applications*, R. Parasuraman and M. Mouloua, eds. Hillsdale, NJ: Erlbaum.
Parsons, H.M.
  1985   Automation and the individual: Comprehensive and comparative views. *Human Factors* 27:99-112.
Phillips, D.
  1995   System failure cited in ship grounding. *Washington Post,* August 11, p. A7.
Planzer, N.
  1995   Briefing to NRC Panel on Human Factors in Air Traffic Control Automation, Washington, DC, August.
Planzer, N., and M. Jenny
  1995   Managing the evolution to free flight. *Journal of Air Traffic Control* March:18-20.
Pritchett, A., and R.J. Hansman
  1993   Preliminary analysis of pilot ratings of "party line" information importance. In *Proceedings of the Seventh International Symposium on Aviation Psychology*, R. Jensen, ed. Columbus: Ohio State University.
Riley, V.
  1994   A theory of operator reliance on automation. Pp. 8-14 in *Human Performance in Automated Systems: Recent Research and Trends*, M. Mouloua and R. Parasuraman, eds. Hillsdale, NJ: Erlbaum.
  1996   Operator reliance on automation: Theory and data. In *Automation and Human Performance: Theory and Applications*, R. Parasuraman, and M. Mouloua, eds. Hillsdale, NJ: Erlbaum.
Riley, V., B. Lyall, and E. Wiener
  1993   An*alytic Methods for Flight-Deck Automation Design and Evaluation. Phase Two Report: Pilot Use of Automation*. Technical Report. Minneapolis, MN: Honeywell Technology Center.
Rouse, W.B.
  1991   *Design for Success: A Human-Centered Approach to Developing Successful Products and Systems*. New York: Wiley.
Rouse, W.B., N. Geddes, and J.M. Hammer
  1990   Computer-aided flight pilots. *IEEE Spectrum* 27:38-41.
Rouse, W.B., and N.M. Morris
  1986   Understanding and enhancing user acceptance of computer technology. *IEEE Transactions on Systems, Man, and Cybernetics* SMC-16:965-973.
RTCA
  1995   *Report of the RTCA Board of Directors' Select Committee on Free Flight*. Washington, DC: RTCA, Incorporated.
Sarter, N., and D.D. Woods
  1992   Pilot interaction with cockpit automation: Operational experience with the flight management system. *International Journal of Aviation* Psychology 2:303-321.
  1994a   Decomposing automation: Autonomy, authority, observability, and perceived animacy. Pp. 22-27 in *Human Performance in Automated Systems: Recent Research and Trends*, M. Mouloua and R. Parasuraman, eds. Hillsdale, NJ: Erlbaum.
  1994b   Pilot interaction with cockpit automation II: An experimental study of pilots' model and awareness of the flight management system. *International Journal of Aviation Psychology* 4:1-28.
  1995a   How in the world did we get ever get into that mode? *Human Factors* 36.

    1995b   *"Strong, Silent, and Out-of-The-Loop": Properties of Advanced (Cockpit) Automation and Their Impact on Human-Automation Interaction.* Technical Report CSEL 95-TR-01. Columbus, OH: Cognitive Systems Engineering Laboratory, Ohio State University.

Satchell, P.
    1993   *Cockpit Monitoring and Alerting Systems.* Aldershot, U.K.: Ashgate.

Sheridan, T.
    1980   Computer control and human alienation. *Technology Review* 10:61-73.
    1996   Speculations on future relations between humans and automation. In *Automation and Human Performance: Theory and Applications.* Hillsdale, NJ: Erlbaum.

Sheridan, T.B.
    1987   Supervisory control. In *Handbook of Human Factors,* G. Salvendy, ed. New York: Wiley.
    1988   Trustworthiness of command and control system. In *Proceedings of the IFAC Conference on Man Machine Systems.* Oulu, Finland.
    1992   *Telerobotics, Automation, and Supervisory Control.* Cambridge, MA: MIT Press.

Singh, I.L., R. Molloy, and R. Parasuraman
    1993   Automation-induced "complacency": Development of the complacency-potential rating scale. *International Journal of Aviation Psychology* 3:111-121.

Sorkin, R.D.
    1988   Why are people turning off our alarms? *Journal of the Acoustical Society of America* 84:1107-1108.

Sorkin, R.D., B.H. Kantowitz, and S.C. Kantowitz
    1988   Likelihood alarm displays. *Human Factors* 30:445-459.

Swets, J.A.
    1992   The science of choosing the right decision threshold in high-stakes diagnostics. *American Psychologist* 47:522-532.

Vakil, S.S., J. Hansman, J. Midkiff, and T. Vancek
    1995   Feedback mechanisms to improve mode awareness in advanced autoflight systems. Pp. 243-248 in *Proceedings of the Eighth International Symposium on Aviation Psychology,* in R. Jensen, ed. Columbus, OH: Ohio State University.

Vakil, S.S., A.H. Midkiff, and R.J. Hansman
    1995   *Mode Awareness Problems in Advanced Autoflight Systems.* Cambridge, MA: MIT Aeronautical Systems Lab. (abstract).

Vortac. O.U.
    1993   Should Hal open the pod bay doors? An argument for modular automation. Pp. 159-163 *Human Factors and Advanced Aviation Technologies,* in D.J. Garland and J.A. Wise, eds. Daytona Beach, FL: Embry-Riddle Aeronautical University Press.

Vortac, O.U., and C.F. Gettys
    1990   *Cognitive Factors in the Use of Flight Progress Strips: Implications for Automation.* Norman, OK: Cognitive Processes Laboratory, University of Oklahoma.

Wald, M.
    1995   Aging control system brings chaos to air travel. *New York Times,* August 20.

Weick, K.E.
    1988   Enacted sensemaking in crisis situations. *Journal of Management Studies* 25:305-317.

Whitfield, D., R.B. Ball, and G. Ord
    1980   Some human factors aspects of computer-aiding concepts for air traffic controllers. *Human Factors* 22:569-580.

Wickens, C.D.
    1992   *Engineering Psychology and Human Performance.* 2nd. ed. New York: Harper Collins.

1994 Designing for situation awareness and trust in automation. In *Proceedings of the IFAC Conference on Integrated Systems Engineering.* Baden-Baden, Germany.

Wickens, C.D., S. Miller, and M. Tham
 1996 The implications of data link for representing pilot request information on 2D or 3D air traffic control displays. *International Journal of Industrial Ergonomics* 606.

Wiener, E.L.
 1977 Controlled flight into terrain. *Human Factors* 19:171-177.
 1985 Beyond the sterile cockpit. *Human Factors* 27:75-90.
 1988 Cockpit automation. Pp. 433-461 in *Human Factors in Aviation*, E.L. Wiener and D.C. Nagel, eds. San Diego: Academic Press.
 1995 Opening remarks on debate on automation and safety. Presented at the annual meeting of Human Factors and Ergonomics Society, San Diego, CA, October.

Wiener, E.L., and R.E. Curry
 1980 Flight-deck automation: Promises and problems. *Ergonomics* 23:995-1011.

Wiener, N.
 1964 *God and Golem, Incorporated.* Cambridge, MA: MIT Press.

Will, R.P.
 1991 True and false dependence on technology: Evaluation with an expert system. *Computers in Human Behavior* 7:171-183.

Wise, J. A., V.D. Hopkin, and M.L. Smith, eds.
 1991 *Automation and Systems Issues in Air Traffic Control.* NATO ASI Series Vol. F73. Berlin: Springer-Verlag.

Woods, D.D.
 1993 The price of flexibility in intelligent interfaces. *Knowledge-Based Systems* 6:1-8.

Appendixes

Appendix A

Aviation and Related Acronyms

AAS	advanced automation system
ABSR	abstract reasoning test
AERA	automated en route air traffic control
AF	Airway Facilities
AFSS	automated flight service stations
ARSR	air route surveillance radar
ARTCC	air route traffic control center
ARTS	automated radar terminal system
ASD	aircraft situation display
ASDE	airport surface detection equipment
ASSET	air safety system enhancement team
ASR	airport surveillance radar
ASRS	aviation safety reporting system
AT	air traffic organization
ATADS	air traffic activity data system
ATC	air traffic control
ATCS	air traffic control specialist
ATCSCC	Air Traffic Control System Command Center
ATOMS	air traffic operations management system
ATTE	air traffic teamwork enhancement
ATWIT	air traffic workload input technique
AUTOLAND	automated landing

CAMI	Civil Aeromedical Institute
CART	controller awareness and resource training
CFC	central flow control
CFIT	controlled flight into terrain
CHIRP	confidential human incident reporting programme
CMAQ	cockpit management attitudes questionnaire
COMPAS	computer-oriented metering planning and advisory system
CRM	crew (formerly cockpit) resource management
CST	coast status
CTAS	center-TRACON automation system
CTI	collegiate training initiative
DA	descent advisor
DBRITE	digital brite, radar indicator tower equipment
DUATS	Direct User Access Terminal System
DYSIM	dynamic simulation
EAS	employee attitude survey
EFAS	en route flight advisory service
ETG	enhanced target generator
FAA	Federal Aviation Administration
FAATC	FAA technical center
FAST	final approach spacing tool
FDIO	flight data input output computer
FDP	flight data processing
FFC	facility flight check
FMS	flight management system
FMAQ	flight management attitudes questionnaire
FSS	flight service station
GCA	ground controlled approach
GNAS	general national airspace system
GPS	global positioning system
GPWS	ground proximity warning system
GS	general service
HFCC	human factors coordinating committee
HOST	computer at en route centers
HSI	horizontal situation indicator
IACO	International Civil Aviation Organization
IFR	instrument flight rules

ILSS	integrated logistics support system
INS	inertial navigation system
IPT	integrated product team
ITS	intelligent transportation systems
ITWS	integrated terminal weather system
LOFT	line oriented flight training
MANPRINT	Manpower and Personnel Integration
MCAT	multiplex controller aptitude test
MCC	maintenance control center
MPS	maintenance processor subsystem
MSAW	minimum safe altitude warning
MTBA	mean time between accidents
MTBF	mean time between failures
M1FC	Model I Full Capacity
NAATS	National Association of Air Traffic Specialists
NADIN	national airspace data interchange network
NAIMS	national airspace information monitoring system
NAS	national airspace system
NASA	National Aeronautics and Space Administration
NASA-TLX	National Aeronautics and Space Administration Task Load Index
NASPAC	national airspace system performance analysis capability
NASSIM	national airspace system simulation model
NATCA	National Air Traffic Controllers Association
NEO	neurotocism, extroversion, openness personality inventory
NMCC	National Maintenance Coordination Center
NORAD	North American Aerospace Defense
NOTAM	notice to airmen
NRP	national route plan
NTSB	National Transportation Safety Board
OCC	operations control center
OKT	occupational knowledge test
OPM	Office of Personnel Management
Ops Net	operations network
ORAT	overall selection rating
OTA	Office of Technology Assessment
PATCO	Professional Air Traffic Controllers Organization
PPI	plan position indicator

PID	proportional-integral-derivative control
PTR	program trouble report
PVD	plan view display
QQPRI	qualitative and quantitative personnel resources inventory
QTP	quality through partnership
RAF	Royal Air Force
RDP	radar data processing
RMS	remote monitoring subsystem
RT	radio telephony
SACHA	Separation and Control Hiring Assessment program
SAINT	system analysis of an integrated network of tasks
SATORI	situation assessment through re-creation of incidents
SFAP	survey feedback action program
SID	standard instrument departure
SIMMOD	airport and airspace simulation model
SMCC	system maintenance control center
SMO	system management office
SWAT	subjective workload assessment technique
TAAM	Total Airport and Airspace Modeller
TAP	terminal area productivity program
TASF	terminal airspace simulation facility
TCAS	traffic alert [and] collision avoidance system
TMA	traffic management advisor
TMU	traffic management unit
TRACON	terminal radar control area
TRADOC	Training and Doctrine Command
VFR	visual flight rules
VOR	VHF omnidirectional range
VSCS	voice switching and control system

Appendix B

Contributors to the Report

Many individuals contributed to the panel's thinking and its drafting of various sections of the report by serving as presenters, advisers, and coordinators of sources of valuable information. The list below acknowledges these contributors and their affiliations.

FAA Liaisons

David Cherry
Lawrence Cole
Mitchell Grossberg
Mark Hofmann

Phyllis Kayten
Carol Manning
Neil Planzer

FAA Advisers and Presenters

Lawrence Bailey
Dana Bain
Dennis Beringer
Robert Blanchard
Brenda Boone
Dana Broach
William Collins
Pamela Della Rocca
Elmer Frazure

Thomas Hilton
O.V. Prinzo
J. Larry Ramirez
Elliott Reid
Mark Rodgers
Mary Sand
Lori Scharf
Hilda Wing
William Young

NASA Ames Presenters

Kevin Corker
Thomas Davis
Heinz Erzberger
Steven Green
Sandra Lozito

Gregory Pisanich
Barry Scott
Barry Sullivan
Sherman Tyler

Other Advisors and Presenters

Jon Alexander	CTA, Incorporated
Richard Bailey	Loral Federal Systems Corporation
Kim Cardosi	Volpe National Transportation Systems Center
Karl Grundmann	National Air Traffic Controllers' Association
John Hansman	Massachusetts Institute of Technology
Anna Olsson	EG&G, Incorporated
William Thomas	CGH, Incorporated

Appendix C

Biographical Sketches

CHRISTOPHER D. WICKENS (Chair) is currently a professor of experimental psychology, head of the Aviation Research Laboratory, and associate director of the Institute of Aviation at the University of Illinois at Urbana-Champaign. He also holds an appointment in the Department of Mechanical and Industrial Engineering and the Beckman Institute of Science and Technology. He received an A.B. degree from Harvard University in 1967 and a Ph.D. from the University of Michigan in 1974 and served as a commissioned officer in the U.S. Navy from 1969 to 1972. He is currently involved in aviation research concerning principles of human attention, perception and cognition, and their relation to display processing, multitask performance, and navigation in complex systems. He is a member and fellow of the Human Factors Society and received the Society's Jerome H. Ely Award in 1981 for the best article in the *Human Factors Journal*, as well as the Paul M. Fitts Award in 1985 for outstanding contributions to the education and training of human factors specialists by the Human Factors Society. In 1993 he received the Franklin Taylor Award from Division 21 of the American Psychological Association. He has also served on the National Research Council's Committee on Human Factors.

CHARLES B. AALFS is a retired air traffic control specialist for the Federal Aviation Administration (FAA). He has over 30 years of experience as an air traffic controller for both the U.S. Navy and the FAA. While with the FAA, he served as an air traffic controller, air traffic automation specialist, air traffic facility officer, air traffic facility manager, air traffic regional office automation specialist and branch manager, and division manager of resource management.

When he retired, he was the manager of the new Southern California TRACON in San Diego, California. As an automation specialist, he was responsible for the software maintenance of the terminal automated radar system called ARTS III and IIIA. He was also the author of many design changes to the ARTS III program, one of which was the design to allow automated handoffs from one ARTS III site to another.

TORA K. BIKSON is a senior scientist in RAND Corporation's Behavioral Sciences Department. She received B.A., M.A., and Ph.D. (1969) degrees in philosophy from the University of Missouri at Columbia and M.A. and Ph.D. (1974) degrees in psychology from the University of California at Los Angeles. Since 1980, her research has investigated properties of advanced information technologies in varied user contexts. Her work emphasizes field research design, intensive case studies, and large-scale cross-sectional studies addressed to the use of computer-based tools in organizational settings. She is a member of Data for Development, a United Nations Secretariat providing scientific guidance on the use of information systems in developing companies, and a technical consultant to the United Nations Advisory Commission on the Coordination of Information Systems. She is a frequent reviewer for professional papers and has authored a number of journal articles, book chapters, and research reports on the implementation of new interactive media. She is a member of the American Academy of Arts and Sciences, the Association for Computing Machinery, the American Psychological Association (fellow), the Computer Professionals for Social Responsibility, and the Society for the Psychological Study of Social Issues. She recently served on a committee of the National Research Council's Computer Science and Telecommunications Board that produced *Information Technology and the Service Society*.

MARVIN S. COHEN is founder and president of Cognitive Technologies, Inc. (CTI) in Arlington, Virginia. His professional interests include experimental research on human reasoning and decision making, elicitation and representation of expert knowledge, training cognitive skills in individuals and teams, development of decision support systems, human-computer interface design, and methods for representing and manipulating uncertainty. His current work at CTI includes experimental research on airline pilot decision-making processes, training decision-making skills under time stress in the ship-based anti-air-warfare environment, training for more effective distributed team decision making in naval air strike warfare, design of interfaces to enhance human performance with automatic target recognition devices, and modeling and training situation-assessment skills of Army battlefield commanders. He has an M.A. in philosophy from the University of Chicago and a Ph.D. in experimental psychology from Harvard University. For 11 years, he was at Decision Science Consortium, Inc., where he was vice president and director of Cognitive Science and Decision Systems. He has taught at George Washington University on the design of human-computer interfaces and has served on a committee of the National Research Council's Air Force Studies Board on tactical battle management.

DIANE DAMOS is an associate professor of human factors at the University of Southern California and president of Damos Research Associates. After receiving her doctorate in aviation psychology from the University of Illinois, she became a member of the faculty of the Department of Industrial Engineering at the State University of New York at Buffalo. Prior to joining the University of Southern California, she was also a member of the faculty of the Department of Psychology at Arizona State University. Her research interests have focused on pilot selection and multiple-task performance, including workload management in advanced automation aircraft. She has authored numerous books and papers and edited *Multiple Task Performance*, which appeared in 1991. She is a member of the editorial board of the *International Journal of Aviation Psychology*.

JAMES DANAHER is the chief of the Operational Factors Division of the Office of Aviation Safety at the National Transportation Safety Board (NTSB) in Washington, D.C. He has more than 35 years work experience in the human factors and safety fields, in both industry and government. Since joining NTSB in 1970, he has served in various supervisory and managerial positions, with special emphasis on human performance issues in flight operations and air traffic control. He has participated in the on-scene phase of numerous accident investigations, in associated public hearings, and in the development of NTSB recommendations for the prevention of future accidents. He is a former naval aviator and holds a commercial pilots' license with single-engine, multi-engine, and instrument ratings. He has an M.S. degree in experimental psychology from Ohio State University and is a graduate of the Federal Executive Institute. He has represented the NTSB at numerous safety meetings, symposia, and seminars and is the author or coauthor of numerous publications.

ROBERT L. HELMREICH is professor of psychology at the University of Texas at Austin. He is also director of the NASA/University of Texas/FAA Aerospace Crew Research Project. He received B.A., M.S., and Ph.D. degrees from Yale University and served as an officer in the U.S. Navy. He studies team performance in many groups, including pilots, astronauts, and surgical teams. He has been involved with the definition and implementation of crew resource management training for nearly 20 years. He is author or editor of 3 books and more than 180 chapters, monographs, and journal articles. He is a fellow of the American Psychological Association and the American Psychological Society. He received the Flight Safety Foundation/Aviation Week and Space Technology Distinguished Service Award for 1994 for his contributions to the development of crew resource management.

V. DAVID HOPKIN is an independent human factors consultant who is based part time at Embry-Riddle Aeronautical University at Daytona Beach, Florida. He was formerly senior principal psychologist at the Royal Air Force Institute of Aviation Medicine at Farnborough and human factors consultant to the United Kingdom Civil Aviation Authority. He has also worked for the International Civil Aviation Organization, NATO, Eurocontrol, the Federal Aviation Administration, and numerous other international and national agencies. He has over 300 publications, including the 1995 *Human Factors in Air Traffic Control*. He has an M.A. in psychology from the University of Aberdeen, Scotland and is a fellow of the Royal Institute of Navigation.

JERRY KIDD is senior adviser for the Committee on Human Factors and its various projects. He received a Ph.D. from Northwestern University in social psychology in 1956; he then joined RAND Corporation to help on a project to simulate air defense operations. He left RAND in late 1956 to join the staff at the Laboratory of Aviation Psychology at Ohio State University. There he worked under Paul Fitts and George Briggs until 1962, when he joined the staff of AAI, Incorporated, north of Baltimore, Maryland. In 1964, he moved to the National Science Foundation as program director for special projects. He joined the faculty of the College of Library and Information Services at the University of Maryland in 1967 and retired in 1992.

TODD T. LaPORTE is professor of political science and formerly associate director of the Institute of Governmental Studies at the University of California, Berkeley. He teaches and publishes in the areas of public administration, organization theory, and technology and politics, with emphasis on the decision-making dynamics of large, complex, and technologically intensive (and hazardous) organizations, and the problems of governance and political legitimacy in a technological society. He is a member of the National Academy of Public Administration, was a research fellow with the Woodrow Wilson International Center of Scholars, and has held visiting research appointments with the Science Center in Berlin and the Max Planck Institute for Social Research in Cologne, Germany. He has a Ph.D. from Stanford University.

ANNE S. MAVOR is study director for the Panel on Human Factors in Air Traffic Control, the Panel on Modeling Human Behavior and Command Decision Making, and the Committee on Human Factors. Her previous work as a National Research Council senior staff officer has included a study of the scientific and technological challenges of virtual reality, a study of emerging needs and opportunities for human factors research, a study of modeling cost and performance of military enlistment, a review of federally sponsored education research activities, and a study to evaluate performance appraisal for merit pay. For the past 25 years her work has concentrated on human factors, cognitive psychology, and information system design. Prior to joining the National Research Council she worked for the Essex Corporation, a human factors research firm, and served as a consultant to the College Board. She has an M.S. in experimental psychology from Purdue University.

JAMES P. McGEE is a senior research associate supporting human factors and related activities in the Division on Education, Labor, and Human Performance of the Commission on Behavioral and Social Sciences and Education. Prior to joining the National Research Council in 1994, he held scientific, technical, and management positions in human factors psychology at IBM, RCA, General Electric, General Dynamics, and United Technologies corporations. He has also instructed courses in applied psychology and general psychology at several colleges. He is a member of the Potomac chapter of the Human Factors and Ergonomics Society and of the American Psychological Association. He has a Ph.D. in experimental psychology from Fordham University.

RAJA PARASURAMAN is professor of psychology and director of the Cognitive Science Laboratory at the Catholic University of America in Washington, D.C. Currently he is also a visiting scientist at the Laboratory of Psychology at the National Institute of Mental Health in Bethesda, Maryland. He has a B.Sc. (Hons.) in electrical engineering from Imperial College, University of London (1972) and an M.Sc. in applied psychology (1973) and Ph.D. in psychology from the University of Aston, Birmingham (1976). Since 1982 he has been at the Catholic University of America, where he has carried out research on attention, aging, automation, cognitive neuroscience, vigilance, and workload. He is a fellow of the American Association for the Advancement of Science, the American Psychological Association (Division 21, Engineering Psychology), the American Psychological Society, the Human Factors and Ergonomics Society, and the Washington Academy of Sciences. He is also a member of the Association of Aviation Psychologists, the Psychonomics Society, the Society for Neuroscience, and the Society for Psychophysiological Research.

JOSEPH O. PITTS retired from the Federal Aviation Administration (FAA) in 1993, after more than 36 years of government service. He is currently employed by the Vitro Corporation, which supports the FAA through its surveillance technical assistance contract. Mr. Pitts supports the integrated terminal weather system (ITWS) program and the air traffic weather division. While employed by the FAA, he held positions as air traffic manager, assistant air traffic manager, branch manager, area manager, and full-performance-level air traffic controller at several air traffic control facilities. In the last 10 years of his tenure with the FAA, he had the responsibility of managing several research engineering and development programs at FAA headquarters; he was very active in both the

FAA's facilities and equipment and research engineering and development budgets.

THOMAS B. SHERIDAN is Ford professor of engineering and applied psychology in the Departments of Mechanical Engineering and Aeronautics and Astronautics and director of the Human-Machine Systems Laboratory at Massachusetts Institute of Technology. He has an S.M. degree from the University of California, a Sc.D. from MIT, and an honorary doctorate from Delft University of Technology, Netherlands. He has served as president of both the Human Factors and Ergonomics Society and the IEEE Systems, Man and Cybernetics Society and is a fellow of both organizations. He has chaired the National Research Council's Committee on Human Factors and has served on numerous other NRC committees. He is senior editor of the MIT Press Journal *Presence: Teleoperators and Virtual Environments* and is a member of the National Academy of Engineering.

PAUL STAGER is professor of psychology at York University, where he has taught since receiving a Ph.D. from Princeton University in 1966. A licensed pilot, his research has been concerned with system evaluation, human error, computer-human interface design, and human performance assessment in complex operational systems, most often within the context of aviation. During the past 20 years, his research has addressed several human factors issues in air traffic control, including the potential impact of bilingual communications on instrument flight operations, the precipitating conditions for operational errors, and the human engineering specifications for an advanced workstation design. Since 1989, he has advised the federal government on all human engineering associated with the development and evaluation of the Canadian automated air traffic system. He was a lecturer at the 1990 NATO Advanced Study Institute on automation and systems issues in air traffic control and, as codirector of the 1992 Advanced Study Institute on the verification and validation of complex human-machine systems, he edited (with J. Wise and D. Hopkin) *Verification and Validation of Complex Systems: Human Factors Issues*.

RICHARD B. STONE retired from Delta Airlines after almost 35 years as a pilot. He served as a line check airman and his last assignment was flying the B 767 extended range to Europe. He has a B.S. from the University of Illinois and an M.S. from the University of New Hampshire. He received his flight training from the U.S. Air Force. During his years as an airline pilot, he also acted as an aircraft accident investigator, represented airline pilots in medical matters, and served as the president of the International Society of Air Safety Investigators. He currently acts as a safety consultant in aviation.

EARL L. WIENER is a professor of management science at the University of

Miami. He received a B.A. in psychology from Duke University and a Ph.D. in psychology an industrial engineering from Ohio State University. He served as a pilot in the U.S. Air Force and the U.S. Army and is rated in fixed-wing and rotary-wing aircraft. Since 1979 he has been active in the aeronautics and cockpit automation research of the NASA Ames Research Center. He is a fellow of the Human Factors Society and the American Psychological Association and currently serves as president of the Human Factors Society. He currently serves on the FAA's Research, Engineering, and Development Advisory Council and the National Research Council's Committee on Human Factors. He is the co-editor (with D. Nagel) of *Human Factors in Aviation* (1988) and *Cockpit Resource Management* (with B. Kanki and R. Helmreich, 1993).

LAURENCE R. YOUNG is professor of aeronautics and astronautics at the Massachusetts Institute of Technology. He received an A.B. in physics at Amherst College, a B.S. in electrical engineering at MIT, an M.S. in electrical engineering, and a Sc.D. in instrumentation at MIT, and a certificat de license in mathematics at the University of Paris. He is also Dryden lecturer at the American Institute of Aeronautics and Astronautics. His research is in the application of control theory to man-vehicle problems, particularly orientation; flight simulators; and space laboratory experimentation on vestibular function. He has worked as an engineer at the Instrumentation Laboratory at MIT and Sperry Gyroscope Co. and as a research assistant at the School of Medicine at the University of Paris. He is a consultant to various industrial and government organizations and has served on the Aeronautics and Space Engineering Board and the Air Force Studies Board of the National Research Council. He has received the Franklin V. Taylor Award in Human Factors from the Institute of Electrical and Electronics Engineers. He is affiliated with the National Academy of Engineering and Institute of Medicine, is a fellow of the Institute of Electrical and Electronics Engineers, the Biomedical Engineering Society, and the American Institute of Aeronautics and Astronautics.

Index

A

Abstract reasoning test, 64, 65
Accountability
  automation and, 137-138
  FAA organizational structure and, 161-162
  leadership issues, 162
Advanced automation system, 215, 249
Age restrictions, 68
Air route traffic control center, 78-79, 82
  responsibilities, 19-21
Air safety system enhancement team, 141
Air traffic activity data system, 156
Air Traffic Control Academy, 55
Air Traffic Control System Command Center, 21, 33
  responsibilities, 49
Air traffic operations management system, 156
Air Traffic Services, 32-33, 52
  in restoration of equipment, 81-82, 185-186
Air traffic teamwork enhancement, 6
  curriculum, 146-147
  limitations, 147-148
  origins and development, 145

Air traffic workload input technique, 209
Airborne automatic conflict warning system, 235
Aircraft identification
  ARTS, 38, 39-41
  automation systems, 254-255
  flight strips for, 36, 42
  tower control resources for, 35-36
Aircraft situation display, 49
Airport design
  future prospects, 22-23
  limits to utilization, 22
Airport operations oversight, 33
Airport surface detection equipment, 35
Airway Facilities
  acquisition and development practices, 187-188, 233
  automation effects, 177-180, 195
  automation trends, 178
  certification activities, 76-77, 80, 183-184
  consumers of services, 76
  employee satisfaction in, 190-191, 196
  future prospects, 179-180, 195
  human factors activities in, 9-10
  human factors research, 191-194

353

maintenance control centers, 78-79, 81, 180-181, 187, 233
monitoring and control operations, 78
organizational structure, 78-79, 82
responsibilities, 76-77, 79-80, 87, 183
restoration of equipment, 81-82, 185-186
staff demographics, 84, 88, 179, 188
staff performance evaluations, 86
staff training, 85-86, 87-88
supervisory control, 181-182
systems approach, 83-84
systems model for assessment, recruitment, and training, 194
teamwork in, 186
workload, 186-187
Airway Facilities specialists, 82
for automation, 83-84, 87, 182-183
equipment, 9-10
performance assessment, 189
responsibilities, 9, 77
staff selection, 84, 188
staffing trends, 178-179
training, 188-189
Anticipatory clearances, 203
Arrivals and departures
aircraft holding procedures, 49
anticipatory clearances, 203
approach sequencing/ghosting, 260-261
challenges for TRACON controllers, 42-43
communications system for, 36-37
constraints to airport efficiency, 21-22
ground traffic management, 35, 37
peak hours, 42
prospects for improving efficiency, 22-23
radar technology, 35-36
tower control responsibilities, 34-35
track deviation alert, 259
TRACON responsibilities, 37-38
ARTS. *See* Automated radar terminal system
Assessment and evaluation
of air traffic control efficiency, 2-3, 157-158

of air traffic control safety, 2-3, 155-157
of cognitive aptitude of candidates for training, 64-67
controller checklists, 56, 57
human engineering criterion measures, 223-224
incident analysis, 201-204
monitoring of automated systems, 277
of personality characteristics of candidates for training, 67-68
physiological measures, 207-209
real-time monitoring, 56
subjective reports, 204-205
workload, 5-6, 205-210
workload drivers, 118-120
See also Performance assessment
Automated radar terminal system (ARTS), 38, 253, 260
failure of, 44
team interaction effects, 149
visual display, 39-42
workload reduction, 123
Automation
accomplishments, 241
accountability and, 137-138
aircraft conflict detection, 259-260
aircraft data block overlap, 256
aircraft guidance systems, 263
aircraft identification systems, 254-255
aircraft operations monitoring, 255-256
Airway Facilities activities, 177-181, 182-183
alerting devices, 258-259
approach sequencing/ghosting, 260-261
authority and autonomy in, 246-247
cockpit, 23-25, 28-29, 242, 251, 262-265, 267
for cognitive support, 178
communication technology, 256
computer assistance concept, 246
computerization and, 182-183, 195
contributors to, 251
cost-benefit analysis, 266-268
current implementation, 250-251, 288
current research, 261-265

INDEX   355

data smoothing, 254
definition, 18, 182-183, 195, 243-244
degradation of situation awareness,
    278-279
efficiency goals, 250
electronic flight strips, 261-262
employee attitudes, 171-172
equipment certification, 184
FAA goals, 249-250
false alarms, 260, 263, 264, 272, 273-
    276
flight crew attitudes, 149-150
flight data management, 253-254
flight service stations, 51-52
flows and slots management, 261
forms of, 244
handoff system, 256
historic failures of, 28-29
human component, 241-242
human performance effects, 13, 242,
    268
implications for controller selection,
    68-69
implications for teamwork, 148-150
implications of, 242-243
information displays, 258
information-sharing, 139-140
interactions between factors affecting
    use of, 279-280
levels of, 244-246, 288
limitations, 241-242
management system for, 152
modernization and, 182-183, 195, 249
navigation aids, 257-258
need for, 248-250
new error forms in response to, 265-
    266, 269-270
overreliance effects, 276-278
perceived reliability, 272
pilot perspective, 23-25
preflight programming, 24
prospects, 247-248, 251
safety concerns, 17-18
safety goals, 249-250
skill degradation effects, 277-278
sources of problems, 267
track deviation signals, 259

user mistrust, 273-276
user monitoring of, 277
user trust, 271-273, 279
vigilance and, 129-130
workload effects, 122-123, 133-134,
    270-271
Automation, human-centered, 12-13
definition, 18, 242
human authority in, 281-282
human empowerment in, 282-283
implementation prospects, 287-288
job satisfaction goals, 282
objectives, 280-281, 284-285
operator awareness in, 283-284
operator trust in, 283
optimal control design, 284
organizational structure for, 284
origins of, 266-267
rationale, 280, 289
supervisory architecture, 285-287
task allocation in, 281
tolerance of nonstandard behavior in,
    284
vs. technology-centered, 265-266
Automation specialist. See GS-2101
    automation specialist
Aviation safety reporting system, 28, 156,
    157, 203

C

Center-TRACON automation system, 12,
    221, 262
Certification and licensure
    air traffic control specialists, 55
    Airway Facilities activities, 76-77, 80,
        183-184
    for full-performance-level controllers,
        71
    oversight, 33
    service, 80
    system/subsystem, 80
Civil Aeromedical Institute, 60, 66, 172
Cockpit resource management, 27
    determinants of success, 143-144
    future requirements, 144-145

implementation in air traffic control, 145
origins and development, 143
teamwork, 6
Cognitive functioning
adaptive flexibility, 98-99
attention processes, 94
attentional resources, 97-98
attitude and, 8, 160-161
automation effects, 13, 24
automation support for, 178
compensation for vulnerabilities of, 4-5
controller error, 103-105, 108
for controller tasks, 92
decision-making, 96, 102-103, 107-108
demands on controllers, 4
expectancy and, 94, 99-100, 105-106
implications for system design, 105-111
implications of proposed automation, 68-69
information monitoring, 97
knowledge-based behavior, 97-98
language of incident analysis, 203-204
long-term memory processes, 95-97, 101-102, 107
management strategies, 96-97
mental models of automated devices, 204-205, 276
organizational mediators, 166
physiological correlates, 207-209
recognition of vulnerability to stress, 142-143
resource allocation, 96-97
response to external events, 92-94
screening of controller candidates, 64-67, 69
for secondary tasks, 206-207
simulator training, 73
situation awareness, 95, 100-101, 106-107
subjective assessments of controller performance, 204-205
task analysis, 92
visual sampling, 99, 105
vulnerabilities in controller tasks, 99-105

work shift rotations and, 131-132
working memory, 94-95, 100, 106
workload assessments, 206-207
workload of vigilance, 125-130
See also Decision-making processes; Information processing
Collegiate training initiative, 70
Collision avoidance systems, 24, 27, 139-140, 149, 264-265, 274
Communication, interpersonal. *See* Interpersonal communications
Communications technology
between air traffic control centers, 21
automated systems, 256
datalink systems, 257
historic failures of, 25-27, 29
national flow control, 50-51
oversight, 33
responsibility for control and maintenance, 77
tower control resources, 36-37
TRACON, 43-44
verbal redundancy design, 106
Computer-oriented metering planning and advisory system, 212
Confidential human incident reporting programme, 203
Conflict management, 146-147
Consumers, 21-22
of Airway Facilities services, 76
Controller awareness and resource training, 145
Controller error
data collection, 222-223
error rate, 59
examples of controller-pilot miscommunication, 138-140
generalizability of human factors studies, 201
implications for systems design, 103-104, 108
incident analysis, 201-202
information processing model, 201-202
predictive modeling, 213
reporting systems for, 203-204
research methodology, 222

INDEX   357

team accountability for
  communications errors, 137-138
technology-centered prevention, 265-266
time on shift as risk factor, 130-131
types of, 104-105
Controller performance
  assessment, 57-59, 163-164
  high-reliability organization context, 154
  implications of automation, 242
  legal liability, 164
  in low workload conditions, 140
  measures of, 223-225
  organizational context variables, 159, 166, 175
  primary-task measures, 206
  recognition of vulnerability to stress, 142-143
  research needs, 225
  secondary-task measures, 206-207
  sleep disruption effects, 131
  subjective assessments, 204-205
  time-on-shift effects, 130-131
  vigilance effects, 125, 127-129
  work-rest schedules and, 130
  work shift rotations and, 131-132
  workload effects, 114-116, 123-124
Controller skills
  adaptive flexibility, 98-99
  cognitive, 4-5, 64-67, 92
  cognitive vulnerabilities, 99-105
  flight plan specialists, 52
  implications of proposed automation, 68-69
  job duties and responsibilities, 62-63
  long-term memory functions for, 95-96, 101-102
  personal qualities, 67-68
  replacement of striking controllers in 1981, 54
  for safety-efficiency balance, 23
Crew resource management, 138
  introduction of, 27
  origins and development, 143
  research findings, 142-143
Critical incident technique, 221

Current air traffic control system
  Airway Facilities operations, 9-10, 179
  automation implementation, 250-251
  baseline system, 30-31
  communications within, 21
  compatibility with new technology, 248-249
  controller aptitude tests, 64-67
  efficiency goals, 22-23
  equipment variation, 177-178, 180-181
  error rate, 59
  flight service stations, 51-52
  flow control, 48-51
  future challenges, 153, 248
  human factors activities in, 31
  human factors design in, 231-235
  job satisfaction in, 168-171
  management model, 153-155
  modeling techniques, 213-214
  national flow control, 48-51
  operations, 19-21
  organizational functioning, 7-8, 52-53, 153-155
  regional differences, 32, 53
  reliability, 2
  safety goals, 21-22
  significant events in development of, 25-30
  stakeholders, 152-153
  stressors, 1, 17
  training of controllers, 3, 55
  work-rest schedules, 130, 131
  work schedules, 6
  workload, 5, 34

D

Datalink, 149
  goals, 257
  modeling responses, 214
  operational implications, 257
Decision-making processes
  automation support for, 178
  collaborative process, 103
  controller vulnerabilities, 102-103, 107-108

recognition of vulnerability to stress, 142-143
system design considerations, 108
team leadership, 142, 155
training considerations, 108
See also Cognitive functioning; Information processing
Deregulation, 27-28
Design process
consideration of local conditions, 239
error-tolerant approach, 103-104
estimates of reliability, 18-19
FAA guide, 187-188
human-centered, 12-13
implications of cognitive vulnerabilities, 105-109
information resources, 10-11
prototyping, 215-216, 238-239, 240
real-time simulations in, 217-218
recommendations, 11-12
regional differences, 234
sequential experimentation protocol, 238-239
teamwork considerations, 148-149
technology-centered approach, 265-266
top-down approach, 239-240
user-centered, 266
user participation, 236-239, 240
See also Human factors design; Systems acquisition and development
Developmental controllers, 55, 60, 70
performance reviews, 57-58
Digital brite, radar indicator tower equipment (DBRITE), 35-36
Disciplinary action, 164
Dynamic simulation, 71, 73-74

E

Efficiency
air traffic control system goals, 21-23
aircraft holding procedures, 49
assessment and evaluation, 2-3, 157-158
automation goals, 249-250
definition, 157
demand for services and, 158
indicators, 157
monetary measures, 158
obstacles to, 22
pilot perspective, 23
policies and procedures, 159-161
prospects for improving, 22-23
safety and, 21-22, 159-160
Employee attitude survey, 8, 168-172, 176, 190-191
En route controllers
assessment, 56
cognitive skills, 4
in FAA organizational structure, 32, 33-34
flight data processing, 253
information resources, 45-46
nonradar areas management, 45
operations, 45
radar resources, 46
responsibilities, 19-21, 45
safety standards, 45
simulation training, 73
staff design, 45
TRACON control and, 48
traffic management activities, 46-48
training program, 70, 71
use of flight strips, 46
Enhanced target generator, 73, 74
Envelope protection, 246-247
Equipment failures
employee attitudes, 171-172
human factors in, 191-192
responsibility for restoration, 81-82, 185-186
small, 185
TRACON response, 44
Equipment maintenance and control
automation of, 80, 180-181
centralized monitoring and control system, 78
certification procedures, 80, 183-184
conceptual trends, 178
human factors research, 192

maintenance control centers for, 78-79, 180-181
responsibility for, 76-77, 183
restoration to service, 81-82, 185-186
workload, 186-187
See also Equipment failures
Expectations, 94, 105-106
effects on perception, 99-100
Eye movement, 208

F

FAA. *See* Federal Aviation Administration
Facility flight check, 156
False alarms
collision avoidance system, 264
flight path conflict detection, 260
ground proximity warning, 263
mistrust of automated systems, 273
threshold setting, 273-276
user response, 272
Federal Aviation Administration (FAA)
acquisition and development practices, 12
in air traffic control system management model, 153-155
automation goals, 249-250
cockpit resource management policy, 144-145
constituents and interested parties, 152-153
controller selection and training, 3-4, 55
databases, 156-157, 163
employee attitude survey, 8, 176, 190-191
flight routing policy, 23, 30
future challenges, 153
human factors design guide, 187-188, 195-196
human factors management, 9, 172-174, 176, 232
human factors policy, 173, 231-235
human factors research, 10, 192-194
job satisfaction in, 168-171
labor-management relations, 164-166

mission, 152, 153
organizational functioning, 7-8
organizational structure, 32-34, 52, 161-163
personnel policies, 163-166, 176
promotion of team training, 6-7
proposed acquisition reforms, 234-235
safety/efficiency assessments, 2-3, 158
safety/efficiency policies and procedures, 159-161
system design philosophy, 2
Feedback control, 244
Field research
applications, 220-221
combined research, 222
limitations, 222
validity, 223
Final approach spacing tool, 212
Flight data input/output computer system, 37, 38
Flight deck operations
automation, 23-25, 242, 251, 262-265, 267
historic failures of, 28-29
leadership style, 142
preflight programming, 24
team training for, 143-145
See also Pilot behavior
Flight level monitoring, 255
Flight management system (FMS), 263
mode confusion, 269-270
principles of operation, 23-24
Flight plans
automated conflict detection, 259-260
automated guidance, 263
automated monitoring, 253-254
daily planning for central flow control, 49
en route adjustments, 47-48
flight service station services, 51-52
HOST system identification, 45-46
routing policy, 23, 30
specialists, 52
tower control responsibilities, 35
visual displays, 121
Flight service stations, 33-34
automated systems, 51-52

number of, 51, 52
reform plans, 52
responsibility for equipment, 77
services of, 51
Flight strips, 36, 42, 46, 53, 253-254
electronic, 261-262
Flow control
central decision-making, 50-51
daily operations, 49-50
determinants of, 48
goals, 48
local/sectoral decision-making, 49-50
resources for, 49
techniques, 49
Full-performance-level controllers, 55, 60
requirements, 71
responsibilities, 56

G

Ghosting, 260-261
Global positioning system, 258
Great circle route, 23
Ground controllers
responsibilities, 19, 35
See also Tower controllers
Ground proximity warning system, 27, 263
Ground traffic
flow control, 49
management pressures, 37
pilot-controller communications, 121
responsibility for, 35
GS-2101 automation specialist, 10, 83-84, 87, 183, 186, 188-189, 196

H

Handoff communications, 21, 35, 36-37, 256
procedure, 43-44
TRACON responsibilities, 37
Hear back problem, 99-100
High-reliability organizations, 154-155, 167

HOST computer system, 45-46, 49, 51, 77
team interaction effects, 149
Human-centered automation. *See* Automation, human-centered
Human factors activities
in Airway Facilities, 9-10, 191-194
costs of automation, 268
in current air traffic control system, 31
FAA management of, 9, 172-175, 176, 232
FAA policy, 173
historic failures of air traffic control system, 25-30
incident analysis, 201-202
modeling, 210, 214
in organizational functioning, 160
research goals, 18
research methodology, 31
research simulations, 218
strategies for research, 10-11, 198-199
training considerations, 144
Human factors design
for Airway Facilities equipment, 187-188
combining research data for, 222-223
contributions of, 228-229, 240
current implementation, 231-235
field studies for, 220
generalizability of research, 200-201, 224
goals, 2
guidelines, 187-188, 195-196
historical development, 227-229
human-machine interface, 187, 219-220
knowledge base, 10-11, 199-200, 228
limitations of research, 200-201
measurement issues, 223-225
modeling techniques for, 210
modes of, 226-227
opportunities for improvement, 110-111
procurement practices, 229-230, 240
professional development, 230-231
prototyping, 215-216
rationale, 227, 228
research literature, 199-200, 201

INDEX

trade-offs, 109
user participation in, 215
See also Automation, human-centered;
Human factors activities

I

Incident analysis, 201-202, 222
Information management
  for central flow control, 49-50
  data collection for incident analysis, 202
  data collection for modeling, 214
  data collection in simulations, 219-220
  at en route centers, 45-46
  flight data, 253-254
  in human-centered automation, 283-284
  human factors research needs, 10-11
  individual reporting behavior, 157
  measurement in complex systems, 223-225
  monitoring activities, 97
  pilot-controller interface, 149
  responsibility for equipment, 77
  safety data collection systems, 156-157, 163, 203-204
  videotaped records, 205
Information processing
  causes of operator error, 201-202
  controller actions, 92-94
  controller error, 103-105
  demands on controllers, 92
  design considerations, 109-110
  long-term memory, 95-97, 101-102
  working memory functions, 94-95, 100
  See also Cognitive functioning; Decision-making processes
Integrated product teams, 234
Interpersonal communications
  air traffic teamwork enhancement program, 146-147
  automation effects, 148-150
  challenges for TRACON controllers, 43
  controller vulnerabilities, 101
  cultural differences, 138-139

employee satisfaction and, 8, 170, 190, 191
examples of controller-pilot interface, 138-140
field studies, 220-221
group process model, 136-137
hear back problem, 99-100
implications of datalink systems, 257
implications of proposed automation, 68-69
individual vs. team accountability, 137-138
leadership style, 142
limits of working memory, 106
nonlinguistic cues, 107
organizational context, 163, 166-167
power relations in, 139
readback, 43
shared assumptions/knowledge in, 101, 139-140, 149
as source of operator error, 202
system redundancy, 106
team-related research, 140-143
for teamwork, 6
visual displays and, 149
willingness to challenge decisions of others, 143, 170
as workload factors, 121
See also Communications technology

J

Job Performance Measurement project, 62
Job satisfaction, 8, 168-171, 176, 190-191, 196

L

Labor relations, 164-166
Leadership
  accountability, 162
  adaptive flexibility, 155
  in high-stress situations, 155
  styles, 142
  team preferences, 142
  See also Management

Legal issues
  controller liability, 164
  equipment certification, 184
Line oriented flight training, 144, 145, 148
Local controllers
  responsibilities, 19, 35
  See also Tower controllers

M

Maintenance control centers, 78-79, 81, 180-181, 187, 233
Management
  acceptance of new technology, 236
  air traffic control system management model, 153-155
  assessment for decision-making, 56-59
  credibility, 160-161
  employee attitudes, 8
  of equipment and systems, 181-182
  FAA human factors activities, 9, 172-174
  FAA labor relations, 164-166
  FAA organizational structure, 32-34, 161-163
  for implementation of automation, 152
  implications of Airway Facilities employee survey, 190-191
  organizational responsibilities, 158
  See also Leadership
Manpower and Personnel Integration (MANPRINT), 230, 231
Military controllers, 21, 70
Minimum safe altitude warning, 27, 38, 42, 259
Model I Full Capacity, 51
Modeling
  analytic, 211
  applications, 210-212
  challenges to, 212-213
  complexity of, 212-213
  current practice, 213-214
  data sources, 214
  for error prediction, 213
  fast-time vs. real-time, 212
  of human factors, 213
  limitations, 214
  for operational support, 212
  outputs, 211
  for policy analysis, 211
  for product acquisition and development, 12
  research needs, 225
  research value, 210
  safety-efficiency interactions, 175
  workload effects, 116-118
  See also Simulators/simulations
Modernization, 18, 249
Multiplex controller aptitude test, 64, 65, 66, 69
Mutual design and implementation, 238

N

NASA Task Load Index (NASA-TLX), 209
National Air Traffic Controllers Association, 164-166
National airspace information monitoring system, 156
National airspace system, 79, 82
National airspace system performance analysis capability, 213
National airspace system simulation model, 213
National data airspace interchange network, 51
National Maintenance Coordination Center, 81
National route plan, 23
Navigations technology
  automation, 257-258
  oversight, 33
New technology
  for airport efficiency, 22-23
  for Airway Facilities operations, 9
  challenges to system-wide introduction, 53
  controller training for, 71-72
  employee attitudes, 171, 191
  FAA research and development structure, 33

INDEX

lack of integration, 177-178, 180-181, 187, 195, 233
organizational functioning and, 8, 174-175
simulation testing, 218
terminology, 18
trends, 178
user acceptance, 235-236, 273-276
user participation in design and implementation, 236-239

O

Occupational knowledge test, 64
Office of Technology Assessment, 161-163
Operational control centers, 178, 193
Operations network, 156
Organizational functioning/structure
acceptance of new technology, 235-236
adaptive flexibility, 155
air traffic control system management model, 153-155
Airway Facilities, 78-79, 82
communications policy, 163
communications style, 166-167
controller performance and, 159, 175
current air traffic control system, 52-53
determinants of, 7
effects of, 7-8
employee satisfaction with, 8, 168-171
FAA structure, 32-34, 161-163
formal context, 158
high-reliability organization, 154-155, 167
for human-centered automation, 284
human factors in, 160
implementation of teamwork concepts, 144
incident analysis, 202
informal context variables, 166-172, 175-176
introduction of new technology, 8, 174-175
management responsibilities, 158

managing human factors activities within FAA, 9, 172-174
as organizational culture, 7
research needs, 8
response to communication of problems, 167
safety outcomes and, 156
subcultures, 167-168
team performance in air traffic control, 135-137
Overseas flights, 21
nonradar areas, 45

P

Perceptual functioning, 69
design considerations, 110
determinants of, 94
display overload, 120-121
expectation effects, 99-100
hear back problem, 99-100
situation awareness, 95, 100-101
visual sampling, 99, 105
Performance assessment
Airway Facilities technician, 84, 86, 189
checklists, 57-58
controller, 3-4
controller selection and training, 55
crew resource management, 142-143
current research efforts, 63, 193-194
employee satisfaction with, 170-171, 191
goals for training program, 74-75
implications of automation, 68-69
for management decision-making, 56-59
minority sensitivity, 63
models for, 3
objective measures for, 58
operational assessment program, 56-57
selection criteria, 60-63
simulators for, 58-59
strategies for research, 10-11
for teams, 163-164
Performance-preference dissociations, 216

Personality traits, 67-68
Physical plant, 77
Physiological stress measurement, 207-209
Pilot behavior
  controller communications, 121
  examples of controller-pilot interface, 138-140
  modeling techniques, 214
  overreliance on automation, 276-278
  recognition of vulnerability to stress, 142-143
  TRACON communications, 43
  See also Flight deck operations
Plan view display radar, 46, 120-121
Preflight actions, 24
Professional Air Traffic Controllers Organization, 70
Prototyping
  applications, 215, 216
  limitations, 216
  for product acquisition and development, 11, 12, 238-239, 240
  research needs, 225

Q

Quality through partnership process, 165-166, 170

R

Radar, 229
  en route center resources, 46
  en route traffic control, 45
  nonradar areas, 45
  responsibility for equipment, 77
  tower control resources, 35-36
  TRACON resources, 38-42
  training, 70, 71
  vigilance effects on use of, 127-128
  visual display, 39-42
  visual sampling, 99
Radar positive control system, 25
Radio telephony, 121

Regional differences, 32, 53
  design implications, 234
  obstacles to performance assessment, 58, 59
  as organizational subculture, 167-168
  in simulations, 59
Reliability
  of air traffic control system, 2
  definition, 18
  measures of, 18
  problems in estimating, 18-19
  trust and, 19
Remote monitoring subsystem, 78
Research methodology, 31
  combining data sources, 222-223
  data collection for incident analysis, 202, 203-204
  efficiency measures, 157
  field studies, 220-222
  generalizability of human factors studies, 200-201, 224
  for human error research, 222-223
  human factors literature, 199-200, 201
  incident analysis, 201-202
  limitations of human factors studies, 201
  measurement in complex systems, 223-225
  modeling techniques, 210-214
  needs for air traffic control research, 225
  prototyping, 215-216
  real-time simulation, 216-220
  requirements, 197
  resources for, 197
  safety analysis, 157
  strategies for human factors studies, 198-199
  teamwork studies, 137
  use of subjective assessments, 204-205
  validity, 197-198, 223-225
  for workload assessment, 205-210
Retirement of Airways Facilities employees, 10, 84, 88, 179

S

Safe Skies for Tomorrow, 161-163
Safety
  air traffic control system goals, 21-22
  analytical procedures, 157
  assessment, 2-3
  automation goals, 249-250
  concerns about automation, 17-18
  cost-effective risk assessment, 159
  data sources, 156-157, 163, 203-204
  efficiency and, 159-160
  high-reliability organizations for, 154-155
  historic failures of air traffic control system, 25-30
  indicators in air traffic control, 155-156
  organizational risk factors, 156
  performance assessments and, 164
  policies and procedures, 159-161
  predictive modeling, 175, 213
  pressures for efficiency and, 21-22
  separation between aircraft, 21
  workload considerations, 133
Security, 33
Separation and control hiring assessment program, 62
Separation between aircraft, 21
  approach sequencing/ghosting, 260-261
  challenges for controllers, 37-38
  en route standards, 45
  need for automation, 248
  in nonradar areas, 45
  technology introduction, 235
  TRACON responsibilities, 37-38
  TRACON standards, 37
  wake vortices, 22, 42
Sequential experimentation protocol, 238-239
SIMMOD, 213
Simulators/simulations
  applications, 217-218
  combined research, 222-223
  constraints on data collection, 219-220
  for controller assessment, 58-59
  current research, 73-74
  for design process, 217-218
  dynamic, 71, 73-74
  fast-time, 212
  features, 211
  fidelity of, 73, 218-219
  for human factors research, 218
  with local features, 59
  part-task training, 73
  rationale, 216-217
  recommendations for utilization, 11
  regional air traffic control system, 217
  research applications, 211
  research needs, 225
  for training, 4, 70, 71, 73, 148
  validity, 223
  See also Modeling
Situation assessment through re-creation of incidents, 4, 120
Situation awareness, 95, 100-101, 106-107, 140
  overreliance on automation, 278-279
Sleep loss, 131
Social learning theory, 72-73
Stacking aircraft, 22
Staff design
  Airway Facilities, 82-84, 188
  en route center, 45
  flexibility in, 44
  flight service stations, 52
  replacement of striking controllers in 1981, 54
  TRACON, 44
Strike of 1981
  outcomes, 28
  replacement workers, 54, 67
Subjective assessments
  in design prototyping, 216
  limitations, 205
  performance-preference dissociations, 216
  research value, 204-205
  for workload assessment, 209-210
Subjective workload assessment technique, 209

Surveillance technology
  oversight, 33
  TRACON resources, 38
System maintenance control center, 79
Systems acquisition and development, 11-12
  for Airways Facilities, 187-188, 233
  controller training, 71-72
  FAA human factors policy, 173
  FAA organizational structure for, 33
  human factors research support for, 173, 174
  incorporation of human factors in, 12, 229-230, 240
  maintenance of predecessor designs, 233
  performance-preference dissociations, 216
  proposed reforms for FAA, 234-235
  standardization in, 181
  user participation in, 12, 236-239
  workload certification, 116

T

Teamwork
  accountability, 137-138
  in air traffic control system, 135-136
  air traffic teamwork enhancement program, 145-148
  in Airway Facilities, 186
  automation effects, 148-150
  communication for, 6
  crew resource management, 142-143
  determinants of, 136
  examples of controller-pilot communications, 138-140
  flight deck, 143-145
  group process model, 136-137
  high workload strategies, 141
  for human factors design, 227-228
  leadership style, 142
  low workload conditions, 140
  performance assessment, 163-164
  research activities, 137, 186
  research findings, 140-143
  significance of, 135, 150
  strategies for improving, 6-7, 150-151
  subcultures, 136
  team members, 135
  training for, 141, 142
Terminal airspace simulation facility, 218
Terminal radar control area (TRACON)
  cognitive skills of controllers, 4
  communications system, 43-44
  crisis management, 44
  en route control and, 48
  equipment failures, 44
  in FAA organizational structure, 32, 33-34
  obstacles to traffic management, 42-43
  physical environment, 44
  radar resources, 38-42, 53
  responsibilities, 19, 34-35, 37-38, 286
  staffing, 44
  use of flight strips, 38-42
Textbooks, 199-200
Time-line analysis, 117
Total Airport and Airspace Modeler, 213
Total systems design, 228
Tower controllers
  communications system, 36-37
  in FAA organizational structure, 32, 33-34
  responsibilities, 19, 34-35, 286
  simulation training, 73
  use of flight strips, 36
  visual resources, 35-36, 53
TRACON. *See* Terminal radar control area
Traffic alert and collision avoidance system, 24, 264-265
  controller-pilot interface and, 139-140, 149
  introduction of, 27
Traffic management advisor, 213, 221
Traffic management coordinators, 49
Train for success philosophy, 70
Training of Airway Facilities technicians, 85-86, 87-88, 178-179, 188-189
Training of controllers, 176
  age limitations, 68

assessment methodology for, 55-56
attrition rate, 66
cognitive screening of candidates for, 64-67
conceptual approach, 70
controller performance related to, 72
course of, 69-70, 71
current practice, 3, 55
current research efforts, 70
current tracking data, 60
decision making, 108
detection tasks in vigilance, 126
goals, 55, 74-75
job-related criteria, 3-4
length of, 71
memory functions, 102, 107
for new equipment, 71-72
opportunities for improvement, 110
performance measures as selection criteria, 60
personal characteristics of trainees, 67-68
prior experience requirements, 70
for radar operations, 70, 71
selection goals, 63
simulators for, 4, 70, 71, 73, 148
situated learning model, 72-73
sources of candidates, 70
subtask training, 73
team for, 70-71
for teamwork, 6-7, 141, 142, 148, 150
Trust, 18, 19
in automation, 271-277, 279
in equipment certification process, 184
false alarm effects, 273-276
human-centered automation objective, 283
management credibility with employees, 160-161
as organizational variable, 171-172

U

User-centered design, 266

V

VHF omnidirectional range, 45
Vigilance
air traffic control and, 127-129
arousal and, 126-127
current understanding, 125, 126-127
definition, 125
implications for automation, 129-130
task factors influencing, 125-126, 133
training effects, 126
workload effects, 127, 133
Visual display
aircraft data block overlap, 256
aircraft flight level, 255
aircraft heading and speed, 255-256
aircraft situation, 49
approach sequencing/ghosting, 260-261
compatability with cognitive processes, 204-205
data smoothing, 254
datalink systems, 257
effects on interpersonal communication, 149
flight path, 121
in human-centered automation, 283-284
informational scope, 258
radar, 39-42, 46
workload factor, 120-121
Visual sampling, 99, 105
Voice switching and control system, 18-19

W

Wake vortices, 22
individual differences in aircraft, 42
Weather
challenges for TRACON controllers, 42-43
constraints to airport efficiency, 22
flight services station services, 51-52
Work schedules
controller performance and, 130-133

current practice, 6, 130, 131, 133
night shift work, 131
potential problem areas, 48
shift changes, 48
shift rotations, 131-132, 133
sleep disruption effects, 131
time on shift as risk factor for error, 130-131

Workload
adaptive strategies, 115, 117-118
airspace load effects, 118-120
in Airway Facilities, 186-187
allocation of cognitive resources, 96-98
assessment measures, 205-206
assessment models, 5-6
automation effects, 122-123, 133-134, 268, 270-271
communications factors, 121
controller performance and, 114-116, 123-124
definitions, 113-114
display factors, 120-121
drivers, 115, 118
extremes of, 5
interaction of factors in, 115, 133
modeling techniques, 214
multitask performance theories, 117, 122
as performance factor, 5
physiological measures, 207-209
primary-task measures, 206
research trends, 113
safety and, 133
secondary-task measures, 206-207
significance of, 5, 112-113
situation awareness and, 100-101, 140
subjective measures, 209-210
system trends, 34
system variation, 113
task load vs. mental workload, 124
team strategies, 141
theoretical models, 114, 116-118
time-line analysis, 117
underload conditions, 112-113, 115, 124, 133, 140, 202
of vigilance, 127-130, 133